T0260410

Green Computing for Sustainable Smart Cities

Information and communication technology and the Internet of Things play key roles in smart city projects. It is challenging to handle the large amount of data generated by the different processes related to land use, the environment, the social and economic milieu, energy consumption, and transportation. This book emphasizes that green computing-based infrastructure initiatives benefit not only the environment but the enterprise as well.

Green Computing for Sustainable Smart Cities: A Data Analytics Applications Perspective covers the need for smart green computing from various engineering disciplines and offers diversified applications for such computing with case studies. The book highlights the sustainable development of smart cities using recent technology and emphasizes advances and cutting-edge techniques throughout. Focused on the different tools, platforms, and techniques associated with smart green computing, this book presents multiple perspectives from academia, industry, and research fields.

The primary audience for this book includes academics, researchers, graduate students, smart city industry practitioners, and city administrators who are engaged in smart cities and related technology.

Green Computing for Sustainable Smart Cities

A Data Analytics Applications Perspective

Edited by

Neha Sharma

Jai Prakash Verma

Sunil Gautam

Valentina Emilia Balas

Saravanan Krishnan

CRC Press
Taylor & Francis Group
Boca Raton London New York

CRC Press is an imprint of the
Taylor & Francis Group, an **informa** business

First edition published 2024
by CRC Press
2385 NW Executive Center Drive, Suite 320, Boca Raton FL 33431

and by CRC Press
4 Park Square, Milton Park, Abingdon, Oxon, OX14 4RN

CRC Press is an imprint of Taylor & Francis Group, LLC

ISBN: 978-1-032-48393-1 (hbk)
ISBN: 978-1-032-48396-2 (pbk)
ISBN: 978-1-003-38881-4 (ebk)

DOI: 10.1201/9781003388814

Typeset in Times
by SPi Technologies India Pvt Ltd (Straive)

Contents

Preface

Green computing in the broader sense covers the practices and procedures of designing, manufacturing, and using computing resources in an environmentally friendly way while maintaining overall computing performance and finally disposing of hardware in a way that reduces its environmental impact. This means (i) reduction in the use of hazardous materials; (ii) maximizing output from the product during its lifetime while minimizing energy consumption; and (iii) the reusability or recyclability and biodegradability of used products and waste. Many corporate organizations are taking initiatives to reduce the harmful impact of their operations on the environment. The United Nations Framework Convention on Climate Change is an international environment treaty whose objective is to stabilize the emission of greenhouse gases in the atmosphere at a level that would prevent dangerous anthropogenic interference with the ecosystem. Sustainable development means developing without damaging the requirements of future generations. That is, meeting human development goals while preserving natural resources and ecosystems on which society depends. This book will provide several important current research directions related to the field of green computing, emphasizing its importance for sustainable development.

The greatest environmental challenge today is global warming, which is caused by carbon emissions. An energy crisis brings green computing, and green computing needs algorithms and mechanisms to be redesigned for energy efficiency. A green computing-based IT infrastructure refers to the study and practice of using computing resources in an efficient, effective, and economic way. Such approaches include virtualization, power management, material recycling, and telecommuting. In broader terms, green computing is also defined as the study of designing, engineering, manufacturing, using, and disposing of computing devices in a way that reduces their environmental impact. It is also known as green information technology (green IT).

This edited book integrates data analytics and data science to provide an understanding of the many applications for smart city projects where green computing can make an impact. In this book we endeavour to provide all the relevant information in one place to enable corporations, academic experts, researchers, students, policymakers, and non-government organizations to understand the ways to achieve the defined goals using data and analytics, the challenges to be faced while using them, and the similarities and fine differences between the data science technologies.

The features of this book include:

1. The need for smart green computing in various engineering disciplines;
2. Diversified applications for smart green computing with case studies;
3. Contributors from different parts of the world;
4. Multiple perspectives from academia, industry, and research fields;
5. The sustainable development of smart cities using recent technology;
6. Emphasizes the advances and cutting-edge technologies throughout;
7. Focuses on different tools, platforms, and techniques;
8. Presents the tools and methods needed for smart green computing.

Editors

Neha Sharma is a data science crusader who advocates for its application for achieving sustainable goals, solving societal, governmental and business problems, as well as promoting the use of open data. She has more than 22 years of experience, is presently working with Tata Consultancy Services, and is a founder secretary of the Society for Data Science. Prior to this she worked as director of the Institute of Pune, which runs post-graduate courses like MCAs and MBAs. She is an alumnus of the College of Engineering and Technology, Bhubaneshwar and she completed her PhD at the Indian Institute of Technology, Dhanbad. She is a senior IEEE member, a secretary of the IEEE Pune Section, and is an ACM distinguished speaker. She is an academic and has organized several national and international conferences and published several research papers. She is the recipient of a "Best PhD Thesis Award" and a "Best Paper Presenter at International Conference Award" at the national level. Recently, she received the Golden Book Award 2023 for her book entitled *Towards Net-Zero Targets: Usage of Data Science for Long-Term Sustainability Pathways*. She has been instrumental in integrating teaching with the current needs of the industry and steering students towards their bright futures.

Jai Prakash Verma is associate professor in the Computer Science and Engineering Department at NIRMA University, where he has been since July 2006. He received his BSc (physics, chemistry, mathematics) and MCA degree from the University of Rajasthan, Jaipur, and a PhD from Charusat University, Changa. His PhD subject area was text data summarization and big data Analytics. His research interests include data mining, big data analytics, and machine learning and he has presented at international conferences and published in journals. He is a recognized PhD supervisor at Nirma University and currently has two students. He is actively involved in conducting various customized training programmes in Big Data Analytics for naval officers at INS Valsura, the Indian Navy, SAC-ISRO, and Ahmedabad scientists.

Sunil Gautam is assistant professor in the Department of Computer Science and Engineering, Institute of Technology, Nirma University, Ahmedabad, India. He did his PhD in Computer Engineering at the Indian Institute of Technology, (Indian School of Mines) Dhanbad, India. He received his MCA degree in Computer Science from Uttar Pradesh Technical University, Lucknow. His main research interest includes wireless sensor mining, network Security, and IoT. He specializes in intrusion detection systems. He has around eight years teaching and research experience, and has contributed a number of research papers to journals and international conference proceedings.

Valentina Emilia Balas is currently full professor in the Department of Automatics and Applied Software at the Faculty of Engineering, "Aurel Vlaicu" University of Arad, Romania. She holds a PhD in Applied Electronics and Telecommunications from the Polytechnic University of Timisoara. She is the author of more than 200 research papers published in refereed journals and presented at international conferences. Her research interests are in intelligent systems, fuzzy control, soft computing, Smart sensors, information fusion, and modelling and simulation. She is editor-in-chief for the *International Journal of Advanced Intelligence Paradigms* and the *International Journal of Computational Systems Engineering*, is an editorial board member of several national and international journals, and is evaluation expert for national and international projects. She served as general chair of the International Workshop Soft Computing and Applications on seven occasions from 2005 to 2016 held in Romania and Hungary. She has participated in many international conferences as organizer, session chair, and as a member of the International Program Committee. She is a member of EUSFLAT and ACM and is a senior member of IEEE. She is also a member of TC – Fuzzy Systems (IEEE CIS), TC – Emergent Technologies (IEEE CIS), and TC – Soft Computing (IEEE SMCS). She was vice president (Awards) of the International Fuzzy Systems Association Council (2013–2015) and is joint secretary of the Governing Council of the Forum for Interdisciplinary Mathematics, a multidisciplinary academic body in India.

Saravanan Krishnan is associate professor, Department of Computer Science & Engineering at the College of Engineering, Guindy, Anna University, Chennai-25, Tamilnadu. He studied ME Software Engineering and has a PhD in Computer Science Engineering. His research interests include cloud Computing, software engineering, Internet of Things, and smart cities. He has presented or published papers at 14 international conferences and in 32 international journals. He has also written 21 book chapters and edited 12 books with international publishers. He has conducted much consultancy work for municipal corporation and smart city schemes. He is an active researcher and academic, as well as a reviewer for many Elsevier journals. He is a fellow of the Institution of Engineers India and a member of the Indian Society for Technical Education, the Indian Science Congress Association, and the Association for Computing Machinery, amongst others.

Contributors

B. Amutha
SRM Institute of Science and
    Technology, KTR Campus
Chennai, India

Mohammed Aqeel
Karnawati University
Gandhinagar, India

Md Azharuddin
Aliah University
Kolkata, India

Vaishali Chourey
Adani University
Ahmedabad, India

Kiyeng P. Chumo
Moi University
Eldoret, Kenya

Santosh Kumar Das
Sarala Birla University
Ranchi, India

Ritwik Duggal
National Institute of Technology
Himachal Pradesh, Hamirpur, India

Sunil Gautam
Nirma University
Ahmedabad, India

Himanshi
National Institute of Technology
Hamirpur, Himachal Pradesh, India

Sonia Joshi
Karnawati University
Gandhinagar, India

Manivel Kandasamy
Karnawati University
Gandhinagar, India

Saravanan Krishnan
Anna University
Chennai, India

Abhishek Kumar
National Institute of Technology
Hamirpur, Himachal Pradesh, India

Ashish Kumar
XIM University
Bhubaneswar, India

Pradheep Kumar
BITS Pilani
Pilani, India

S. Sathish Kumar
Institute of Engineering and
    Technology
Hyderabad, India

Vikash Kumar
Sarala Birla University
Ranchi, India

Kaushal Nileshbhai Maniyar
Nirma University
Ahmedabad, India

Riya Mehta
Adani University
Ahmedabad, India

Kithinji Muriungi
Purdue University
W. Lafayette, Indiana, USA

S. Muthukaruppasamy
Velammal Institute of Technology
Panchetti, India

Ramesh Ram Naik
Nirma University
Ahmedabad, India

J. Jesu Vedha Nayahi
Anna University Regional Campus
Tirunelveli, India

Rohit Pachlor
MIT ADT University
Pune, India

Prashant Pansare
Rubiscape Labs
Pune, Maharashtra, India

Keyurkumar Patel
Rashtriya Raksha University
Gandhinagar, India

Sanjay Patel
Nirma University
Ahmedabad, India

Alok R. Prusty
DGT, NSTI
Bhubaneswar, Odisha, India

Sk Md Abidar Rahaman
Aliah University
Kolkata, India

J. Ramkumar
SRM Institute of Science and
    Technology
KTR Campus, Chennai, India

N. Renugadevi
Indian Institute of Information
    Technology
Tiruchirappalli, India

Siddhartha Roy
University of Engineering and
    Management
Kolkata, India

Sandeep Sahu
Vellore Institute of Technology
Bhopal, India

Niomi Samani
Karnawati University
Gandhinagar, India

S. Saravanan
Anna University
Chennai, India

Raju Shanmugam
Karnawati University
Gandhinagar, India

Neha Sharma
TCS Pune
Pune, India

Nagendra Pratap Singh
National Institute of Technology
Jalandhar, India

Akankshya Subhadarshini
College of Engineering and
    Technology
Bhubaneswar, India

C. M. Naga Sudha
Anna University MIT Campus
Chennai, India

Rohit Thakur
National Institute of Technology
Hamirpur, Himachal Pradesh, India

G. Arun Sampaul Thomas
Institute of Engineering and Technology
Hyderabad, India

Tejas Uttare
Karnawati University
Gandhinagar, India

Jai Prakash Verma
Nirma University
Ahmedabad, India

Ravi Verma
VIT Bhopal University
Madhya Pradesh, India

Mark Wall
Tata Consultancy Services
London, United Kingdom

1 Introduction to Green Computing for Smart Cities

Ravi Verma
VIT Bhopal University, Madhya Pradesh, India

Sunil Gautam
Nirma University, Ahmedabad, India

Sandeep Sahu
Vellore Institute of Technology, Bhopal, India

1.1 INTRODUCTION

Digitalization is growing all around the world, and today most people have multiple devices to process their requests; every device will consume a certain amount of energy. It would be a big challenge for computer device developers and designers to make energy-efficient designs and machines that can operate with the user's requirements in an efficient way regarding energy utilization; therefore, it is recommended that developers design green energy-based technological devices, such as computer systems, tablets, and other wearable devices. To sustain green engagement, everyone makes a great effort to prepare for a green energy-based solution that uses a variety of computer resources to address every issue related to current energy use. These issues motivate us to recycle and re-engineer various computing devices [1].

In other words, we can say that green computing will be used in order to bring about an environment that can use computing resources according to the green computing framework; accordingly we can also say the future of computing technology is all about green computing so that it will be a part of major research and analysis for designing, developing, and then recycling of various digital devices using the concept of green re-engineering principles that will result in a reduction in pollution and other environmental hazards. We need to work equally for maintaining green computing along with all the recommended computing devices and resources. In 1992 the U.S. Environmental Protection Agency (EPA) started a number of information technology programmes that motivated green computing and resources in order to improve information technology energy efficiency. Clean computing practices will include a variety of green development strategies to improve green computing with the efficiency principle if we had energy consumption through different types of digital devices, as well as advancements in disposal and recycling technology that will

DOI: 10.1201/9781003388814-1

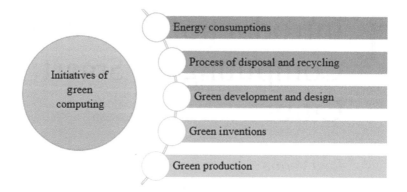

FIGURE 1.1   Initiatives of green computing.

allow us to call for recycling whenever necessary through the following green computing initiatives (as shown in Figure 1.1).

- **Energy consumption**: The green computing paradigm entails a distinct procedure to reduce the energy consumption of various computing technologies and equipment. By using this procedure, we may enhance the exposure of green computing activities.
- **The process of disposal and recycling**: The process of disposal and recycling of all the unwanted or damaged digital circuits and much other equipment needs to be disposed of in an eco-friendly environment.
- **Green development and design**: The manufacturing industries should follow the rules and principles of green computing to manage design; development functions should be processed under specific guidelines of the green computing environment in which industries will deploy the latest energy harvesting techniques.
- **Green inventions**: Green inventions include the process of recycling and reutilizing computing resources using greener ways to reproduce computing peripherals such as the computer and its various equipment. In today's unhappy scenario, we find that industries are focused on the minimization of wastage during the implementation of innovations in the manufacturing of future computing possibilities [2].
- **Green production**: The goal of the most recent technological developments was to maximize the use of energy in computing processes by manufacturing digital equipment in a more environmentally friendly manner. This was achieved by manufacturing and recycling digital equipment, which reduced the amount of energy wasted by the devices [3].

1.1.1   THE SMART CITY

Multiple definitions have been presented for smart city initiatives that are adequate. According to IBM, a smart city is one that can primarily enable an optimization mechanism for all the interconnected and available energy resources so that it can be understood by everyone and in real time. However, we can also say a smart city uses a kind of dynamic framework that combines information and resources together in a most optimized way to enhance efficiency and reduce the energy losses.

1.1.2 HISTORY OF THE SMART CITY

From the mid-1960s to the 1970s the idea of a smart city came into existence; before this initiative, everyone was working with their own database technology to gather information from multiple sources in which, due to heavy resource engagement, energy was wasted to a high extent and the source of energy was not effectively managed. In order to address the needs for a better solution to effectively utilize and reduce the wastage of energy resources smart city initiatives have taken place as the primary solution for the entire first-generation computing industries.

In first-generation computing requirements, we considered that various day-to-day life needs to be addressed along with various activities that improve the comfort of people and enhance the utilization of energy resources. After considering various aspects of citizen requirements, the extended part of smart city initiatives considered a second generation in which developers are concentrating on several other smart cities innovations that address various real-time issues to improve communication, development, and the transformation of various application domains using smart city development projects [4].

The third generation of the computing domain was developed by the city of Vienna, which increased the number of users of smart city technology by connecting various people through several communication platforms as a social community and that completely takes away control from technology leaders and shifts it towards the engagement of citizens.

1.1.3 MOTIVATION

A smart city project needs strong support from different types of energy modes and communication sensors to collect instructions, as data, to make various decisions to complete online ongoing activities and their execution. Such data are very useful to manage efficiency at the operational level and helps cities to smartly manage the functional cycle of various computing elements [5] such as hospitals and educational institutes. All the connected resources are the primary assets of smart city projects because the information gathered is used to control and analyse the various decisions and functions related to citizens, forecasting, education, other utilities, and the transformation of several other domains.

The growing advancement in the technological development of various domains motivates us to work and improve the energy sectors. We need to focus on renewable energy areas and technological development to extend the benefits at a large scale.

Due to continuous changes in users' requirements for their comfort with the use of computing technology, smart city projects encourage us to focus on renewable energy sectors and dedicatedly work with the integration of green energy computation and the IoT, the cloud, and blockchain technology to make communication more reliable and secure.

1.1.4 OBJECTIVES

The concept of green computing in smart cities will play a major role in the future. There is various research going on in this area, and several research-based projects also address various energy-renewable issues. The major objectives of this chapter are:

- To be a guide to all learners and readers; it is specially designed for beginners who are new to this computing area;
- To define and address various green computing issues and also to focus on the green computing framework for smart city projects;
- To address the various challenges of smart city projects and their various features;
- To defining smart city projects and technological development.

1.1.5 ORGANIZATION OF THE CHAPTER

Section 1.2 describes smart city project features and their impact on technological development. Section 1.3 is focused on smart city description and its correlation with the green computing paradigm. Section 1.4 describes how green computing is useful for smart city projects: through a literature review we found energy utilization and its impactful uses is necessary for future computing technology. Section 1.5 details technology enhancement that can combine the internet, the IoT, and the cloud for the betterment of smart city projects and its implementation for making the future of computing technology greener. Section 1.6 highlights the smart city framework used for smart finance, administration, and smart lifestyles, and which makes life easier using smart concepts such as various smart wearable devices. Section 1.7 details the classification of reconciled energy and its role in smart city applications, and describes the way to reconcile energy for its effective and efficient utilization. Section 1.8 defines the role of green infrastructure and buildings to make life smarter and easy. Section 1.9 considers the concepts of renewable energy in the form of solar to make computation and energy utilization more prominent. Section 1.10 considers various technological developments in smart city projects, since a smart city is a combination of multiple paradigms that needs the cooperation of various techniques and tools so that smarter services can be extended from one technology to another to improve services for daily activities. Section 1.11 describes the process cycle or consequences of different procedures in smart city projects. Section 1.12 defines the various challenges that a smart city project will face during and after implementation. Section 1.13 concludes the chapter.

1.2 FEATURES OF THE SMART CITY PROJECT

With the combination of IoT technology, artificial intelligence techniques, and machine learning concepts it is now possible to adopt smart city deployment for multiple applications, such as implementing the smart parking facility using different platforms, which would help citizens regarding the parking space and also help them to make digital payment. Similarly we can also consider the smart traffic system where we can control traffic using the smart city concept. Optimized traffic lights will use the concept of machine learning in order to optimize the best route for users. Apart from this it will also guide while driving to the location and will help one to reduce the overall time due to heavy traffic load. Similarly, we can also work smartly in order to maintain energy resources. Whenever the system finds an empty row it will turn the light on. As a result, it has the ability to control energy waste and make better use of the energy resources that are already accessible, and only acquire and implement a smart city-based framework for a variety of applications. Smart city methodology is also useful to manage the side effects of natural disasters, such as earthquakes. Air pollution management is also effected by various Internet-enabled sensors, through which we can measure the different parameters to maintain the quality of the environment and observe various challenges in the near future for the purpose of waste management. In a similar way smart city methodologies also allow the management of safety parameters and various logs based on the monitoring and tracking of crime activities through various Internet-enabled sensors for advanced warning of various unwanted events [6].

In the consideration of smart city initiatives, smart infrastructure such as buildings can also participate to perform real-time management and various other structural monitoring systems. This will give the advantage of measuring future requirements based on the need, for example, for using intelligent sensors. A smart building is sufficient to monitor the leakage of a piping system that has been installed in the building and for other infrastructural parameters. In an inconclusive way we can say that a smart city technology is the combined solution for various manufacturing and farming requirements for the purpose of managing energy efficiently and effectively. To date a smart city would be the best solution to jointly achieve multiple day-to-day solutions for the entire citizen domain.

1.3 SMART CITY DESCRIPTION

A smart city utilizes a variety of tools and technologies, collectively known as information and communication technology, to enhance the efficient exchange of information within the public system. This approach aims to deliver government services of high quality and meet the daily needs of citizens. Consequently, smart city initiatives prioritize the fulfilment of citizens' requirements, taking into account the numerous factors that contribute to a more convenient and advantageous lifestyle.

Smart city projects explore diverse communication methods to facilitate the exchange of essential data between different locations. The concept of smart city projects enables us to optimize the utilization of various government facilities while minimizing energy consumption. In summary, the comprehensive implementation of smart city projects aims to address multiple issues associated with the various components of such projects [7, 8].

The implementation of smart city technology aims to optimize the utilization of energy resources and streamline the flow of information in a more efficient manner.

This involves integrating various entities from the Internet of Things and other internet-based applications to facilitate citizen engagement. By leveraging available networking resources according to individual requirements, the smart city programme minimizes energy wastage and delivers highly efficient outcomes.

The primary goal of the smart city programme is to enhance operational efficiency and stimulate economic development by implementing advanced smart technologies and data analysis tools. This initiative also aims to improve communication and service delivery for citizens. A key aspect of this programme is exploring the utilization and accessibility of these technologies. This statement delineates the essence of smart city initiatives, wherein real-time tools are employed to effectively manage energy resources. By leveraging the availability of these tools, energy measurement is conducted, ultimately resulting in the generation of guaranteed outcomes aimed at minimizing and reducing wastage of energy resources. The implementation of such measures is expected to yield significant positive effects.

The most listed objectives that can be determined using smart cities are:

- The technology networks around the citizen, using the smarter infrastructure;
- The entire paradigm needs environmental initiatives to be taken in the process of deploying smart city mechanisms for a variety of functional areas;
- The smarter city will have its own highly effective and efficient functional public transport that will help everyone to make their life easier and more comfortable;
- The smart city concept is founded on robust infrastructure and ambitious plans that are designed to effectively handle numerous operations simultaneously;
- With the assistance of intelligent urban centres, individuals may now access a variety of services in a real-time urban setting, effectively using available resources and obtaining desired outcomes in response to specific requests.

The success rate of a smart city project hinges on the collaboration between public and private sectors to execute a series of tasks and functions aimed at establishing and sustaining an intelligent environment. This involvement will undoubtedly facilitate optimal efficiency in the reuse and recycling of resources, particularly in the management of energy resources. Furthermore, the implementation of a data analyst is crucial for the effective utilization of the information derived from various sensors within a smart city setting. By analysing these diverse sets of information, the analyst can make informed decisions regarding the management of future services. This systematic approach greatly facilitates the identification of appropriate solutions for each input received within the intelligent environment [9, 10].

1.4   GREEN TECHNOLOGY FOR SMART CITY PROJECTS

Globally, we can see the challenges has been growing on socially and environmentally you noted to make life more easier simpler and comfortable therefore multiple cities globally redirecting their development approach to make city smarter using believe green energy concepts, with the help of smart city projects the major objective behind the green smart city projects are increases the number of production units how long with more sufficient efficiency and minimum wastage of available energy of the system, with the help of smart city projects the green technology enhance the overall capacity of every equipment's that we have been used the process of major functioning that have been deployed in order to manage smart requirements for this system with the aim of turning normal city into smarter will play an important role in order to make city advance through various smart projects which brings user friendly platform to its inhabitants in order to serve them a quality of life through offering various opportunity to have a safe, healthy and quality spending of life through green computing and smart city projects [11]. On a global scale, there has been a noticeable increase in the challenges faced in the realms of social and environmental aspects. These challenges are aimed at enhancing the ease, simplicity, and comfort of human life. Consequently, numerous cities around the world are altering their development strategies to embrace the concept of smart cities, which prioritize the utilization of green energy. This transition is being facilitated through the implementation of smart city projects. The primary objective of green smart city projects is to increase the number of production units while improving efficiency and minimizing energy wastage. Through the implementation of smart city projects, green technology enhances the overall capacity of equipment used in major functions, thereby managing the smart requirements of the system. The transformation of a regular city into a smarter one through various smart projects plays a crucial role in advancing the city and providing its inhabitants with a user-friendly platform.

This section considers the various aspects of climate change and its impact on smart cities along with its expected functions which include various green motives such as smart city project innovations, smart governance systems, and smart financial department programmes. As the latest report defines it, the smart city project considers three major roles for managing the greener future of industrial requirements through smart city projects such as smart buildings, smart lighting, and intelligent mobility, which also functions with the object of the development of various sustainable plans in order to conduct multiple programmes for managing the entire development framework through smart city projects [12].

Currently, there is a widely recognized trend of global population growth, which is paralleled by a corresponding decrease in available energy resources. This phenomenon has significant implications for the increasing cost of energy, as well as the hindrance it poses to the development of cities in terms of global climate change and air pollution. Consequently, there is a pressing need to transform traditional cities into smart cities through the integration of advanced technologies. Recent research indicates that a substantial portion of cities worldwide, approximately 70%, heavily rely on greenhouse gas emissions, while the remaining 30% of energy consumption is distributed across various urban areas [13].

As stated in [14], the worldwide demand for carbon dioxide absorption surpassed 32 gigatons in the year 2016. This figure, which represents a 50% rise from the levels seen in 1990, serves as a crucial determinant of the elevated concentration of CO_2 in the Earth's atmosphere. The smart city projects are definitely capable of improving citizens' lives because they serve various types of opportunities for managing the quality of life through culture and social growth in a meaningful, efficient, safe, and stimulated environment [15].

1.5   TECHNOLOGICAL ENHANCEMENT

Today, achieving the technical goal is possible because of the advancements of cutting-edge technology, which change the functions of the city and services through the smart deployment of strategies. The movement of this system encompasses intelligent materials, efficient transportation solutions, interactive communication technology, product recycling processes, and the Internet of Things (IoT). This integration facilitates the delivery of optimal citizen services directly to their residences and all operating sites.

According to research conducted by Navigant, the value of smart city development is projected to reach 25.3 billion euros by the year 2023. This growth trend has been consistently observed in previous studies on smart city development projects. It is anticipated that approximately 85% of urban projects will be classified as smart city projects, while the remaining projects will focus on rural development. This progress aims to ensure that a wide range of services are readily available to all citizens, taking into account their specific requirements and dependencies on various factors. These projects will prioritize the digitalization of everyday life needs and aim to improve the efficiency of cities. During the process of segmentation, smart city initiatives will take into account several advantages and factors that contribute to their development, such as:

- The innovation of smart grids, which can exchange information for the best of services on a demand basis; here the project primarily focuses on the management of energy-based resources, where the energy requirements would be manageable [16].
- Focussing on the safety and control mechanism, where it will consider all the smart-led solutions for managing the traffic and maintaining the pollution criteria for a specific area of the communication system.
- The proposed solution revolves around a smart mobility system that incorporates innovative features in vehicles. These vehicles are equipped with smart technology that enables the utilization of natural energy sources, specifically battery-based services. This trend is gaining momentum in the bike and vehicle industry, as there is a growing emphasis on enhancing battery life and optimizing its management. By reducing reliance on traditional fuels, the aim is to improve the cost efficiency of vehicles [16].
- Smart vehicle technology is also dedicated to the future prediction of various events that can be caused anytime. Natural disasters will be predicted in advance so that the government can take decisions for the management and life of each and every citizen of the country.
- Through a smart city project, we can gather all the data together and perform an analytical study based on the different datasets and patterns to find major solutions and so to take the decisions that would definitely enhance the productivity of the entire business cycle, which would be better for the nation [17].

In relation to the recent statement issued by the former CEO of Google, who unfortunately passed away due to an accident involving smart project technology, it is possible to establish a connection and effectively manage multiple datasets. This enables improved analysis and filtering techniques, facilitating the identification of various patterns that can greatly assist in the formulation of strategies, decision-making processes, and other pertinent aspects related to both business operations and the overall well-being of a nation's citizens. These terminologies can be categorized as "big data," encompassing a range of functions aimed at filtering diverse data elements and obtaining optimal outcomes in the form of results, which subsequently inform decision-making processes.

1.6 SMART CITY FRAMEWORK

The smart cities framework brings a systematic and efficient environment to its users for providing a quality-of-life solution with analysis of various real-time activities in a more accurate and specific way. Smart devices enable us to gather real-time information through various connected assets of the smart city framework to improve decision-making capacity. In addition, now with smart cities allowing its citizens to connect and communicate through their assets, using mobile devices and other infrastructure-based sensors, by integrating various devices and smart city tools together will reduce the overall communication cost and improve quality of life through sustainable technology [18]:

- **Smart finance and economy**: The smart economy depends on the various business decision and working frameworks along with major marketing policies and interaction integrations that can be managed smartly through the smart mobility concept to enhance productivity and its greener impact over the entire business growth.

In order to optimize the utilization of available energy resources and enhance mobility, various city projects have been initiated. These projects are designed based on an ICT framework that incorporates multiple factors such as digital technology, sustainable energy, green computing initiatives, and efficient business empowerment plans. The underlying concept of these projects revolves around the utilization of renewable energies to maximize energy regeneration and improve overall energy efficiency.

- **Smart administration (governance)**: Smart governance offers smart administrative functioning through smart and intelligent decision-making policy to enhance productive output over various aspects in a transparent manner so that everyone can see the various political strategies and forecasting approaches for the betterment of business and the lifestyles of citizens [19].

In particular, the development of a smart green project will also enhance multiple other possibilities. What is the participation of citizens in politics as well as in the day-to-day communication practices, since the smart city project also provides an innovation in the form of tools to suggest a digital democracy, openly available governments, the maximum empowerment of citizenship, and likewise multiple digital data democracy practices. This is also happening in rural as well as in urban areas. Digital information technology tools are facilitating the inclusion of multiple options in the responsibilities of society and in the context of digital democracy. Presently, citizens can readily engage in political processes and governmental regulations through various connections provided by social media. This phenomenon is remarkable, as it has the potential to significantly increase civic participation in a transparent manner, if implemented effectively. Digital democracy also plays an important role as a mediator between government agencies and the public domain itself, where the open governance environment also enhances the accessibility feature of civilizations for various categories of data and processes. One of the major objectives behind a smart city project is to make simpler and facilitate a user-friendly communication environment to all citizens and government administrations so as to lead most things in a more administrative and transparent manner. Similarly we can find nowadays that open governance policy will also enhance productivity in urban areas by reaching out to consider maximum requirements and address it through the legitimate services offered by smart city projects [20].

The initiative for citizen empowerment through the smart city project will also be responsible for contributing actively to the process of decision-makng and the various changes that need to be implanted in the city itself. By way of creating a wonderful opportunity for the participation of citizens is a responsible contribution to the city and its services. In the context of a smart city project, individuals within the public domain have the ability to readily communicate their ideas, therefore maximizing the potential for transforming their innovative concepts into tangible realities. Furthermore, this facilitates enhanced accessibility to government services as shown in Figure 1.2.

- **Availability**: The smart city project also plays an important role by taking advantage of highly available and sustainable smart technology and devices throughout the process cycle, where smart communication and correlation information technology infrastructure make communication easy to handle using smart mobility protocols and devices. In addition, it will also enhance the accessibility of services across multiple locations in an efficient way.
- **Smart paradigm**: This results in the improvement in the specifications of various direct and indirect parameters of environmental conditions such as air pollution, protection, and the management of all components of the system.

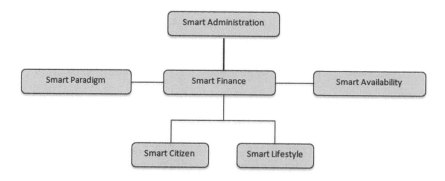

FIGURE 1.2   Smart city framework modelling.

- **Smart lifestyle**: This is the combination of various essential day-to-day life needs. Smarter technological development will ensure the availability of the best of services through various quality parameters in order to find a better solution-demand basis which considers various situations like health condition, personnel safety, security, education advancement, and smart living concepts, which help one every time and in multiple ways to obtain major benefits as per the requirements.

1.7   THE CLASSIFICATION OF RECONCILED ENERGY

Using multiple technologies the process of transformation of cities into smart cities will concern multiple areas of intervention that will address the solution to utilizing energy resources and improving the efficiency of existing infrastructure, such as buildings, with the introduction of renewable energy sources, overall large-scale development, along with various mobility plans. The aforementioned domains represent the most relevant factors to be taken into account in order to successfully address the environmental requirements associated with the reduction of energy consumption in the context of waste management.

The combination of environmental needs, financial goals, and societal needs will combine multiple objective searches for energy consumption management and the elimination of cost-effective advanced technology. Financially, it will lower the costs for all local residents. According to the International Renewable Energy Agency, there is a significant opportunity in the field of energy utilization. By 2023, 9.8 million people may be able to find work in the renewable energy sector [21].

Since the renewable energy programme directly supports various social and economic goals, the increment in the employment level can be recognized as a central component of the global energy management system. It was discovered, according to [22], that the solar energy management systems industry employed 3.1 million people in 2016, accounting for up to 12% of the workforce. The majority of this employment was in China, India, and the United States. When we talk about jobs in the USA, the solar industry increased 19 times more and faster than the overall economy of the globe, which increased jobs by around 25% extra from the previous year in the same way window installation technology also contributed a 7% increment in the wind employment sector, which also raised jobs by up to 1.2 billion. Similarly in various other countries like Brazil, India, and China there has been an increased percentage of jobs in the sector of bioenergy, where the use of biofuel has added 2.2 billion jobs. In conclusion, we may estimate that by the year 2030, there will be about 24 million jobs in the renewable energy sector worldwide. For biogas, that number is 0.8 million, while for 2030 it is 1.4 million.

1.8   GREEN INFRASTRUCTURE AND BUILDINGS

As numerous reports have indicated, building construction will also play a significant role in the advancement of economic growth in every country, as the Internet of Things (IoT) and the concept of a "smart city" work together to manage multiple requirements simultaneously. This will also address infrastructure and building requirements more prominently than traditional, non-smarter buildings, as we can currently manage multiple things using various sensors and utilize energy resources. Today, about 40% of the energy used for consumption is by homes. Public and private offices, markets, and shops also contribute equally to the use of energy. These can be broadly categorized as: 43% goes towards homes, 44% goes towards industry, and the remaining 13% goes towards infrastructure-based buildings. In this way, we can say that residential homes use 67% of the total energy consumption, and 15% goes towards lighting and various other electronic devices to produce some better solution [22].

From multiple research studies, it has been clear that smarter buildings have some potential for the improvement of energy utilization and efficiency by the end of this year which will definitely improve cost efficiency and management throughout the multiple business process and lifetime [23]. In addition to this wind power shield energy production will also enhance the capacity of buildings to utilize their roof and install solar shields in order to generate a huge amount of energy that can be utilized for multiple day-to-day activities so that in total 40% of the electricity demand of European countries could be satisfied by having these solar shields until the end of 2025. Since in Europe especially most people like to stay in their home it is necessary to design a building in a proper way so that adequate construction can provide a better energy-efficient solution as per needs and requirements. So in this way we can manage and maintain the available natural energy using some advanced smart city project technique in construction so that the building can have in-house capability and the potential to utilize and reuse the available energy resources in a better and efficient manner [24].

In buildings, like offices and schools, we find that various types of equipment are present that will engage them and consume a significant amount of energy each day. By using a smarter method, we can determine when the buildings with a few sense-based connections will utilise energy, such as all the fans only being able to run when someone is present; otherwise, no lights and fans will be on. In this way, we can enable a high amount of available energy to be safe and reduce wastage with an efficient energy utilization policy through a smart green city project.

For managing energy consumption effectively we can say that we can manage all installed equipment for every house by introducing a smarter plan in order to obtain, distribute, utilize, and effectively transform the energy from one form to another. If we successfully promote this programme then definitely common electrical equipment will be used in a very effective and efficient manner and a higher number of offices will also take part in the process of managing the available energy resources. At the same time we can take advantage of such a renewable energy solution for future betterment.

1.9 SMART SOLAR GRID

In the 21st century, we can say that the installation of solar devices along with solar panels on the roof will definitely promote energy management programmes in a more effective way with the objective of making the future greener. Solar panels are a simple way for homeowners to manage their energy supplies. The government is supporting this initiative by providing various facilities, such as installation and additional cost benefits, to the public, making it possible for everyone to install this smart, green-based solution.

All such initiatives will support energy consumption for urban and rural areas by deploying smart city concepts so that the grid-based solar technology will also smartly increase the efficiency for public lighting purposes which consecutively replace the outdated and highly consuming electronic lamps with the solution of green energy distribution that will take care of all areas of functioning that has been replaced successfully by the green energy solution schemes.

It has come to our attention that the city's power management calls for a smart solution that can operate in a dynamic manner, handle requests in real time, and manage both types of energy distribution on the basis of four devices. Additionally, for the best systems, the smart solution must be able to maintain and balance the supply and demand ratio continuously as a point of reference. The Italian Energy Authority first proposed a concept for distributing smart energy systems with limited functionality, which we called smart electricity counters. These systems were installed in various categories between 2001 and 2016, but a new kind of extended feature smart energy distribution system is required, one that can offer smart counters that operate more intelligently.

In comparison to the previous one, it should be much more user-friendly. Additionally, each user should have a few options to choose from, allowing them to manage energy with smart grid-based solutions in accordance with their needs and capacity. Finally, they can generate a report detailing their energy usage for a given period of time.

Therefore we need a smart grid-based solution that can work in multiple directions on multiple transactive grids. So that we can manage the active and passive transactive grid at the same time, it should be available for a range of communication that offers multiple other devices to take part in the process within a house, outside of the house, or any specific zone of communication. Additionally, you must have the opportunity to purchase whatever particular equipment you need in order to queue up a certain quantity of energy at a specified time.

A more secure system can be used to close the process through the wearable microbit, which is great for supporting communication processes. The process of verifying the authenticity of individual contracts using the blockchain-based technique to use the available energy resources more effectively and efficiently is another way that this approach truly sets it apart from other traditional energy supplies. Traditional energy supplies will offer an energy supply process through private buyers from the governments or any private limited company that may belong to the same domain or some other network hub's set process.

Smart intelligent mobility is one of the major features of smart city projects through which multiple cities can offer the best services regarding the transformation process that would be supported by large structural investment as well as low-cost initiatives that definitely support the social involvement process and public awareness. The intelligent smart mobility system will consider various city concerns regarding their services:

- The advancement and efficiency of the entire system in order to improve public transport and integrate modern technology with electric vehicles will definitely improve the green energy utilization process and the importance of smart city projects, as now one can think about electric as well as hydrogen motors to run driverless vehicles.

- The promotion of electric and hybrid vehicles will motivate citizens to use smart energy-efficient solutions for transport with support.
- It will also motivate and enhance the sharing and polling concepts of bike and vehicle policies to overcome the previous traditional listed issues.

1.10 TECHNOLOGICAL DEVELOPMENT IN THE SMART CITY

In order to make life easier, the smart city concept uses multiple tools, the user interface, and the segmentation of networks. We can also use the IoT so that all people can connect through the process in a real-time environment to deliver the data and other interchanges of information for any specific functioning. Therefore, the IoT also plays a crucial role, which is smart city sensitivity using multiple devices which are interconnected to each other for exchanging the data which help with smart city deployment in a broader way to process information.

The IoT encompasses a wide range of items that would be associated with a smart city. It takes into account everything from cars to home appliances. Additionally, own-street sensor-based operations can frequently be conducted through IoT technology. In this way, IoT technology gathers data from all the sensors and processes it to optimize the outcome and guarantee multiple aspects of the city's sufficient intelligence. With such an idea, the daily lives of numerous citizens have been made easier and better, so everyone wants to be involved in the development of smart cities in order to contribute to multiple aspects of the city.

Nowadays much IoT technology deploys the concept of cutting-edge computation that will ensure the relevant and important data are exchanged across the communication network. Apart from this, the network system will also enhance the smart city development in IoT operations by deploying secure service technology through multiple protocols in order to eliminate hacking activities and managing data loss issues so that when the smart city deals with the data its authentication and authorization process will be taken care of by the network itself to ensure protection, security, and privacy. Now if everyone takes part in the service criteria of smart cities, where data are very crucial. It should not be used in some unauthorized way so that we need to work on monitoring and control policy. In order to ensure security, they must also be deployed carefully so that the data in the cloud network can be maintained for smart cities.

With the help of IoT technology, the smart city can pursue the following services as a basic requirement:

- **Involvement of application interfaces**: These will deploy multiple application interfaces so that different users can use their individual interface in order to choose smart city services in a more prominent and user-friendly way.
- **Artificial intelligence**: In order to make the machine more interactive, date machines came into existence simultaneously and carried out their tasks based on human actions. Additionally, the concept of machine learning will undoubtedly improve citizens' lives by utilizing the smart city project. By acting as an AI-based solution, this solves a variety of problems

related to medical science, business operations, finance, and the fulfilment of daily life. In essence, this involves taking into account the various facets of humankind in order to engage with the machine.

- **Cloud services and the IoT**: The combination of the IoT with cloud services definitely improves the storage part of the data as well as the security monitoring and controlling of the entire database in a real-time scenario by having these cloud computing services in a smart city project easily collecting the data from multiple devices and managing them over the cloud network, which is especially dedicated for managing and monitoring services while the communication is taking place.
- **Communication through dashboards**: The dashboard is a part of the communication and interaction with environmental elements that can be combined on a single platform in order to operate multiple things in a user-friendly way since the dashboard considers multi-operations and functions in the form of options which will encourage many citizens to access their data in functions according to requirements.
- **Integrating machine learning techniques and tools**: How long will the IT-based solution framework allow the machine learning concept to play a critical role? Embedded programming is necessary to make the machine more interactive. The machine learning concept will bring about an embedded programming interface, enabling the machine to make decisions and assist with daily needs and business needs. These days, many offers include wearable devices, so having the machine learning concept will enable all of the devices to take you through tasks like measurements, pulse count, oxygen level, and heartbeat ratio while also providing multiple medical equipment types that will gather patient data and send it to doctors. As a result, all such decision-making processes will be addressed.
- **Machine-to-machine interactions**: Now we are working in an environment where machines can communicate with humans. They can also communicate with other network elements using the concept of human–computer interaction, therefore machine-to-machine interaction will also improve the lifestyle of citizens because multiple machines can take care of multiple requirements simultaneously. Now you can work on multiple tasks with efficiency and produce reliable output that would be fruitful.
- **Networking**: The networking concept is also very important since all of the exchanging scenario is taken care of by the network itself. Therefore, the network supports a point-to-point and multipoint connectivity scenario with a different type of smart city functioning. By having a perfect networking topology and design we can deploy this smart city framework in a more efficient way. With minimum requirements and a reduced configuration audit we can deploy economically feasible solutions and best practices.

1.11 THE WORKING PROCESS OF SMART CITIES

Smart cities basically integrate various tools and technology such as the IoT to improve quality of life parameters as well as enable economic growth around the world. This uses the following four steps as a life cycle procedure for maintaining smart cities initiatives.

1. **Collection and correlation**: Various smart Internet-enabled sensors are connected together to trace real-time data and activities to correlate various sensor-enabled devices for multiple actions in a real-time environment.
2. **Analysis**: Analysing various real-time parameters for finding detailed insights for taking future business and other decisions helps smart city technology with its operations and services.
3. **Communication**: The data which have been collected and analysed through previous phases are communicated to the decision makers for further decisions and expansion of operations and services.
4. **Action**: Take action to manage internet-enabled assets, operations, and services so that a smart city can cooperate with citizens who reside in such an environment.

The smart cities framework brings a systematic and efficient environment to its users for providing a quality-of-life solution with analysis of various real-time activities in a more accurate and specific way. The smart devices enable the gathering of real-time information through various connected assets of the smart city framework to improve decision-making capacity. In addition, smart cities now allow their citizens to connect and communicate through their assets using mobile devices and other infrastructure-based sensors. By integrating various devices and smart city tools together will reduce the overall communication cost and improve quality of life through sustainable technology (Figure 1.3).

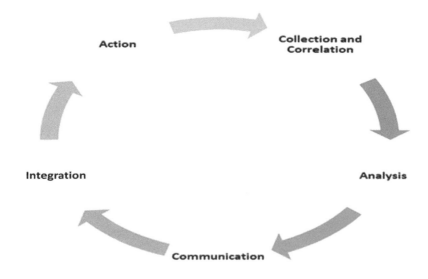

FIGURE 1.3 Working procedure followed by a smart city.

1.12 SMART CITY CHALLENGES

Smart cities are one of the most successful and growing technologies used and adopted by one and all today. Behind their big success there is a big chain of challenges that need to be overcome. Various problem-solving approaches are recommended, such as governments taking initiatives to allow for a major participation of national citizens. Along with this there is a need for positive participation of public and privates sectors to align and deploy the technology with residents so that they can contribute to smart city projects.

The smart cities project should be scaled up with transparency and availability features using a number of open and mobile-based data portals that can be accessed in a more user-friendly manner so that everyone can perform their personal tasks and operations as per their individual requirements. Since smart city technology also offers various financial services, it is required that the most secure and safest system for collecting and managing data safely is installed and that data are protected from outside unauthorized users and hackers. The management of privacy and security will be the biggest challenge to smart city service developers and providers. Another big challenge is to manage millions of smart city users that have all been connected through IoT devices and application interfaces. Such an environment demands the maintenance of all ongoing activities while the demand for service is increasing. Smart cities also need to address the gap in managing social factors that mitigate the social needs of citizens and that offer a sense of understanding in between multiple social communities.

To manage the safety and security measures all around the clock we need to integrate all the security solutions together so that we can eliminate breaches and gaps by deploying protocols and guidelines designed and developed by different network supervisors, administrators, manufacturers, database administrators, and cloud security solution managers to enhance the security of services to its users, which will consequently increase the trust of its users to engage with smart city services for all their essential requirements. The following measures of security management need to be maintained to overcome security and privacy challenges:

- **Availability**: This facility consists of the mechanism in which data should be available based on demand with secure data access services to perform real-time operations through various online digital tools and techniques provided by smart city infrastructure-based frameworks.
- **Integrity**: This facility considers the benefits of finding the required data and services with accuracy features to gain the benefit of high-level security that prevents the data being manipulated by outside attackers or unauthorized users.
- **Confidentiality**: The concept of confidentiality concerns the mechanism of keeping sensitive data safe from unauthorized users or attackers so that the data cannot be misused at any point of operation by means of using firewalls and other security measures.

- **Accountability**: Accountability is achieved through monitoring and tracing various ongoing interactions and activities so that the system manages the transaction and interaction of historical records for accessing and manipulating data that someone can call upon at any time when required. The file logs are used to maintain access records of data to resolve issues relating to unwanted accessing practices.

1.13   CONCLUSION

This chapter has described a systematic study of green computing processes using smart city features, applications, and frameworks. To find various types of services for people, smart city projects target the basic needs of citizens with the integration of artificial intelligence, machine learning, and IoT technology. Smart city initiatives address solutions for various types of challenges and difficulties.

The chapter should be helpful for beginners as well as experts who seek to expand their knowledge and for future research and development.

REFERENCES

[1] T. Yigitcanlar, M. Kamruzzaman, M. Foth, J. Sabatini-Marques, E. da Costa, and G. Ioppolo, "Can cities become smart without being sustainable? A systematic review of the literature," *Sustain. Cities Soc.*, vol. 45, pp. 348–365, 2019.

[2] V. Ravi, S. K. Pramod, J. Neelesh, and A. Chetan, "Enhancing MANET communication services through blockchain technology," In *AIP Conference Proceedings* 2597, 030003, 2022, doi: 10.1063/5.0115134

[3] *The 3rd Smart City Master Plan 2019 2023*, Korean Ministry of Land, Seoul, South Korea, Jul. 2019.

[4] J. Desdemoustier, N. Crutzen, and R. Giffinger, "Municipalities' understanding of the smart city concept: An exploratory analysis in belgium," *Technol. Forecasting Soc. Change*, vol. 142, pp. 129–141, 2019.

[5] G. F. Camboim, P. A. Zawislak, and N. A. Pufal, "Driving elements to make cities smarter: Evidences from European projects," *Technol. Forecasting Soc. Change*, vol. 142, pp. 154–167, 2019.

[6] D. Sikora-Fernandez, "Smarter cities in post-socialist country: Example of poland," *Cities*, vol. 78, pp. 52–59, 2018.

[7] F. T. Neves, M. de Castro Neto, and M. Aparicio, "The impacts of open data initiatives on smart cities: A framework for evaluation and monitoring," *Cities*, vol. 106, 2020, Art. no. 102860.

[8] T. Yigitcanlar, N. Kankanamge, L. Butler, K. Vella, and K. Desouza, "Smart cities down under: Performance of Australian local government areas," In *Queensland University of Technology*, Brisbane, QLD, Australia, Tech. Rep., Feb. 2020.

[9] T. Yigitcanlar, M. Kamruzzaman, L. Buys, and S. Perveen, "Smart cities of the sunshine state: Status of Queensland's local government areas," In *Queensland University of Technology*, Brisbane, QLD, Australia, Tech. Rep., Mar. 2020.

[10] Fast Company. *Smart Cities*. Accessed: Mar. 15, 2019. [Online]. Available: https://www.fastcompany.com/section/smart-cities

[11] J. Gibson. *CITIE: A Resource for City Leadership*. London, UK: Nesta. Accessed: Aug. 7, 2015. [Online]. Available: https://www.nesta.org.uk/report/citie-a-resource-for-city-leadership/

[12] I. B. School. *Smart Cities Guide*. Barcelona, Spain: Univ. Navarra. Accessed: Mar. 15, 2019. [Online]. Available: https://www.iese.edu/library/guide-smart-cities/

[13] Juniper Research. Accessed: Mar. 15, 2019. [Online]. Available: https://www.juniper research.com/document-library/white-papers/smartcities-on-the-faster-track-to-success

[14] A juntament de Barcelona. *22 Barcelona Plan*. Accessed: Mar. 29, 2020. [Online]. Available: https://www.scribd.com/document/361584416/Dossier-22-Castellano-p-pdf

[15] J. Yang, Y. Kwon, and D. Kim, "Regional smart city development focus: The South Korean national strategic Smart City program," *IEEE Access*, vol. 9, pp. 7193–7210, 2021, doi: 10.1109/ACCESS.2020.3047139

[16] L. A. Tawalbeh, A. Basalamah, R. Mehmood, and H. Tawalbeh, "Greener and smarter phones for future cities: Characterizing the impact of GPS signal strength on power consumption," *IEEE Access*, vol. 4, pp. 858–868, 2016, doi: 10.1109/ACCESS.2016.2532745

[17] Y. Lim, J. Edelenbos, and A. Gianoli, "Identifying the results of smart city development: Findings from systematic literature review," *Cities*, vol. 95, Dec. 2019, Art. no. 102397.

[18] S. Goldsmith (16 September 2021). "As the chorus of dumb city advocates increases, how do we define the truly Smart City?" datasmart.ash.harvard.edu. Retrieved: 27 Aug. 2022.

[19] L. Qi, et al., "Privacy-aware data fusion and prediction for Smart City services in edge computing environment," in *2022 IEEE International Conferences on Internet of Things (iThings) and IEEE Green Computing & Communications (GreenCom) and IEEE Cyber, Physical & Social Computing (CPSCom) and IEEE Smart Data (SmartData) and IEEE Congress on Cybermatics (Cybermatics)*, Espoo, Finland, 2022, pp. 9–16, doi: 10.1109/iThings-GreenCom-CPSCom-SmartData-Cybermatics55523.2022.00043

[20] T. Li, Y. Xiao, and L. Song, "Integrating future smart home operation platform with demand side management via deep reinforcement learning," *IEEE Transactions on Green Communications and Networking*, vol. 5, no. 2, pp. 921–933, June 2021, doi: 10.1109/ TGCN.2021.3073979

[21] N. Kumar, R. Chaudhry, O. Kaiwartya, and N. Kumar, "ChaseMe: A heuristic scheme for electric vehicles mobility management on charging stations in a Smart City SCENARIO," *IEEE Transactions on Intelligent Transportation Systems*, vol. 23, no. 9, pp. 16048–16058, 2022, doi: 10.1109/TITS.2022.3147685

[22] S. Douch, M. R. Abid, K. Zine-Dine, D. Bouzidi, and D. Benhaddou, "Edge computing technology enablers: A systematic lecture study," *IEEE Access*, vol. 10, pp. 69264–69302, 2022, doi: 10.1109/ACCESS.2022.3183634

[23] P. Agarwal, R. Kumar, and P. Agarwal, "IoT based framework for smart campus: COVID-19 readiness," in *2020 Fourth World Conference on Smart Trends in Systems, Security and Sustainability (WorldS4)*, London, UK, 2020, pp. 539–542, doi: 10.1109/ WorldS450073.2020.9210382

[24] Z. Tan, H. Qu, J. Zhao, S. Zhou, and W. Wang, "UAV-aided edge/fog computing in smart IoT community for social augmented reality," in *IEEE Internet of Things Journal*, vol. 7, no. 6, pp. 4872–4884, 2020, doi: 10.1109/JIOT.2020.2971325

2 A Review of Green IT, Computing-Enabled, Sustainability-Based Cities

Nurturing Living Standards

G. Arun Sampaul Thomas
Institute of Engineering and Technology, Hyderabad, India

S. Muthukaruppasamy
Velammal Institute of Technology, Panchetti, India

S. Sathish Kumar
Institute of Engineering and Technology, Hyderabad, India

Saravanan Krishnan
Anna University, Chennai, India

2.1 INTRODUCTION

A sustainable smart city is an inventive place that makes use of information and communication technology as well as other modern tools to raise standards of living, increase the effectiveness of urban service-based actions, and boost effectiveness, while also addressing the cultural, social, environmental, and economic needs of both the present and the future. Over 300 worldwide experts were involved in a multi-stakeholder process used by UNECE (United Nations Economic Commission for Europe) and ITU (International Telecommunication Union) to develop the definition of a smart city [1, 2].

A "smart" city is one that makes efforts to improve its sustainability, effectiveness, inclusiveness, and pleasantness. The main ambition of a smart urban environment is to raise the standard of living for its residents. The tight relationship between sustainability and quality of life, resource efficiency, governance, and sustainability are typically the secondary goals of smart cities. As computing grows more prevalent, "green computing" has emerged as the subject that worries businesses and governments across the globe the most over the past ten years. The term "green computing" signifies the use of defensible business practices in the creation, disposal, use, production,

DOI: 10.1201/9781003388814-2

and of diverse resources used in information technology (IT). Green computing is simply the use of IT in a sustainable way to decrease the impact on the atmosphere. Implementing energy-efficient servers, peripherals, and CPUs is one way to promote green computing. Another is to dispose of e-waste properly.

2.2   THE IMPORTANCE OF GREEN COMPUTING AND ITS APPLICATIONS

Reduced utilization of IT services, combined with an increase in energy consumption and carbon emissions, emphasizes the need for green technology and its methodologies. By 2030, it is predicted that the energy consumption of the IT industry will increase by 50%, with a corresponding 26% increase in emissions. Given these trends, it is crucial to concentrate on a sustainable digital future as countries all over the world begin their digital transitions.

As technology enters every aspect of daily life, study into green computing must consider the distinctive needs of every trade. Cloud data centers using cloud computing makes it possible to store enormous amounts of data by virtualizing physical resources. Working remotely is more efficient as a result, but e-pollution has increased due to a year's worth of work from home [3, 4].

The power consumption of cloud computing is significantly reduced because of the absence of servers and cooling systems. Green cloud computing aims to make these numbers even lower. The field, which is restricted by network costs, is rising because of the usage of compression techniques over long distances in the increased use of virtual machine migration. Although this seems positive, there are still obstacles to overcome.

The development of smartphones will eventually render desktop PCs obsolete. Due to the proliferation of mobile applications, the cost of computation on phones has increased, necessitating a sustainable approach to their design and production. Each component included in the CPU and RAM, which combine stationary and active power, is different. Software technology such as WiFi and GPS require tail electricity. The first is dependent on the device and receives its power consumption rates from its insulating capabilities, whilst the latter has an impact on the power consumption of the device, for example, by rapidly depleting the battery [5, 6].

To save energy and extend the life of mobile phones, analysis reports evaluate a device's power depletion. They highlight hardware plus software solutions for green phone computing, utilizing state of charge and code analysis techniques and accounting for variables such as battery age and charging and discharging rates. Other initiatives include improving code generators, resolving energy bugs, and educating developers.

However, this business demands high accuracy, and current technologies typically produce inaccurate assessments. A network of sentient things called the Internet of Things (IoT) eliminates the need for human interaction by exchanging the knowledge they receive from their surroundings. IoT facilitates this process via machine learning, artificial intelligence, and even cloud services. To put it briefly, it requires the collaboration of communication methods and supporting technologies. Examples include wireless sensor networks, radio frequency identification, and green M2M communication. By leveraging infrastructure, such as eco-friendly tags and algorithms, such as regular reporting techniques, these solutions enable a green IoT. It also faces numerous architectural and technical difficulties, though.

These examples make it simple to deduce common practices that are critical to enabling and improving green computing. Let's start with virtualization, which enables the creation of virtual machines from many physical systems, greatly decreasing the need for hardware and energy. Another important factor is algorithm efficiency, because bad code design significantly increases the device's overall energy usage [7, 8].

A slack reduction algorithm, a simulation-based algorithm for latency forecasting of surge labour production, and the reckoning rate based algorithm, which chooses a server from a pool of potential servers to reduce the load balancing and power consumption. Additionally, it uses recyclable glass and aluminium in its products, which also have energy-saving characteristics. In a similar vein, Wipro has often reaffirmed its dedication to environmental sustainability. Google, which started Errand APIs epochs ago and uses salvaged water to track its amenities, carries the flame forward.

The Indian government has taken the lead by launching the "Green IT Initiative" within the Department of Electronics and Information Technology. Under the same roof, the division began working on developing green technologies and smart building solutions with minimal carbon emissions. To do this, it has partnered with the Center for Development of Advanced Computing (CDAC), in Chennai and Bangalore, on the design and development of systems and solutions for smart cities using the IoT, as well as with CDAC, Chennai, Bangalore, Hyderabad, and Trivandrum.

"Green computing," sometimes known as "green IT," is the study and use of ecologically friendly computing. Green computing, to put it simply, is the technique of minimizing the environmental impact of technology by effectively exploiting its resources. In general, green computing entails:

1. **Green use**: is the use of reserves in a technique that minimizes the use of precarious substances.
2. **Green design**: is the process of creating products and services that are environmentally friendly.
3. **Green disposal**: is recycling electronic garbage with little to no environmental impact.
4. **Green engineering**: is the identification and creation of novel goods that lessen or do away with the use or production of precarious materials in engineering [9].

2.3   STUDIES RELATED TO GREEN COMPUTING

The future of computing is frequently described as "green computing." The following are some essential reasons why using green computing can help your company to develop and define its operations:

1. **Environmental statistics**: According to the Organisation for Economic Co-operation and Development, the United States emitted 6,169,592.14 tonnes of CO_2 in 1990. In 2011, that figure had risen to 666,570,000 tonnes. Furthermore, the amount of generated e-waste is increasing. For instance,

in 2010, there were 252,574 tonnes of primary garbage produced, of which 9,447 tonnes were hazardous trash and 19,714 tonnes were from industry manufacturing. This demonstrates the extent of environmental contamination and the impending demise of green growth [10].

2. **Energy consumption**: The US Department of Energy estimates that, in 2010, 1.1–1.5% of the world's energy was consumed by data hubs. A normal office PC may use around 90 watts when on, which comprises 40 watts for a common LCD display. This is fascinating new information [9].

3. **Electronic waste**: Discarded electronic devices, including televisions, laptops, and mobile phones, are included in the category of "e-waste." According to Wikipedia, "approximately 50 million metric tonnes of e-waste are created each year, of which 15–20% is reutilized, with the remaining amount going straight to landfills."

2.3.1 TOP FIVE GREEN COMPUTING IMPLEMENTATION STEPS

The following steps must be taken, whether you're an individual or a corporation, to certify the environmentally friendly depletion of workstations and gadgets.

1. **Designing energy-efficient data centres**: air control, warmth retrieval, and the electrical arrangement of data centres are all part of energy-efficient data centre architecture, which significantly lowers energy consumption. Additionally, on-site power cohort and spare warmth recycling are features of contemporary data centre designs.

2. **Using products with the Energy Star label**: An Energy Star badge rates an appliance's energy efficiency, such as a TV, from one to ten stars. The efficiency increases as the number of stars increases. Consequently, purchasing an appliance is a green choice.

3. **Adhere to these fundamental computer ethics**:
   (a) In your operating system, enable the PC power management feature.
   (b) When not in use, turn off your computer.
   (c) Energy is not saved by screen savers. Turn off your monitor and stay away from screen savers.
   (d) Use LED monitors rather than LCD monitors.
   (e) Whenever you leave your computer on, use sleep mode.
   (f) Always use PC power supplies that are 80-plus approved.

4. **E-waste recycling**: Recycling or reusing e-waste, such as outdated computers and monitors, is referred to as "e-waste recycling." Always donate to charities and non-profits or recycle it through society or remote services.

5. **Telecommunication**: It is an organization in which workers use the internet, phone, and email to do business while working from home. To implement green computing, "teleconferencing," "telework," and "telepresence" must be frequently used interchangeably. Numerous benefits are provided by these technologies, including a decrease in greenhouse gas emissions when connected to portable devices, improved job contentment, and decreased expenditure on heating, cooling, and lighting in offices.

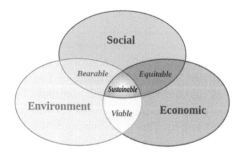

FIGURE 2.1   Environmental sustainability.

2.3.2   GREEN COMPUTING ADVANTAGES

In addition to ensuring social and environmental sustainability, as depicted in Figure 2.1, green computing also ensures the economic sustainability required for human survival.

1. **Better resource utilization**: Green computing is an effective strategy for making efficient use of resources like office space, data centres, computers, heat, light, and electricity.
2. **Cost savings**: Because green computing makes optimal use of resources, overall operating costs are significantly reduced.
3. **Improved corporate and social image**: By adhering to legal and regulatory requirements, green computing helps firms to improve their corporate image. Additionally, it is a great approach to satisfying staff and customer aspirations towards sustainability.

2.3.3   GREEN ADOPTION OBSTACLES

Even though green computing is primarily seen as the way things will be done in the future, there are still several problems that prevent its widespread use. As an illustration, consider the "privacy complications" that result from recycling obsolete computers, the expensive equipment needed to comply with green data region enterprise, and engineering laws. Working on green computing benefits both the individual and the industry. The green IT movement should be accelerated, and individuals and businesses should work together and separately to do their part to save the planet by computing more efficiently [11].

2.4   GREEN, IT-BASED COMPUTING IS AN UPCOMING TREND

The prioritization of reducing the usage of hazardous chemicals and improving energy efficiency in IT systems is indicative of a commitment to environmental sustainability. It seeks to cut back on energy use, thereby containing global warming. Green IT is the practice of ecologically sustainable computing, with the goal of reducing the adverse effects of IT activities on the environment through environmentally friendly computer enterprises, fabrication, depletion, and discarding. The proposal, development, and supervision of computers used by both individuals and corporations together constitute IT. By developing, manufacturing, running, and disposing of various computers and other computer-related products in an environmentally responsible manner, green IT helps to curtail the harmful effects of IT operations on the atmosphere.

The components of green IT are:

1. Reducing the environmental impact of the use of IT.
2. Encouraging the biodegradability or recycling of obsolete products and manufacturing waste.
3. Making use of IT to address environmental problems, green IT is a highly clever way to cut a company's IT costs by over 50%. It is also constantly updated and sustainable.

Green computing is another name for green IT [12]. The green IT system is constructed so as not to harm the environment but to conserve as many natural resources as possible. This is the study and application of IT or computing that is environmentally sustainable. The goal of sustainable computing, a branch of mathematics and computer science, is to use a variety of approaches and strategies to maximize all societal, economic, and environmental resources.

A wide range of policies, practices, and projects that cover the full spectrum of IT use are included in sustainable computing. In this sense, "sustainability" refers to the capacity to generate enough energy to sustain the world's biological systems while utilizing the processing power of computers to handle massive amounts of data.

Going green has several benefits, some of which include:

1. **To lessen e-waste**: The technological world of today is flooded with gadgets like computers and mobile phones. When these gadgets cease to be required, they turn into e-waste and are thrown away using inappropriate removal methods that might distress much life that resides on the planet. Therefore, decreasing e-waste is essential.
2. **To lower operational and spending costs**: When adopting ineffective procedures, there are many additional costs and expenditures that rise dramatically. Going green will help to reduce these extra costs to some extent by using suitable, effective solutions.
3. **To make better use of resources and energy**: Each organization or corporation strives to maximize its earnings in every way by making the best use of the resources that are currently available.

Several strategies for green computing are:

1. **Product durability**: The main contribution this makes to green IT is to extend the equipment's lifespan.

2. **Recyclable materials**: To keep dangerous substances like lead, mercury, and hexavalent chromium out of landfills and to replace equipment that would otherwise need to be created again, all obsolete materials and parts should be properly recycled. This will save energy and reduce carbon emissions.

3. **Power administration**: You can change your power settings to save energy by using a computer system's Power Management Configuration, which includes the ability to turn off, on, or sleep the display. From a single location, you can create, change, and delete power schemes for users. When users are not at their desks, you can reduce the cost of power. Specific power schemes can be created and applied to users via configurations.

4. **Terminal servers**: Users at a terminal connect to a central server, but even though the real computing is done on the server, the terminal is where the end user interacts with the operating system. The phrase "thin client" is sometimes used to refer to terminal servers.

5. **Virtualization**: The practice of having consecutively two or more rational computer techniques on a single set of hardware, which is an abstraction of computer resources.

6. **Working remotely**: Green computing efforts frequently use teleconferencing and telepresence technologies.

The benefits are a rise in employee satisfaction, a reduction in travel-related greenhouse gas discharges, and higher profit margins because of fewer expenses on workspaces, warmth, and so on.

Benefits that a company could experience because of using green IT include: [13]

1. Cost-cutting;
2. Improved branding;
3. Risk administration;
4. Making use of resources;
5. Computers with longer lifespans;
6. Reducing current exposure to chemicals and nerve damage in laptops;
7. Environmental stewardship;
8. Enhanced recovery from disasters;
9. Enhanced corporate reputation;
10. Reduced risk to workers' health [14].

Green computing examples include renewable energy sources. Sustainable power sources do not use petroleum products. They are unreservedly accessible, amicable with the earth, and create almost no contamination. Apple is constructing new corporate headquarters, and intends to use wind turbine technology to control a sizeable portion of the structure, while Google has already created a server farm that is wind powered. Elective energy sources are not only available to large enterprises or the wind. Property owners have long had access to solar energy. It is currently feasible for property owners to install solar-powered lights, water radiators, and wind generators to meet at least some of their energy needs. Other well-known green energy sources include geothermal and hydroelectric power [15].

2.5   CREATING A GREEN CITY IN FOUR STEPS

A sustainable city protects the environment and combats global warming. We must start making changes right away because our planet is facing numerous issues like urbanization, globalization, and climate change. Cities presently produce 70% of the world's trash and deplete up to 80% of its energy. In underdeveloped nations, sprawl has the most damaging consequences for biodiversity. For instance, less than 2% of the Atlantic woods of coastal Brazil are still present within Sao Paulo's metropolitan boundaries, and thousands of species from this area of high endemism have reportedly been pushed to extinction, the majority of which have never even been named by taxonomists [16]. Four steps to create a green city are:

1. **Reduce, reuse, and recycle**: Figure 2.2 shows how food waste that is not recycled or used again is causing landfills to overflow. Methane, a greenhouse gas that is more damaging to the environment than CO_2, is produced as it decomposes. Due to inadequate and improper storage, about 70% of the food supply is wasted. Food waste could be limited and distributed through charities and food banks by a responsible global citizen. To reduce waste, we should practise the three Rs: reduce, reuse, and recycle.

FIGURE 2.2   Green, red, and yellow dustbins to be used in smart and green computing-based cities.

2. **Effective use of public transport**: Private vehicles like cars are mostly responsible for the transportation sector's emissions. Many vehicles pollute the air, emit hazardous gases, and raise the risk of respiratory illnesses. The best strategy to cut emissions is to use public transportation more frequently. To promote walking, bicycling, and using public transportation, cities should build bike lanes, bridges, and walkways.

3. **More neighbourhood parks**: Urban areas are losing their green spaces; 75% of the world's population is anticipated to reside in urban areas by the year 2050. This causes more CO_2 and pollution, which is bad for our health. Green spaces are the cornerstones of a sustainable city. The presence of trees, plants, and green areas is necessary for improved air quality and general wellness. In parks, children can play while adults can relax, take a fresh breath, and reconnect with nature. Additionally, public green areas can enhance general well-being and mental health [14, 17].

4. **Make your building green**: We may make our buildings green in a variety of ways. Drinking-water and wastewater quality must be improved, while water usage must be decreased. It is necessary to stay away from substances and materials that emit harmful or toxic emissions. Make sure everyone using the building follows the mantra "Think reduce, reuse, recycle."

2.6   THE IMPACT OF SMART CITIES ON SUSTAINABLE DEVELOPMENT

Smart cities have come to light as a viable remedy for the environmental problems brought on by increased urbanization. It is believed that they are required for a long-term sustainable future. Smart technology and solutions are used in smart cities to support socioeconomic development and improve the quality of the whole lifecycle. In common, cities present great prospects for economic growth (80% of global GDP is produced there), profession progression (urban residents earn three times more than their counterparts in rural areas), and sustainability. [1].

Since cities have more than 80% of the world's GDP, and are essential to the global economy. The bulk of the energy produced on the planet is also used by our cities, and most of them were not built with sustainability in mind. Additionally, especially in emerging countries, cities are considerably growing in population, and most experts forecast an average increase in world temperatures. However, sustainability encompasses more than just environmental concerns; it also considers other aspects that people value and require, such as employment, a high standard of living, good health, and an excellent education [18].

2.7   GREENER CITIES

There is no single environmental solution that can be used in every city in the world, because the challenges, needs, and opportunities facing every city are different. There is a serious problem that needs to be solved globally concerning where cities acquire their energy from and how they use it, as seen in Figure 2.3.

FIGURE 2.3   Green structure building in Barcelona, Spain.

The most important ways in which cities can reduce their impact on the environment are:

- Increase system efficiency, such as in transportation, and change citizen behaviour to reduce energy and resource use;
- Recycle and use leftover energy and resources;
- Use cleaner energy sources.

While there is not a single technological panacea, cities may significantly lower their carbon emissions by exploitation of a diversity of tactics. A few examples follow.

2.7.1   IoT FACILITATES SUSTAINABLE SMART CITY DEVELOPMENT

Smart cities use IoT gadgets like sensors, cameras, lighting, and metres to gather and analyse data. Then, these data are applied to progress civic conveniences, infrastructure, and amenities. By offering real-time data on city operations, the IoT, together with other digital technologies like cloud computing or open data, helps link various city investors, enhances public participation, and delivers new and improved amenities.

Beyond the apparent differences in economic and technological capabilities, certain nations may lack the research capacity needed to contextualize programmes like smart cities. Lacking their strategy detail, developing countries are compelled to follow frameworks developed and tried in more established nations, which is not necessarily the best course of action given their unique set of circumstances.

2.7.2   RENEWABLE ENERGY

Although many cities could produce cleaner energy within their borders, it is highly improbable that most could ever be considered "self-sufficient" in that area given the amount of energy they use. The good news is that this is not necessary since fresh energy can be attained from other sources, such as offshore windmill-based farms.

Creating electricity from the sun, heating buildings and water with solar thermal panels, using wood from nearby forests to produce low-carbon energy for heating, geothermal energy are all instances of clean ways to produce energy in cities. The expenditure of sustainable energy can vary greatly from one locality to another; therefore, affordability is a major concern.

2.7.3 STREETS AND STRUCTURES

In cities, buildings frequently consume most energy and produce most carbon emissions. There are several intriguing instances of how other cities have handled this issue.

For instance, we are aware that building energy-efficient structures (insulating walls, windows, and roofs, as well as utilizing energy-efficient lighting and heating systems) is frequently more economical and beneficial than producing green energy. A building that is so well insulated that it hardly needs to be heated is the "German Passive House" design.

Small actions, like painting roofs white or planting trees, can lower city temperatures by up to 2 °C, which considerably lowers city energy usage, as New York and Singapore have demonstrated. Warm water for heating is distributed to buildings in Scandinavian and Eastern European nations via insulated pipes buried beneath the streets. The heat can be produced by highly efficient power plants that simultaneously produce electricity and heat or by reclaiming heat from industries like breweries, bakeries, and distilleries. Glasgow is installing intelligent lamps in several regions of the Scotland. LED lights can be used to replace older yellow sodium streetlights, improving safety while lowering emissions.

2.7.4 TRANSPORT METHODS

A city's transportation infrastructure is essential to its smooth operation, yet it can also cause gridlock, poor air quality, and gas emissions. Although it would be ideal to have fewer private vehicles on the road, some towns have developed innovative solutions. One of the biggest car-sharing organizations in the UK is in Edinburgh. Members of this organization pay to join and use a car only when they need it. Copenhagen promotes cycling to work by giving cyclists priority at traffic lights, resulting in a "green wave." High-quality bus and underground train systems have been built in cities like London and Singapore, and low-emission zones have been established where merely e-vehicles are permissible.

2.7.5 SERVICES AND GOODS

It's easy to overlook the fact that everything we buy at the store has a carbon footprint, including the food we eat, the vehicles we drive, the concrete in our buildings, and the buildings themselves. When items, like the fresh food at the store, travel thousands of miles to get to us, their carbon footprint is frequently higher. Locally produced goods may be more sustainable and have a smaller carbon footprint.

2.8 SMART CITIES, SUSTAINABILITY, AND MAKING THE URBAN ATTRACTIVE

Nearly 75% of the world's greenhouse gas (GHG) emissions are attributed to cities; the biggest contributors are commuting, construction, electricity, and waste disposal. At the same time, less than 2% of the Earth's surface is covered by cities. Improving the sustainable development of cities has been a cherished objective consistently [19].

However, recent events have made a significant impact on cities. As a result of lockdowns, less individuals commuted and people grew accustomed to working from home. The pandemic forced the middle classes out of cities and into the countryside in Western nations, and many people abruptly refused to live a life in built-up areas and realized they preferred the fresh air. In European countries, office occupancy rates have stabilized at 45% following COVID. Property owners are contemplating what to do with city centres because businesses are still finding it difficult to draw employees back into offices. The question is that offices should be converted into apartment buildings and rely on fresh greenfield projects.

2.8.1 CHALLENGES TO GREENING CITIES

Making cities greener has historically been difficult due to problems with basic budgeting, which frequently results in most of the funding going to necessary services like policing, education, and sanitation. Political obstacles must constantly be overcome. Additionally, COVID-19 had an impact on city budgets for green projects and delayed actual completion dates because funds were required for other crucial urban services. Most governors of cities across the globe were compelled to sign a petition organized by the C40 Cities Climate Leadership Group that demanded that green stimulus expenditure be given top priority in the wake of the pandemic. At the Smart City Expo World Congress (SCEWC) event in November 2021, which was attended by over 30,000 experts from urban innovation systems in more than 120 nations, the need was underlined. According to SCEWC director Ugo Valenti, "Our 10th anniversary edition comes at a moment when cities need to be together even more, and this summit gathers to restart a much-needed drive for sustainable urban change." To decarbonize public transportation and energy, enhance the estimation of global warming, and promote more sustainable waste omission, the initiative will grant funds to cities including Johannesburg, Jakarta, Kuala Lumpur, Lima, Lagos, Bogota, and Mexico City [20].

2.8.2 PROJECTS UNDERWAY

The modernization of urban infrastructure is a top priority for governments in rising markets around the world. Leading the way are initiatives like Egypt's "New Cairo," also known as the greenfield Administrative Capital for Urban Development, which emphasizes renewable energy and employs IoT technologies to conserve power. The goal of the city is to provide 15 square metres of green space per resident, and it intends to build a central "green river" that is twice the size of Central Park in New York and to consist of both open water and planted vegetation [21].

One hundred smart, sustainable cities are to be built across India as part of the ambitious Smart Cities Mission. Approximately 31% of India's population currently resides in cities, which also account for 63% of the country's GDP. By 2030, these numbers are expected to increase to 40% and 75%, respectively. India is looking at smarter ways to boost efficiency and improve living standards in light of the country's rapid urbanization. Urban space projects, smart water and solar projects, smart road and urban mobility programmes, and other initiatives are all currently being enacted in Indian cities [21].

These kinds of project initiatives show a deep commitment to improving the quality of life and sustainability in metropolitan areas where people live and work. Extreme weather occurrences, an increase in waste, air pollution, and even things like social upheaval must all be managed by cities. Therefore, there is pressure on city authorities all over the world to discover fresh, creative solutions that enhance safety, security, air quality, and general wellbeing index rating to draw people into cities.

2.8.3 TECHNOLOGY MATTERS IN MAKING CITIES MORE ATTRACTIVE

Designing and deploying services that will make cities more sustainable and draw people into them requires the use of IoT, AI, and big data analytics. IoT and big data analytics are employed in forward-thinking cities to monitor and optimize real-time energy use and advance sustainable energy agendas. According to Gartner [6], by 2028, commercial intelligent buildings will have more than four billion IoT-connected devices, thanks to smart infrastructure based on 5G and high-efficiency Wi-Fi 6 or 6E. Power, trash, and water will all be supported by intelligent utilities in these structures. According to Deloitte Insights, these kinds of transformative energy initiatives could help develop circular economies in cities by increasing the generation of energy from renewable sources [21, 22].

2.8.4 EXAMPLES OF CITIES PURSUING SUSTAINABILITY

Digital technology is being used by forward-thinking cities to cut down on airborne pollutants produce by energy use. Amsterdam has installed smart energy metres around the city to drive lower usage, and Beijing has used IoT sensors to track pollution sources and adjust activity as necessary, resulting in a 21% reduction in airborne pollutants. Current initiatives include technology-driven plans to address climate change, energy efficiency, and smart transportation infrastructure. Barcelona now has more than 20,000 active IoT sensors in place that collect data on temperature, air quality, and mobility. Using Sentilo, the city's cloud-based, open-source software platform, the data is analysed and presented. Singapore is the pioneer in Asia for sustainable urban infrastructure projects. By 2030, 80% of its structures are expected to be eco-friendly, according to its green building masterplan, which is a component of Singapore's strategy, which calls for Super Low Energy designation for 80% of new structures and an 80% increase in energy efficiency for green structures. The plan also emphasizes traffic control and waste management procedures that can improve the sustainability and appeal of the city.

2.8.5 New Solutions to Ongoing Challenges

Cities and governments will face a wide range of difficulties in the year 2022 including problems with the economy, society, and the environment. Cities may generate real, quantifiable benefits that increase their sustainability and appeal to residents by implementing strategies and solutions based on digital technologies like data analytics, IoT, cloud computing, and AI. Indicators of quality of life, such as sustainability and health, can be improved by 10–30% with the implementation of technology-driven smart city programmes. The "Green Reboot" report of the Worldwide Finance Corporation predicts that if cities in 21 emerging countries give climate-smart growth top priority in their post-COVID recovery strategies [23] by 2030 they might generate up to $7 trillion in investments and 144 million new employment places. Hopefully, these kinds of projects will inspire people to return to cities and ignite a global process of continual, sustainable improvement in urban settings [17].

Most businesses have not given much attention to how to reduce their environmental impact. According to Stanford University, it is possible to save between 17 and 74% of the energy used to power each individual workstation. Even when we unplug and turn off our electronics, technology continues to produce a significant output [13].

Some ways that businesses and departments can contribute to sustainable IT are:

- **Relocate (and gather) servers**. To save cooling and energy costs, make the most of the space in your data centre. If possible, move your servers to colder regions to cut GHG emissions by 8%.
- **Adhere to pre-eminent practices for data centres**.
  (a) Automate the controls for outdoor cooling, security, and lighting;
  (b) Cool only as much as is necessary;
  (c) Arrange the aisles into hot and cool sections;
  (d) Strive to achieve a power use effectiveness;
  (e) Disconnect and delete zombie servers.
- **Toggle to the cloud**. Economy of scale tends to make cloud energy more effective.
- **Utilize cutting-edge IT**. Because of their large size and excessive heat output, legacy frameworks frequently demand more electricity and more cooling.
- **Enable users to use and buy computers with high ratings for energy efficiency**. Organizations like TCO Certified and Energy Star evaluate and certify equipment for its sustainability and efficiency.
- **Provide budget increases or rebates to teams who advance sustainability**. Unions may allow some teams and departments to work virtually.
- **Allow each team to choose the best options for meeting their needs**. According to the Stanford study, research teams would save more energy when they are given the freedom to choose their options rather than being forced to do something.

2.9 ELITE SUSTAINABLE SMART CITIES AND THEIR CRITERIA

Table 2.1 lists the many worldwide indices that were taken into consideration when ranking sustainable smart cities [1, 24].

Ten elite smart cities [1] were selected based on each city's standing in:

- Environment stability;
- Societal and economic growth;
- Air purity;
- Energy growth with renewable sources;
- Orthodox lifestyle;
- Waste omission;
- Water purity;
- Digitally linked markets;
- Environment, social, and governance functionality;
- Agile ecological system.

1. **Copenhagen, Denmark**: Copenhagen is the smartest city on our list, with the strongest plan for leveraging machinery to make the city greener and increase the standard of living. The city makes use of GPS in buses, wireless records from cell gadgets, and sensors in sewers to empower the city's status in real time. Copenhagen aims to become a carbon neutral metropolis by 2025. Denmark wants to be fossil fuel free by 2050. Sustainable shipping alone saved more than a third of all fossil fuels used in shipping.

2. **Oslo, Norway**: More than 70% of automobiles in Oslo are electric powered, with intentions to eliminate access to non-electric automobiles by 2025. According to the metropolis of Oslo, initiatives towards a sustainable smart city entail:
   - Zero-emission building sites;
   - Upgrading domiciles with sensors for construction management systems;
   - Establishing waste omission and an energy conservation system.

TABLE 2.1
Sustainable Smart Cities Worldwide Index and Their Metrics

Index	Metric
IMD Smart City Index	The technological solutions of cities in five core areas are mobility, activities, opportunities, health and safety, and management [14, 25]
Global Condition of Air	Air purity
Energy	Migration to renewable energy sources
Mercer Quality of Living City Ranking	Living standards, social climate, the political, the economic climate, the socio-cultural climate, socio-medical factors, schools and education, public transportation, leisure activities
Waste Management	Waste estimation in the regions
Water Condition of Sustainable Towns	Water purity
RobecoSAM Nation Stability Ranking	Environmental, social, and governance criteria serve to motivate the technological growth ethically

3. **Zurich, Switzerland**: The mayor of Zurich proposed "Smart City Zurich" in 2018 to meet forthcoming demands, promote renovation, and make Zurich a smart city. Focus topics include integrated public transportation, digital infrastructure upgrading, and effort to invite businesses and residents to participate in the creation of Smart City Zurich.
4. **London, United Kingdom**: London is a global powerhouse for clean-tech, government technology, digital health, and education technology. It is one of the world's most connected cities, with one of the foremost progressed open transportation systems, and the objective is to create the most intelligent city.
5. **Stockholm, Sweden**: Stockholm is a global and worthwhile city. In conjunction with the academic and commercial sectors, the city is continually creating and testing savvy economical arrangements in both present and future domiciles. Stockholm's objective is to achieve carbon neutrality by 2040. Smart technologies are used to make the best use of clean air, efficient energy conservation, and irrigation.
6. **Singapore**: Singapore is on track to be the Asian pioneer in shrewd and sustainable domiciles. It aims to make 80% of its buildings eco-friendly by 2030.
7. **Amsterdam, the Netherlands**: The Amsterdam Smart City programme kicked off the city's environmental measures in 2009. The initiative employs technological advancements to attain sustainable city targets. The metropolis intends to reduce carbon dioxide emissions by 50% by 2025.
8. **Sydney, Australia**: Sydney's framework serves as the foundation for the resourceful "Sydney 2030" initiative. It envisions the city as a thriving and irrepressible one to be converted into an entirely renewable-energy-based metropolis. This will lower the city's emissions by approximately 20 tons per annum, which is equivalent to the electrical consumption of 4,000 households.
9. **New York City, USA**: When it comes to genuine environmental measures, New York is a city to be admired. Regulations such as the 2019 NYC Structures Emission Act, also known as the Climate Call Act, establish carbon discharge limitations on construction sites. With 8,000 enterprises, the city is ready to prime the smart cities mutiny.
10. **Munich, Germany**: Munich is adopting modern smart city technologies. The objective is to use breakthrough data-driven technology to reduce the use of fossil fuels. Munich has increased its initiatives by establishing a smart city innovation lab. Technologists and city officials will collaborate to develop intelligent and sustainable alternatives for a smarter Munich.

2.10   CONCLUSION

The true cost of smart technology is how it is applied to improve people's lives. Smart cities take advantage of technological advancements and data analysis to improve the standard of life for residents. Thus residents in the world's brightest cities are amidst the joyous people on the globe. They commit to live sustainably while putting heart and soul into everyday activities. The list of sustainable smart towns is a by-product of people's passion for both the environment and technology. To tackle the problems of global warming and climate change in an ethical manner, firms must embrace green IT methods that make sense from both a moral and a business stance. Technology has shown promise in addressing today's most serious issues from rural ecology to world conflicts. As a result, it is crucial to use it in harmony with the atmosphere and provide societally oriented welfare. The effort over global warming requires green and sustainable IT-based cities. The goal is to improve everyone's quality of life and contribute to the progress of the future metropolis.

REFERENCES

[1] https://www.disruptive-technologies.com/blog/the-top-20-sustainable-smart-cities-in-the-world

[2] https://www.acecloudhosting.com/whitepapers/green-computing/

[3] Shuja, J., Gani, A., & Ahmad, R. W., Ahmed, A. I. A., Siddiqa, A., Nisar, K., Khan, S., & Zomaya, A. (2017). Greening emerging IT technologies: Techniques and practices. *Journal of Internet Services and Applications*, 8, 9. https://doi.org/10.1186/s13174-017-0060-5

[4] Moghaddam, R. F., Mohamadpour, M., & Kazemi, M. (2020). A review on green computing-based smart cities and sustainability. *Sustainable Cities and Society*, 54, 101995. https://doi.org/10.1016/j.scs.2019.101995

[5] Chowdhury, S. N., Kuhikar, K. M., & Agnihotri, A. (n.d.). Green computing: An overview with reference to India. *International Journal of Electrical, Electronics and Computer Systems (IJEECS)*, 3(2). http://www.irdindia.in/journal_ijeecs/pdf/vol3_iss2/5.pdf

[6] Al-Obaisat, R. F., & Elci, A. (2021). Green computing and smart cities: A review of the state-of-the-art, challenges, and opportunities. *Journal of Cleaner Production*, 312, 127799. https://doi.org/10.1016/j.jclepro.2021.127799

[7] Ohri, A. (2020). *Green Cloud Computing: A Beginner's Guide*. Jigsaw Academy. https://www.jigsawacademy.com/blogs/cloud-computing/green-cloud-computing/

[8] Gholami, R., Gholami, S., & Zhou, Q. (2021). A review on smart city green computing technologies and sustainability. *Environmental Science and Pollution Research*, 28(19), 23602–23615. https://doi.org/10.1007/s11356-021-13860-2

[9] https://deepanshugahlaut.medium.com/the-ultimate-guide-to-green-computing-73e30ba2a485

[10] https://stats.oecd.org/Index.aspx?DataSetCode=air_ghg

[11] https://www.cyberpeace.org/green-computing-what-is-the-significance-in-21st-century/

[12] https://medium.com/@dhawa39/the-emerging-trend-of-green-computing-86d2e4edb838

[13] https://www.bmc.com/blogs/sustainable-it/

[14] Arun Sampaul Thomas, G., & Harold Robinson, Y. (2020). "Real-Time Health System (RTHS) Centered Internet of Things (IoT) in healthcare industry: Benefits, use cases and advancements in 2020," *Springer's Multimedia Technologies in the Internet of Things Environment (Scopus Indexed)*, 29 September 2020, ISSN: 978-981-15-7965-3.

[15] https://medium.com/@rmsrn.85/green-computing-and-its-advantages-disadvantages-examples-b13f42d65cab

[16] https://medium.com/age-of-awareness/5-ways-to-make-a-green-city-d26989c42535

[17] https://www.weforum.org/agenda/2022/04/global-urbanization-material-consumption/

[18] https://www.bbc.co.uk/bitesize/articles/zdqt7nb

[19] https://www.orange-business.com/en/blogs/smart-cities-sustainability-and-making-urban-attractive-again

[20] https://www.smartcityexpo.com/2021-highlights/

[21] Arun Sampaul Thomas, G., & Harold Robinson, Y. (2020). "IoT, Big Data, blockchain and machine learning besides its transmutation with modern technological applications," *Springer Book Chapter-Internet of Things and Big Data Applications - Part of the Intelligent Systems Reference Library Book Series* (ISRL, volume 180), 25 February 2020, pp. 47–63, ISSN: 978-3-030-39118-8.

[22] https://www.axians.com/news/to-remain-attractive-the-cities-of-the-future-will-be-efficient/

[23] https://www.un.org/sustainabledevelopment/wp-content/uploads/2022/03/2021-Report.pdf

[24] Naeem, M. A., & Akhtar, N. (2021). The role of green information technology (IT) in sustainable smart cities: A systematic review. *Journal of Cleaner Production*, 314, 127974. https://doi.org/10.1016/j.jclepro.2021.127974

[25] Singh, R., Singh, R., Singh, J., & Jha, V. (2021). Green computing for sustainable smart cities: A review of challenges, solutions, and future directions. *Sustainable Cities and Society*, 70, 102902. https://doi.org/10.1016/j.scs.2021.102902

3 The Impact of Green Smart Cities on Socio-Economic Growth

Vaishali Chourey and Riya Mehta
Adani University, Ahmedabad, India

Neha Sharma
TCS Pune, Pune, India

Sunil Gautam
Nirma University, Ahmedabad, India

3.1 INTRODUCTION: SMART CITIES

Smart cities are synonymous with an integrated living standard with a controlled organization of the functioning of the city and homogeneously creating sustainable environments for all stakeholders of the city. Smart cities have contributed to smart urbanization with solutions for sustainable living. Smart cities are a hub for residential properties, public amenities, institutions of purpose, industry, business, and enterprises. An interconnection among the entities of the city defines the processes to manage it. Today, in order to make living sustainable in cities, there is a need for a clear vision and support of harmoniously moderated activities. Smart cities comprehend such functions of the city with technology to bring efficiency and translate them into digitally administrative ones [1]. Smart cities architecturally are intelligent and digitally managed. The knowledge base is thereby created and shared amongst the stakeholders of the smart city. The data is dynamically created and channeled to the right placeholder for smart governance. Although both the terms "smart governance" and "smart city" are used interchangeably, yet there stands a procedural distinction, as depicted in Figure 3.1, where smart governance is seen to be a top-down approach. The initiation of governance and using technology for it is the key approach to smart governance. The policies and rules enable schemes and projects to be deployed with the use of technology. A smart city is a bottom-up approach where the technically aided tools, deployed all around the city, generate data to understand the requirements of the city [2]. Data allow the selection of the best policies to allow governance driven by the data. The changing needs of the stakeholders with time flexibly bring about the best solution in governance. Smart governance is envisioned under the umbrella of the smart city in today's conditions. "Smart" in the context of

DOI: 10.1201/9781003388814-3

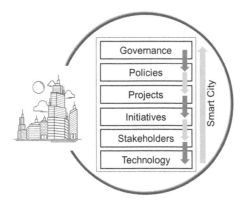

FIGURE 3.1 Smart city functional paradigm.

this chapter has a perspective with a definition in terms of intelligence, digitization, technology, quality of life, and sustainability [3]:

- Intelligence in competencies that need to be explored from the existing human resources of the city.
- Digitization and technology come together for automation and control.
- Quality of life improves with economic development accompanied with innovations and initiatives from private and non-profit organizations.
- Sustainability is in terms of being future ready with optimal use of perishable resources and using renewable resources to manage the functioning of the city. Energy efficiency and smart logistics and transport are perfect examples.

Smart cities are applications of information and communication technology (ICT) that allow services to be digitally managed amongst the entities of the smart city [4]. The services are offered by way of the cohesive contributions of public and private institutions rendering facilities in their specialized domains. For example, a city has a mature system of public services with municipal corporations, but the infrastructure development and transport are managed by private organizations. Thus, an agreement between the central, state, and private organizations bring together smart solutions for sustainable living.

Cities face challenges in catering to the needs of the increasing population and the higher complexity of city management. Under these circumstances, city management is streamlined with the use of technology for better services and a greener ecosystem. The city must be developed with the available resources, preserving the culture and ethnicity of a region. Smart cities dwell on technology for solutions to public issues like traffic, pollution, waste management, skill employment, and population crowding. Smart cities aim at improving livelihood, whereas green smart cities attempt to make living sustainable by preserving the climate.

The remainder of this chapter is organized as follows. Section 3.2 enlists the functionalities of smart cities that are the overall administrative requirements. Section 3.3 is a UN perspective of the smart city with regard to Sustainable Development Goals

(SDGs). This section also describes the distinct pillars of smart city initiatives. Section 3.4 summarizes the socio-economic growth indices. Section 3.5 concludes the chapter.

3.2 DRIVERS FOR GROWTH IN THE SMART CITY

The ever-increasing population and full-paced urbanization have raised questions regarding the quality of life of the inhabitants of cities. Smart cities is a concept that is a solution for sustainable living. The city needs to be securely planned for critical functionalities to effect the solution as a smart city. Schematic planning for the smart city includes certain services covering the amenities and administration of the city. The functioning of services is the key area to focus technology on. The use of ICT and the internet for managing the chores of the city and gathering the necessary data to take decisions is the core of the smart city. A smart city is one that makes use of digital technology to connect its residents, protect them, and improve their lives. Sensors connected to the Internet of Things (IoT), video cameras, social media, and other inputs function as a nervous system, providing the city operator and inhabitants with continuous feedback so that they may make educated decisions.

A futuristic smart city gathers economic development across all segments of society – government or societal. To achieve this growth, a specially planned city infrastructure is drafted and data amongst the components are shared. Such a formulation enables looking into deprived components and making them self-sustainable. To identify the components an understanding with respect to the functioning of the city must be framed. The basic components of a smart city integrate solutions for upgrading and innovation for sustainable economic growth from the sectors of education, clean and green energy, health, ICT, infrastructure, mobility and transportation, safety and surveillance, urban development, water management/conservation, and so on. As shown in Figure 3.2, these domains contribute to the measurable factors of sustainability, like carbon emission reduction. Various organizations, government or private, moderate the conceptual building of the smart city and facilitate projects to realize them.

The manner in which city leaders develop and provide public services is being reshaped as a result of rapid urbanization, shifting demographics, and the rapid growth of emerging technologies. The following is a list of the important themes that

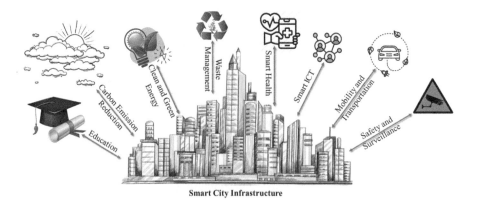

Smart City Infrastructure

FIGURE 3.2   Smart city functional domains.

are driving the need for innovative approaches to the operations of cities and the delivery of their services:

- Skilled manpower and economic development: Cities need to both recruit and keep individuals who are talented and skilled if they are to maintain their competitive edge. Because of their robust innovation ecosystems and extensive industrial expertise, smart cities typically have a significant advantage over other cities. Such settings entice new businesses and young people, which in turn stimulate additional economic expansion, job creation, and innovative thinking. This also encourages additional expansion in several other industries, such as the entertainment and real estate industries. The analysts at IDC Government Insights are of the opinion that ongoing economic growth will be beneficial to smart cities [5].
- City infrastructure and resources: A congested city with an increasing population blocks the growth of a physical infrastructure that is sufficient to meet demands. The most visible impact of this can be seen as traffic jams on the roads. Also, with capacity-constrained schools, hospitals, and other resources, the infrastructure fails to scale with the population. The only solution is to manage resources with technology and cater to future needs substantially.
- Climate change and energy efficiency: Take for instance "climate departure", which suggests that the coldest days now will become the hottest for a city. Studies prophecy that the earth will witness climate departure in 2047 with some cities facing irrevocably hotter climates. Smart cities are an important part of making energy use more efficient and lowering carbon emissions. There are many ways to make cities more efficient, such as with better street lighting, green buildings, more people using public transportation, and less traffic.
- Digital divide, trends, and accessibility: This trend forces city leaders to come up with new ways to provide services while keeping the old ways open, such as call centers and office hours. Cities have to meet the needs of both older or low-income people who may not have much access to technology and younger people who are always online.
- Technology for data explosion: New information regarding city operations can be obtained from surveillance footage captured by video cameras, transponders that collect tolls from highways, and sensors that are fixed to bridges, parking spots, water pipes, streetlights, and garbage cans. These data keep flooding in, and require specialized machinery to process them and make precise decisions. The use of machine learning techniques and IoT opens areas of research in this field.

Innovative technology for developing cities that are egalitarian, safe, healthy, and collaborative raise the quality of life while simultaneously improving business operations and the delivery of public services in order to produce and provide a good impact on society. This can only be achieved with public services that are enhanced with private organizations collaborating for smart solutions. With an array of services, upcoming projects, and model solutions in the smart city initiatives, some deliverables expected in each category are summarized in Table 3.1.

TABLE 3.1

Sectors, Deliverables, and Opportunities in Smart City Projects

Sector	Expected Deliverables	Opportunities	Challenges
Smart Healthcare [6]	Affordable services, developing and delivering vaccines, provisioning advanced medical treatments, e-health and m-health initiatives, innovation in medical diagnosis, skill development and training for community health care manpower.	Software and hardware innovation in diagnosis, treatment, consultation and awareness regime.	Equal access to healthcare services in population at rural/remote locations.
Smart Governance [7]	Citizen and community engagement in government project interfaces, operational coordination synchronization, and inter-departmental data sharing for planning and policy implementation, improved routine tasks and operations, disaster management, structured dissemination of generic information to the public in the native language, and creating public awareness in regional languages for greater public involvement. Data-driven governance in a sustainable and inclusive manner.	Mobile and web-based solution for administration and governance, community and social engagement platform development, native and vernacular language support for public engagement, policy information dissemination.	Lack of awareness and training, IT and network facilities not accessible to all, legacy practices still need to accelerate technology adoption.
Smart Waste Management [8]	Use of technology for waste management, clean technology adoption, carbon level reduction, treating pollution of all kinds, strategizing waste reduction, reusing waste resources like plastic, and evolving economic and sustainable solutions for energy efficiency (like solar and alternate fuels).	IoT-based smart solutions for waste management, sustainable hardware and app-based implementations, public–private collaboration.	Lack of motivation and awareness, native drives required for local positions.
Water Resource Management	Sustainable water projects to conserve water resources, provide safe and clean drinking water, community-owned water projects, disposal of wastewater from industries, conserving rivers and lakes for a clean environment, and preventing diseases that are water-borne.	Smart solutions for the distribution of clean drinking water for all and wastewater management projects.	Most unattended issues in government projects.
Smart Infrastructure [9]	Resilient infrastructure for industries, residential properties, roads, and public amenities like open spaces, libraries, parks, and theaters. Smart parking.	Innovation in building, construction designs, and materials.	Population density and existing colonization poses serious challenges.

(Continued)

TABLE 3.1 (Continued)
Sectors, Deliverables, and Opportunities in Smart City Projects

Sector	Expected Deliverables	Opportunities	Challenges
Safety and Surveillance [10]	Preventing crimes, secure and dependable community development, internet or cloud-based solutions for surveillance and disaster management, traffic management, smart tracking solutions, cyber and network security solutions.	Monitoring and surveillance strategies, city mobility guides, and routing solutions.	Technology not available and accessible to public domains.
Supporting Start-ups for a Cause [11]	Sustainable urban and rural planning, smart asset management, sustainable grids and grooming the city with ease of livelihood. Digital solutions.	Technological innovation and research for financial independence.	Persistent business still is rare even with funded start-ups, scaling, and implementation level failures.
Smart Education [12]	Education in a technology blended mode, accessible to all and skill development projects for enhancing employability.	Technology-enabled pedagogy and free open courseware for skill enhancement in vernacular languages.	Training and facilitation challenges to public schools.

The process of transformation for any city to its "smart" version produces opportunities for all cadres of society to contribute to. Research and development suit prospective universities to develop entrepreneurial strengths in the vast domain knowledge areas. Similarly, organizations collaborate to devise solutions in conjunction with government offices for city administration and governance.

3.3 GROWTH: SUSTAINABLE DEVELOPMENT GOALS IN PERSPECTIVE

The concept has a strong foundation with the SDGs defined by the UN [13]. Some research aligns with the Public Private Partnership (PPP) for accomplishing these sustainability goals. A fruitful example is the Indian adoption of infrastructure projects with PPP in the 1990s (Ministry of Finance 2014) for economic liberalization [14, 15]. The PPP procurement led to a profitable incentivized private sector growth with abundant opportunities developed for skilled human resources. Many initiatives grow with the sustainability issues taking command over traditional governance. A mapping of SDGs onto projects, policies, and schemes for financial support and PPP is visible in all segments of government administration (see Table 3.2). The SDGs correlate with all the targets of Goals 11 and 16. But, taking into account the targets defined under the SDGs, a synchronized mapping can be drawn with the domains of a smart city.

TABLE 3.2

Mapping SDGs to Smart City Domains

Smart City Domain	SDGs and Targets
Smart Healthcare	Goal 3 – Target 3.7, 3.8
Smart Governance	Goal 1 – Target 1.4, 1.5
Smart Waste Management	Goal 12 – Target 12.4, 12.5
Water Resource Management	Goal 6 – Target 6.1, 6.2, 6.5, 6.6
Smart Infrastructure	Goal 7 – Target 7.1, 7.a; Goal 9 – Target 9.1–9.3
Safety and Surveillance	Goal 3 – Target 3.6, 3 (a–d)
Supporting Start-Ups for a Cause	Goal 8 – Target 8.3, 8.6, 8.10
Smart Education	Goal 4 – Targets 4.1, 4.3, 4.4

The SDGs have been pursued progressively at a global level. The projects and policies are modulated to implicitly imbibe the conceptual framework as in the SDGs and their respective targets. This opens venues of work, skill deployment, and entrepreneurship opportunities. The PPP model also enhances strategic development in socio-economic growth. However, a clear view of the economic perspective is discussed in detail in the next section.

3.3.1 Socio-Economic Factors

The socio-economic factors are concerned with the interaction between social factors and economic factors. The smart city emphasizes urban sustainable development with economic and societal growth. Thus, there are several economic and social factors that affect smart cities. The economic factors include growth, employment, healthcare, transportation, infrastructure, and urbanization. The social factors include security, equality, awareness, participation, and city management. The interaction of economic and social factors is shown in Figure 3.3. The socio-economic factors are the quality of life, a smart eco-system, residual waste management, residential growth, occupational profile, educational profile, and collaboration.

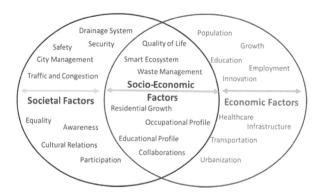

FIGURE 3.3　Socio-economic factors affecting the smart city.

3.3.2 QUALITY OF LIFE

The World Health Organization defines quality of life as "a person's view of their place in life in relation to their objectives, aspirations, standards, and concerns in the context of the culture and value systems in which they live." Thus, public safety, time and convenience, public health, social connectedness and civic participation, jobs, cost of living, and other factors can all be considered aspects of quality of life [16]. Public safety in the context of a smart city includes a decrease in occurrences of assault, burglary, auto theft, and robbery as well as a decrease in fatalities from homicide, fires, and traffic. Residents in smart cities should be allowed freedom of movement and peace of mind in order to ensure public safety [10].

Additionally, statistical analysis that can predict crime, identify gunshots, and offer home protection and intelligent monitoring can be used to create real-time crime mapping. Response times from law enforcement agencies can also be accelerated through smart systems. The everyday commute of residents of smart cities has been made more convenient and time efficient. By considering variables like the density of a smart city, resident travel patterns, infrastructure, and other considerations, smart mobility applications can help to lessen traffic congestion and shorten commute times. Through digital signage or mobile apps that provide real-time information about delays, smart transit systems can also serve to streamline the experience for users. Additionally, the intelligent synchronization of traffic signals has the potential to shorten inhabitants' typical travel times (Figure 3.4).

Numerous applications can aid in the monitoring, prevention, and treatment of chronic illnesses like diabetes or cardiovascular disease for the public health of people living in smart cities [6]. Digital gadgets are used in remote patient monitoring systems to take critical readings, which are subsequently transmitted to doctors for evaluation elsewhere. The system can promptly provide medicine and identify any early intervention needs, reducing patient problems and hospital stays. A reduction in disability-adjusted life years can be achieved through improving child and

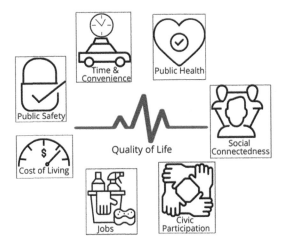

FIGURE 3.4 Quality of life with smart economic perspectives.

maternal health and spreading information about sanitation, vaccines, and adhering to antiretroviral medication regimens.

In order to improve quality of life, social connection must be strengthened. The resident may feel more connected to the local community and the government by using digital channels for communication and in-person involvement [17]. Additionally, the two-way channels of communication can improve the government's responsiveness to the citizens of the smart city. As a result, these systems not only disseminate information but also provide venues for citizens to gather data, voice their concerns, or provide input on planning-related topics.

3.3.3 A Smart Ecosystem

The idea of an "ecosystem," where each component is interconnected to create an appealing whole, is exemplified by smart cities (Figure 3.5). Organizational silos need to make way for a problem-solving ecosystem that is backed by technological solutions that help address a wide range of issues and opportunities. These smart cities are not only capable of providing inhabitants with higher quality services more effectively, but they can also save operating costs and free up resources for further value-added initiatives [18].

In a nutshell, the strength of the ecosystem is the determining factor in whether or not smart cities are able to make use of the opportunities presented by digital innovation [19]. When picking their partners, municipal officials have two considerations that they need to give priority to from a technological point of view. They are required to develop partnerships with technology vendors who are able to deliver a platform that can gather data from all of the key organizations across the city's ecosystem, make sense of that data, and communicate the insights to the users through an intuitive interface. And they need to connect with vendors who can demonstrate that they have success stories throughout the ecosystem. This will allow them to apply their knowledge of industry-specific business models and skills to incorporate analytical insights into the appropriate business processes.

Within the ecosystem of a smart city, there are four distinct categories of value creators. When people think of a smart city, the first thing that typically comes to mind are the services that are offered by municipal and government agencies. Some examples of these services are smart parking, smart water management, and smart

FIGURE 3.5   A smart ecosystem.

lighting. In point of fact, the smart city is home to three additional value providers and users: businesses, communities, and inhabitants. Each of these groups works together to make the city a success. Services that use and create information to produce outcomes for a company's stakeholders are something that companies and other organizations can develop. Enterprises such as Uber and NextDoor, which facilitate the exchange of information, and Google, which assists with the planning of commutes and traffic, are examples of "smart" businesses. Communities are like small smart cities, but they have requirements that are specific to their area. Housing developments/neighborhoods, commercial districts, airports, cargo ports, multi-dwelling units or apartment complexes, university campuses, office parks, airports, and even individual "smart" structures are all examples of possible smart communities. They have requirements for intelligent services that can be adapted precisely for the stakeholders in their organization. Residents of the smart city, as well as individual residents, can also function as smart service providers. A person who lives close to a street intersection that is known to be hazardous can point a camera at the intersection and stream live video of the intersection to traffic planners and police. When specific periods of the year come around, residents install air quality measurement sensors on their homes to keep track of the levels of pollution and pollen in the air, and they share the data they collect with the other people who live in the community. Residents have the option of making these smart services temporary or permanent, as well as choosing whether or not to pay for them.

3.3.4 RESIDUAL WASTE MANAGEMENT

Governments across the world are starting to put more emphasis on waste management. As the global population expands at an exponential rate, urban areas will become increasingly congested, leading to a rise in public litter. An estimated annual worldwide garbage production is 2.3 billion tons and, by 2050, that number is predicted to rise to 3.4 billion tons. Awareness of waste patterns is crucial for optimizing management techniques and population control in the long run. Residual waste management in the future will rely heavily on IoT devices and big data technology (Figure 3.6).

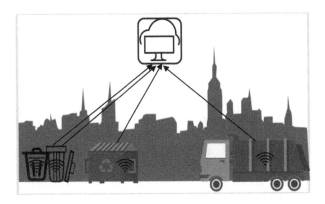

FIGURE 3.6 Residual waste management information gathering.

At this time, the smart bin is the most well-known piece of equipment associated with Smart Waste Management. Using IoT-powered sensors, the smart bin tracks how often garbage cans are emptied and promotes eco-friendly waste management practices [8].

Public space trash and bin overflow continue to be an issue in cities around the world as urban settings and people continue to rise. With its built-in sensors, the smart bin is able to detect when a bin is overflowing and send an alert to nearby municipal or private monitoring systems [16]. This allows for more efficient bin collection routes, which in turn leads to less garbage on the streets. Smart bins have a GPS module and radio frequency identification (RFID) tags that allow garbage trucks to be tracked. RFID tags on the truck and the trash cans read each other and send a signal to the cloud once the truck reaches the designated area. The information will be stored in the cloud and then transmitted to the user's app and the control panel [20]. Waste management firms can save money on gas and labor hours by using data-driven route planning. Numerous major metropolitan areas, like Singapore, Dubai, and Hong Kong, have already implemented this technological advancement. The public can be educated about recycling's significance with the help of smart bin technology. When people are detected by sensors, the screens provide recycling education to encourage proper behavior. Additionally, by keeping track of garbage collection schedules, both individuals and municipalities can learn more about trash management, which in turn can lead to more effective efforts to reduce garbage production and promote sustainable results. The value of having a smart bin in the home will rise as people learn more about the benefits of recycling and reducing food waste. Governments and local councils might use this information to keep tabs on recycling rates and punish non-recyclers with limits or fines, for instance.

There are other apps that may be used to monitor the cleanliness of a community on a daily basis, and users can communicate with municipal officials to share their concerns about local problems. In addition, citizens are encouraged to offer comments to government officials regarding how they believe waste management should be improved. Furthermore, the user can also arrange for the garbage truck to pick up trash at his or her location for a fee. New York City's garbage management system is among the most complex in all of North America. Nearly 15,300 pounds of trash are produced everyday by just the half a million people that go through Times Square. As part of New York City's largest public space recycling project, 30 big belly smart garbage and recycling stations were installed in Times Square in March 2013. These units have features such as trash compaction, continuous monitoring of the fill level, and automatic reminders to call for collection. As a result, the intelligent stations boosted overall waste capacity by approximately 200% while decreasing collection frequency per bin by 50%. After seeing such positive results, the city decided to deploy a total of 197 smart stations.

The Hague, a Dutch city, started putting in underground trash cans in 2009 to accommodate the growing amount of garbage produced there. There are currently 6,100 units in the city, all of which are buried beneath the pavement with only the top of the bin protruding above ground at waist height. Approximately 3,500 of these underground bins have sensors installed, enabling waste management workers to track the progress of each bin's contents remotely and create optimized emptying schedules. The success of these subsurface containers led to their inclusion in a study on New York City's Zero Waste Design guidelines in 2017.

3.3.5 RESIDENTIAL GROWTH

The center of domesticity in a city is found in its residential areas. These must be a top priority of the administration if we are to improve the quality of life in a city [7]. A smart city may level up residential regions by distributing food stores and medical facilities fairly around the city and quickly filling in the gaps where they are missing. Another way a future smart city might improve life for its citizens is by making the region as clean as possible by installing trash cans at regular intervals and educating the locals on the need for civic responsibility. Civic sense can be instilled in the community through civic education, civic responsibility awareness, and training on how to be a model citizen. Regular offenders face strict penalties/fines. It is crucial to develop a successful strategy for altering people's deeply ingrained civic worldview, whether through rigorous education or a system of rigid, unyielding governance. On-street spitting, urinating, and defecating must be rigorously prohibited. The care of stray animals, including cows and dogs, must be provided at a nearby animal shelter, preventing the risk of animal waste on the highways and possibly resulting in more employment and adoption opportunities. It takes time and effort to build the perfect neighborhood in a city, and both parties must make an effort for the agreement to succeed.

In addition, water supply and waste disposal; heat energy consumption for heating and ventilation, hot water supply, and electric energy consumption; consumption of renewable and secondary energy resources; the comfort of the environment (heat and humidity, air, light, and acoustic modes) is made possible by a major improvement in the performance quality of the environment's life support systems [21]. The implementation of an optimal management plan based on a statistical model of the thermal and physical properties of a building as a comprehensive energy system is made possible by the employment of cutting-edge IT technology for self-learning. Due to urbanization, more people are relocating to urban areas. By utilizing the capabilities of cutting-edge technology to remove inefficiencies, architects and developers may make a significant contribution to sustainability. Fortunately, the technology needed to support the rise of smart cities will also draw in tech-savvy citizens [22].

Installing a network of smart pipes is an option if we want to incorporate smart city technologies into our home's plumbing and water supply. Data such as how much water runs through a pipe or the periods when water is turned on can be sensed and categorized by smart pipes. A sensor can therefore forecast the complete lifetime of a pipe piece. If it discovers a leak or other emergency right away, it can even inform us, as modern technology may be used to improve every element of building design and administration. Smart glass is a unique kind of glass that has the ability to measure how much sunlight is entering it. The lighting system in the building is then given instructions. Based on the amount of sunshine coming through the glass, smart glass automatically adjusts the light levels. There is no need to light a room that is already receiving plenty of sunlight in order to save energy and increase utility costs. Smartphones today are far more sophisticated than older analogue phones. Smartphones are all-in-one tools that enable us to track goods, place restaurant orders, and carry out a variety of other essential functions. Residents can open doors, enter buildings, and switch on air conditioners and televisions using their phones with a mobile access control system. The technology can also provide ease to building workers and tenants to control access to the property from anywhere. It also permits visitors who require property access to make two-way video calls to users and gives homeowners the option to grant one-time delivery credentials to couriers and virtual keys to regular guests.

3.3.6 OCCUPATIONAL PROFILES

One of the major initiatives of Prime Minister Narendra Modi is the "Smart Cities Mission," which will use cutting-edge solutions to transform India's urban environment and develop cities that are more inclusive, livable, and conducive to economic progress (Figure 3.7).

The Smart Cities program is divided into two main parts: area-based development, where real estate investors and construction firms play a significant role in building new townships or renovating existing spaces in areas between 50 and 500 acres; and ICT-based pan-city solutions [23], where possible interventions could be in the areas of energy management, water management, or mobility and transport management. For the pan-city interventions to deliver creative, scalable solutions at low prices, infrastructure and IT business skills would be required. The program uses a variety of strategies, including public–private partnerships, best practices in urban planning, digital and information technologies, and policy change, to alter the urban environment. For the next seven to ten years, public and private investments of $30–40 billion are expected to be made. This level of added expenditure and the requirement to utilize existing resources would necessitate business expertise, expert competency, and familiarity with cutting-edge technologies. Small and medium-sized construction and IT enterprises would be eligible to obtain contracts relating to the implementation of the scheme because the entire program will be divided into a number of projects. There will be considerable demand for service providers with the necessary topic knowledge and experience.

Professionals and students in this field should anticipate tremendous prospects as system and network integration services, programming, data analytics, IT consulting,

FIGURE 3.7 Occupational profiles in smart city execution.

and other related services will be in high demand [24]. IT solutions companies, network service providers and integrators, data analysis companies, infrastructure developers, real estate companies, construction companies, solar energy companies, transport engineering companies, and project and program management companies are some examples of the services or businesses that could directly benefit from the smart cities mission [25, 26]. Startups, innovation hubs, and accelerator programs have given the economy a significant push in the right direction. There will be a considerable demand for knowledge-based services like consultancy and advisory services. The program has resulted in the creation of enormous new prospects, and it is urgently necessary to take advantage of these by engaging in entrepreneurship in the aforementioned industries. Businesses may make a strong argument for participating actively in smart city initiatives based on direct and ancillary benefits such as ease of doing business, stronger relationships with the government, and new sources of revenue. For example, when the city hall was created in Barcelona District, new employment opportunities were generated: more than 55,000 employment places with over 1,500 new companies and new institutions, primarily in ICT and media businesses [27].

3.3.7 EDUCATIONAL PROFILES

Citizens need to have a firm grasp of the advantages and revolutionary potential of smart cities to make the most of the opportunities presented by them. Because individuals are typically reluctant to change, educational activities aimed at gaining the support and participation of city residents are essential to the success of any smart city development project (Figure 3.8).

Smart cities require a level of knowledge and technology advancement that can only be provided by a modern/future education. To make education more engaging, collaborative, and focused on building talents and abilities that help us address societal issues, we need to make a deliberate attitude shift. This type of education needs to be adaptable to the various learning styles of its participants by giving them more control over their time spent studying and the pace at which they complete assignments. We can only meet these demands if we adjust both the curriculum and the methods of instruction to reflect the new circumstances.

FIGURE 3.8 Knowledge and education management.

Thus, for cities to prosper, it will be necessary for their accountable local authorities, civil societies, and commercial organizations to streamline their educational system and policies, learning tools, and resources to meet the following trends, skills, and learning needs. There needs to be a decentralization of education so that schools are not the only places where people can acquire knowledge. Despite their centrality to the learning process, they must make room for the participation of stakeholders from the business, parental, government, and civil society sectors. It is up to the students themselves to serve as a teaching resource for their other classmates. The curriculum should also promote interdisciplinary student exchanges between different schools.

Students have a better opportunity of collaborating with problem-solvers and innovators who have the ability to transform their knowledge and abilities into action with this instructional approach. Students require hands-on experience with the transformation process and with shaping ideas into solutions that can be applied in real-world settings. Building a "smart" city relies heavily on "smart" education. "A model of learning tailored to new generations of digital natives," smart education aims to do just that [12]. Smart education is an alternative to the more traditional classroom teaching style that emphasizes student participation and teacher flexibility by using interactive, collaborative, and visual elements. An integral aspect of what makes a city "smart" is its commitment to and support of high-quality education at all levels, from kindergarten through graduate school. A key component of smart cities is "education programmes creating graduates with modern knowledge, practical skills, and collaborative attitudes." Different technologies were tried out in the classroom throughout the last century, with mixed results [28]. Many of these resources centered on the instructor and were made to complement antiquated pedagogical practices that failed to capture the attention of the students of the digital native generation. These days, more strategic use of digital technology is essential to the growth of a smart education ecosystem. Educators can make better use of classroom devices by adopting new approaches that emphasize student and teacher–student interaction.

The goal of smart education is to create a learning environment in which students are actively engaged in their own education and are given the tools and support necessary to become independent and lifelong learners [29]. In order to take charge of their own education, students must be intrinsically motivated and equipped with the knowledge, values, and abilities necessary to do so. Teachers, on the other hand, will play the role of guides, pointing the way and making themselves available to students in order to foster an attitude of gratitude toward the learning process and a sense that their needs are being met. The learning environment, including its equipment, data, resources, and physical location, should be tailored to each individual student's requirements and designed to foster the development of his or her individual strengths. In this case, two things are at play. One is that students should emerge from the educational system with a strong desire to continue their study outside the traditional classroom. Second, smart cities are those that center around technology and experience daily, multifaceted expansion. Blended learning systems incorporate digital tools like online discussion forums, quizzes, and simulations to provide students with a variety of ways to engage with course material. In order for smart cities to function and thrive, they need residents who can apply, control, and disseminate the necessary circuitry, data, and skills.

For the purpose of sustainability, the citizen living in a smart city should understand the complexities involved in this type of urban setup and be willing to actively contribute in encoding, decoding, and loading signals to the systems so that the various data analytics professionals and computing devices can study the algorithms and thus make consistent decisions in the urbanization journey of a smart city, thereby improving the services of the city.

3.3.8 COLLABORATION

The term "collaboration" refers to "a process or collection of activities in which two or more agents work together to achieve common goals" [30]. According to research, "smart cities" are characterized as "cities with smart collaboration" from an ideal-typical point of view, with the emphasis placed on governance. The term "smart collaboration" refers to a platform that enables government agencies, non-profit organizations, and corporate companies to collaborate effectively (Figure 3.9) [11].

The primary purpose of the cooperation is to build an ecosystem in which thought leaders, innovators, consultants, analysts, networkers, government, and commercial and public organizations may join together to work together. Several facets of cooperation have been investigated, including team collaboration and collaborative learning [31–33]. Collaboration can take place on many different levels; it can take place between organizations; it can take place across industries; or it can take place through the interaction between the government and its citizens. The improvement of department-to-department cooperation and integration has emerged as a top priority for administrations all around the world [34]. External collaboration involves governmental and non-governmental partners, such as businesses, non-profit organizations, and civic groups, or the involvement of both governmental and non-governmental partners, such as businesses, non-profit organizations, civic groups, or individuals. Taking into account the organizational changes, the goals of smart city initiatives are

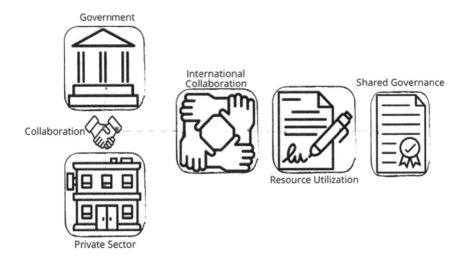

FIGURE 3.9 Collaboration and partnerships.

to improve the provision of information and services, as well as the efficiency and effectiveness of public administration, as well as aspects that promote smart governance to encourage greater collaboration between stakeholders [35]. Other advantages of collaborative cooperation are increased information sharing between departments and agencies, more efficient use of resources, and a more involved citizenry in the formulation of public policy. Information sharing optimizes and minimizes the hurdles to increasing government performance. It also enhances situation awareness, which supports well-informed, collaborative decision-making and coordinated actions. A collaborative government is able to apply collective intelligence in order to develop novel solutions to problems, and it is also able to give shared governance, which ultimately helps to nurture citizens' trust and confidence in their own governments [30]. One of the distinguishing features of smart governance in comparison to e-government is its emphasis on the collaborative environment [36]. The collaboration should establish sustainable smart cities that are physically and digitally safe, respectful of the natural environment, improve the quality of lives, develop a competitive economy, and are linked with the SDGs.

3.4 A SOCIO-ECONOMY INDEX FOR THE SMART CITY

In light of the many new developments taking place in smart cities, a number of institutions, including educational institutions and private consulting firms, are conducting studies to evaluate and rank smart cities. Smart cities are evaluated on a yearly basis and their rankings are published in news and online media. Although some city governments have made use of such rankings, it is unclear how widely accepted these rankings are by both other city governments and residents of the cities in which they are used. In addition, there is still skepticism regarding the rankings, specifically regarding their objectivity, accuracy, reliability, and the methods of measurement and comparison that they use. At this time, there is no international organization whose primary mission is to produce a form of smart city index and ranking that is universally agreed upon and accepted by all parties. Some new indexes have been developed, but their scope and usefulness are restricted, whereas others have been in existence for several decades.

The most important smart city indexes and rankings are the AT Kearney Global Cities Index (GCI), the IMD-SUTD Smart City Index, the Mori-Foundation Global Power City Index, the Smart Eco-City Index, and the Smart Cities Index Report, as well as the Cities of the Future Index and the IESE Cities in Motion Ranking (CIMI). The AT Kearney GCI is one of the smart city indexes that has been around the longest and was first introduced in the year 2008. The GCI evaluates smart cities based on a set of 29 indicators that are divided among 5 categories: human capital (30%), business activity (30%), cultural experience (15%), information exchange (15%), and political engagement (10%). The IMD-SUTD index takes into account the economic, social, and technological aspects of smart cities. There are five primary factors that need to be considered for this assessment: health and safety; mobility; activities; opportunities; and governance. The Global Power City Index ranked the world's smart cities based on their economy, research and development, livability, cultural interactions, accessibility, and environments. The index used 70 different indicators to determine the rankings. The Smart Eco-City Index evaluates smart cities based on their performance in seven distinct categories: sustainability; transport and mobility; governance; digitalization; innovation economy; living standard; and the perception of industry experts. The Cities of the Future Index conducted an evaluation of smart cities based on environmental sustainability, digital life, business technology infrastructure, and innovations in mobility. The evaluation of cities by CIMI is based on nine criteria and 101 indicators, including governance, human capital, mobility and transportation, social cohesion, international security, economic development, environmental protection, urban planning, and technological advancement.

The socio-economic factors and the index that is related to them are still absent from each of the indexes and rankings that have been presented so far. The current research elaborates on socio-economic factors to be measured, evaluated, and compared in the growth of smart cities using seven key criteria including quality of life, smart eco-system, waste management, residential growth, occupational growth, educational profile, and collaboration.

3.5 CONCLUSION

The results from the various studies and our summary shows that socio-economic growth apparently depends on the seven factors identified. These factors are decisive to indexing the growth with their attributes namely: **Q**uality of Life, **S**mart Ecosystem, **C**ollaboration, **O**ccupational Profile, **R**esidual Waste Management, **R**esidential Growth, and **E**ducational Profile. The proposed aggregate index is called "**Q – SCOR^{2}E**."

Figure 3.10 suggests that the proposed index has an influence demarcation based on the stakeholders who bring the impact. For instance, quality of life can only be measured with citizen participation within a region. So, the impact is measured by

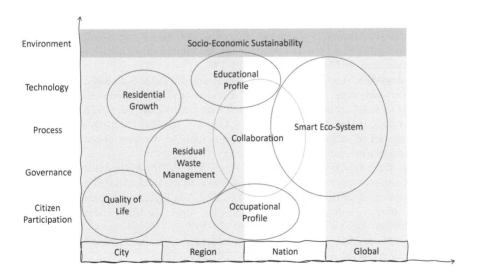

FIGURE 3.10   Growth index: Q – SCOR^{2}E.

the degree of urbanization within a city. Similarly, the stakeholders are governance, process, technology, and environment, which are also the beneficiaries of growth.

To implement Q – SCOR²E, the process needs to be defined and applied with the metrics outlined in the sections above. The scope of this chapter is limited to introducing the concept and concluding the evaluation of socio-economic growth with a methodology. The methodology evolves with the established metrics and so the validation is not so critical for the suggested strategy.

Finally, we conclude with the seven attributes of socio-economic growth applied to smart cities. With the advent of technology utilized by city governance, the evaluation and application are easy. There is a need to establish a process to timely assess the same and work on the weaker attributes and mark the growth amongst the developed cities globally.

REFERENCES

[1] Das, D. (2020). In pursuit of being smart? A critical analysis of India's smart cities endeavor. *Urban Geography, 41*(1), 55–78.

[2] Patrão, C., Moura, P., de Almeida, A. T. (2020). Review of Smart City assessment tools. *Smart Cities, 3*(4), 1117–1132. https://doi.org/10.3390/smartcities3040055

[3] Dameri, R. P. (2013). Searching for smart city definition: A comprehensive proposal. *International Journal of Computers & Technology, 11*(5), 2544–2551.

[4] Cisco Report, Smart Cities and Internet of Everything – The Foundation for Delivering Next-Generation Citizen Services, sponsored by *Cisco*. 2013.

[5] Sharifi, A. (2022). Smart City indicators: Towards exploring potential linkages to disaster resilience abilities. *APN, 12*(1), 76–90.

[6] Al-Azzam, M. K., & Alazzam, M. B. (2019). Smart city and smart-health framework, challenges and opportunities. *International Journal of Advanced Computer Science and Applications*, 10(2), 171–176.

[7] Susanti, R., Soetomo, S., Buchori, I., & Brotosunaryo, P. M. (2016). Smart growth, smart city and density: In search of the appropriate indicator for residential density in Indonesia. *Procedia-Social and Behavioral Sciences, 227*, 194–201.

[8] Marques, P., Manfroi, D., Deitos, E., Cegoni, J., Castilhos, R., Rochol, J., & Kunst, R. (2019). An IoT-based smart cities infrastructure architecture applied to a waste management scenario. *Ad Hoc Networks, 87*, 200–208.

[9] Nitoslawski, S. A., Galle, N. J., Van Den Bosch, C. K., & Steenberg, J. W. (2019). Smarter ecosystems for smarter cities? A review of trends, technologies, and turning points for smart urban forestry. *Sustainable Cities and Society, 51*, 101770.

[10] Alsamhi, S. H., Ma, O., Ansari, M. S., & Gupta, S. K. (2019). Collaboration of drone and internet of public safety things in smart cities: An overview of qos and network performance optimization. *Drones, 3*(1), 13.

[11] Pardo, T. A., Gil-Garcia, J. R., & Luna-Reyes, L. F. (2010). Collaborative governance and cross-boundary information sharing: envisioning a networked and IT-enabled public administration. In Rosemary O'Leary, David M. Van Slyke, and Soonhee Kim (eds), *The Future of Public Administration Around the World: The Minnowbrook Perspective*. Washington, DC: Georgetown University, 129–139.

[12] Singh, H., & Miah, S. J. (2020). Smart education literature: A theoretical analysis. *Education and Information Technologies*, 25, 3299–3328.

[13] Statistics, U. N. (2019). Global indicator framework for the sustainable development goals and targets of the 2030 agenda for sustainable development. *Developmental Science and Sustainable Development Goals for Children and Youth*, 439.

[14] Ministry of Finance (2014). *Public Private Partnerships: India Database*. New Delhi: Department of Economic Affairs, Ministry of Finance, Government of India.

[15] Patil, N. A., Tharun, D., & Laishram, B. (2016). "Infrastructure development through PPPs in India: Criteria for sustainability assessment." *Journal of Environmental Planning and Management 59*(4), 708–729.

[16] Onoda, H. (2020). Smart approaches to waste management for post-COVID-19 smart cities in Japan. *IET Smart Cities*, *2*(2), 89–94.

[17] Hassankhani, M., Alidadi, M., Sharifi, A., & Azhdari, A. (2021). Smart city and crisis management: Lessons for the COVID-19 pandemic. *International Journal of Environmental Research and Public Health*, *18*(15), 7736.

[18] Khang, A., Rani, S., & Sivaraman, A. K. (Eds.). (2022). *AI-Centric Smart City Ecosystems: Technologies, Design and Implementation*. Florida: CRC Press.

[19] De Guimarães, J. C. F., Severo, E. A., Júnior, L. A. F., Da Costa, W. P. L. B., & Salmoria, F. T. Governance and quality of life in smart cities: Towards sustainable development goals. *Journal of Cleaner Production*, *253*, 119926 (2020).

[20] Ali, T., Irfan, M., Alwadie, A. S., & Glowacz, A. (2020). IoT-based smart waste bin monitoring and municipal solid waste management system for smart cities. *Arabian Journal for Science and Engineering*, *45*(12), 10185–10198.

[21] Suakanto, S., Supangkat, S. H., & Saragih, R. (2013, June). Smart City dashboard for integrating various data of sensor networks. In *International Conference on ICT for Smart Society* (pp. 1–5). IEEE.

[22] Washburn, D., Sindhu, U., Balaouras, S., Dines, R. A., Hayes, N., & Nelson, L. E. (2009). Helping CIOs understand "smart city" initiatives. *Growth*, *17*(2), 1–17.

[23] Fennell, S., Kaur, P., Jhunjhunwala, A., Narayanan, D., Loyola, C., Bedi, J., & Singh, Y. (2018). Examining linkages between smart villages and smart cities: Learning from rural youth accessing the internet in India. *Telecommunications Policy*, *42*(10), 810–823.

[24] Trencher, G. (2019). Towards the smart city 2.0: Empirical evidence of using smartness as a tool for tackling social challenges. *Technological Forecasting and Social Change*, *142*, 117–128.

[25] Bakıcı, T., Almirall, E., & Wareham, J. (2013). A smart city initiative: The case of Barcelona. *Journal of the Knowledge Economy*, *4*(2), 135–148.

[26] Caragliu, A., & Del Bo, C. F. (2019). Smart innovative cities: The impact of Smart City policies on urban innovation. *Technological Forecasting and Social Change*, *142*, 373–383.

[27] Batlle, J, Majo, A, Ventura, J. L., Vila, M., & Ponti, I. (2011). Interviews, Barcelona City Council.

[28] Zhu, Z.T., Yu, M. H., & Riezebos, P. (2016) A research framework of smart education. *Smart Learning Environments*, *3*, 4.

[29] Batagan, L. (2011). Indicators for economic and social development of future smart city. *Journal of Applied Quantitative Methods*, *6*(3), 27–34.

[30] Chun, S. A., Luna-Reyes, L. F., & Sandoval-Almazán, R. (2012). Collaborative e-government. In *Transforming Government: People, Process and Policy*.

[31] Cheng, X., Fu, S., & Druckenmiller, D. (2016). Trust development in globally distributed collaboration: A case of US and Chinese mixed teams. *Journal of Management Information Systems*, *33*(4), 978–1007.

[32] Cheng, X., Fu, S., Sun, J., Han, Y., Shen, J., & Zarifis, A. (2016). Investigating individual trust in semi-virtual collaboration of multicultural and unicultural teams. *Computers in Human Behavior*, *62*, 267–276.

[33] Cheng, X., Yin, G., Azadegan, A., & Kolfschoten, G. (2016). Trust evolvement in hybrid team collaboration: A longitudinal case study. *Group Decision and Negotiation*, *25*(2), 267–288.

[34] Alhusban, M. (2015). The practicality of public service integration. *Electronic Journal of e-Government*, *13*(2), 94–109.

[35] Alawadhi, S., Aldama-Nalda, A., Chourabi, H., Gil-Garcia, J. R., Leung, S., Mellouli, S., & Walker, S. (2012, September). Building understanding of smart city initiatives. In *International Conference on Electronic Government* (pp. 40–53). Springer, Berlin, Heidelberg.

[36] Scholl, H. J., & Scholl, M. C. (2014). Smart governance: A roadmap for research and practice. In *IConference 2014 proceedings*, University of Illinois at Urbana.

4 A Study of the Future Generation of Smart Cities Using Green Technology

Siddhartha Roy
University of Engineering and Management, Kolkata, India

4.1 INTRODUCTION

The rapid development and revolution in the field of Information and communication technology (ICT) and the Internet of Things (IoT) is to provide proper coordination among various devices such as cars, smartphones, electronic gadgets, and laptops easier. A smart city consists of smart governance, a smart economy, and a quality lifestyle through strong human capital [1]. It is the novel advancement of a technology to raise effectiveness, reduce expenses, and enhance the quality of life of individual people. Digital communication is now used to develop a smart city to connect everything around the city through the sharing of digital information instantly to any part of the country through various electronic handheld devices. ICT and the IoT are deployed in a smart, sustainable society to improve effective inter-city and intra-city communication, which will meet the requirements of the present generation for many aspects such as economic, socio-cultural, and interpersonal ones. The objective of a smart city is to connect government, industries, and researchers to produce a sustainable green city. The latest digital technologies such as AI, wireless sensor networks (WSNs), the IoT, and cloud computing are used to connect each other to share information efficiently. Several IoT edge devices and WSNs are used for acquiring, processing, and transferring information to produce superior-quality images which are always desirable for smart cities. This study focuses on technological innovation to develop the smart city [2]. In this chapter, a detailed study is conducted on the implications for various use cases, such as renewable energy, smart agriculture, smart transport, and smart healthcare, which are the main focusing areas for a recognition of sustainable development strategies and programs within the existing judicial framework for the smart city (Figure 4.1).

Green computing, a part of green technology, is extensively used in an emerging smart city. The goal of green computing is to reduce greenhouse gas emissions and maximize renewable energy sources, so providing sustainable computing. Green computing encompasses the analysis, design, and implementation of computer

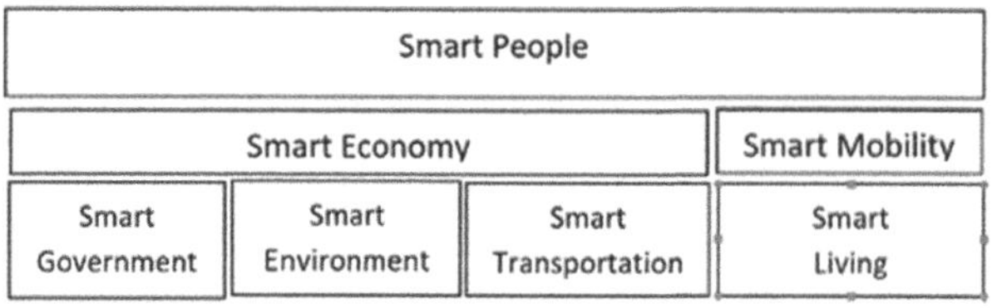

Smart People			
Smart Economy			Smart Mobility
Smart Government	Smart Environment	Smart Transportation	Smart Living

FIGURE 4.1 To become smart, six smart city indicators are crucial: smart economy, smart environment, smart government, smart living, smart mobility, and smart people.

systems aimed at minimizing their environmental impact. This objective is accomplished through the efficient utilization of hardware resources via virtualization technology. By conserving resources, green computing aims to reduce the overall environmental footprint. One approach involves optimizing energy consumption by minimizing the number of servers required for computations.

The development of green computing faces certain challenges, primarily the lack of awareness and knowledge within society and the absence of economic incentives to implement such technologies. However, as smart cities continue to grow, the importance of green computing will become increasingly vital. A smart city infrastructure heavily relies on the IoT and big data, forming the fundamental pillars of green computing. Additionally, data analytics plays a crucial role in facilitating informed decision-making within smart cities.

While smart city computing is still in its early stages, emerging solutions are addressing the pressing issues arising from urbanization, population growth, social inequalities, and environmental concerns. The future of smart cities lies in leveraging green computing technologies to harness the power of technology while responsibly utilizing our resources to ensure sustainable living.

The smart cities of the future are anticipated to enhance sustainability, be energyefficient, and be environmentally friendly; green technology will play a crucial role in achieving this goal. The integration of green technology in a smart city infrastructure will help cities reduce their carbon footprint, improve air quality, and create more livable, resilient communities.

Green technology can be used in various aspects of smart city development, including transportation, energy, water management, waste management, and building construction. As an illustration, the utilization of renewable energy sources like solar, wind, and geothermal power can be employed to generate electricity for various purposes, while energy-efficient building materials and smart building management systems can reduce consumption of energy.

Carbon emissions can be drastically reduced by incorporating smart transportation systems, such as electric vehicles and bike-sharing programs, which in turn improve air quality. Additionally, intelligent waste collection systems and recycling programs can support waste reduction andmitigate the environmental consequences in cities. Moreover, smart water management systemscan optimize water consumption and minimize water wastage.

Green technology can also be integrated into urban agriculture and food production to promote sustainable food systems; these advancements can enhance access to fresh, nourishing food and foster the development of urban green spaces, such as parks and green roofs, and can also be used to improve air quality and provide natural habitats for wildlife.

In this chapter, I will study how various digital technologies are used to make a smart city. In Section 4.2 I give a brief literature survey. I then go on to discuss how smart cities are made smarter with the various latest technology in Section 4.3. Some use cases are discussed in Section 4.4. In Section 4.5 I discuss some issues related to the smart system. Section 4.6 ends with some concluding remarks.

4.2 LITERATURE SURVEY

Since the Industrial Revolution a rise in carbon dioxide in the atmosphere has been the cause of serious temperature change that could be fatal for the human race. In 1920, the term "global village" was introduced after radio communication (one-way data communication) was invented. The surge in global population, along with escalating air pollution and the rising costs of diverse energy sources, pose significant challenges for cities to overcome in order to ensure their sustainability and resilience. A rise of smart cities and smart mobility based on green technology is the only solution for sustainability in the future. Owing to the inadequate energy efficiency of transportation and buildings, contemporary cities contribute approximately 70% of global greenhouse gas emissions and consume over 60% of the world's energy. Moreover, the existing mobility systems, reliant on fossil fuels, not only account for more than 25% of polluting emissions but also fail to meet the demands of urban areas effectively. Nowadays most electronic and electric instruments have sensors to recognize the world around them. These sensors provide us with data about their surroundings and how energy is being consumed. Every year 12.6 million people die with increasing illnesses and health damage due to various environmental forms of pollution, such as soil, water and air pollution, ultraviolet radiation, and chemical exposure. Chithaluru et al. discuss the increase in the value of renewable energy storage [3]. Kumar et al. discuss the role of the IoT in forming existing and future smart cities [4]. The IoT together with remote sensing are used to generate big data and detect the environment, as discussed by Huang et al. [5]concerning the significance of environmental sustainability and the concept of smart cities. The rapid proliferation of IoT devices across various sectors, including healthcare, industry, and corporate domains, has revolutionized conventional practices and paved the way for innovative advancements; such practices are discussed by Nguyen et al. [6]. How data analytics use a large number of sensors connected together is discussed by Lloret et al. [7] and Gao et al. [8], focusing on building a sustainable and green smart city using IoT wearable sensors [8]. Bani-Saleh et al. analyze a huge quantity of data from sustainable development [9].

Green technology plays a crucial role in reducing carbon emissions, minimizing waste, preserving freshwater, and consuming less energy compared to traditional technologies. In the context of smart cities, sustainability is prioritized to foster engagement, encourage positive consumer behavior and energy conservation, and leverage renewable energy sources like solar and wind power to preserve natural resources. Researchers, such as Al-Kiyumi et al., have developed a conceptual framework that integrates the IoT and AI-enabled smart city systems to enhance stability and sustainability in distribution systems. The significance of IoT technology as a key component of smart cities has been thoroughly examined by these experts [10]. These specialists not only focus on implementing green and sustainable communication techniques within green infrastructure but also address performance and energy consumption challenges associated with IoT sensor devices during the transmission of smart city data.

The smart city has the ability to monitor, analyze, and automate activities and maintain social equilibrium and quality of life among people in realtime through the IoT, sensors, and cloud computing, which are applied in different areas such as transportation, healthcare services, energy, and waste management. Numerous researchers are actively working to enhance the performance of public services through advancements in technology, knowledge, and economic development. In Singapore, in the Intelligent Island project, with the help of ICT and the IoT, work and life are being transformed [11, 12]. Taiwan has developed the "e-Taoyuan" and "u-Taoyuan" systems to develop an e-governance system which is used to improve the quality of life among citizens [13]. A knowledge city extensively embraces the concept of a smart city to foster the growth and empowerment of its citizens in adapting to technological advancements. By embracing a smart city approach, there are ample opportunities for the city to unlock its human potential and foster a culture of creativity and innovation in people's lives.

4.3   GREEN TECHNOLOGIES FOR THE SMART CITY

Green technologies can play an essential role in enabling a smart city that is sustainable, energy-efficient, and environmentally friendly.

4.3.1   AI-DRIVEN APPROACHES FOR THE SMART CITY

In the past few decades, computational intelligence-based technologies have played a crucial role in facilitating intelligent decision-making processes across various sectors, including healthcare, agriculture, education, environment and waste management, mobility and smart transportation, risk management, and security. A significant influence on AI adoption in smart cities has been observed [14]. Figure 4.2 shows the actual and predicted revenue generation in the case of a smart city.

Particularly in the aftermath of the pandemic, the healthcare industry has experienced significant impact from AI-based algorithms such as artificial neural networks and convolutional neural networks in medical image processing. In the context of smart cities, AI plays a vital role in analyzing and monitoring data generated by businesses, and aiding in decisions regarding the optimal utilization of renewable energy sources. It also enables city planners to monitor and reduce energy waste effectively. AI-based security cameras are employed to analyze real-time images, detect anomalies, and promptly address any arising situations. Additionally, AI is utilized for detecting CO_2 levels in the environment, facilitating measurements of energy consumption and pollution levels. Smart cities leverage AI technology for CO_2 detection, which subsequently informs decision-making processes related to transportation.

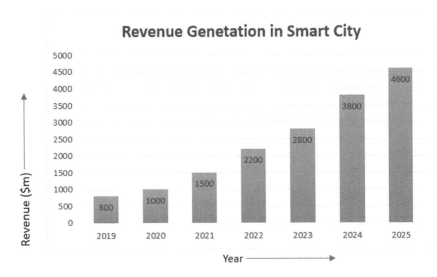

FIGURE 4.2 The adoption of AI in smart cities results in a significant increase following the pandemic.

4.3.2 SMART SENSOR NETWORKS

WSNs comprise sensor nodes that are strategically and extensively deployed to communicate, monitor, and analyze information remotely. These sensor nodes are capable of collecting data on environmental conditions and transmitting them wirelessly to a central base station within a smart city. WSNs find applications in diverse smart city scenarios, especially in areas that are challenging for human access, as the sensor nodes can be deployed in such locations. The gradually reducing cost of WSNs is enabling them to be used in smart home appliances such as controlling room temperature. It can also be used for congestion avoidance, route planning, and safer driving. WSNs play a crucial role in smart cities by enabling the provision of clean drinking water through water quality monitoring and pipeline spillage detection in water distribution networks. This monitoring system helps to rapidly identify and mitigate wastewater due to spillages. WSNs also find application in monitoring natural disasters, hazards, pollution, and potential biological or chemical attacks, particularly in war conditions. The U.S. Geological Survey has undertaken a long-term (ten-year) initiative to study and develop countermeasures against threats arising from natural disasters. The purpose of designing new technologies is to monitor networks and ensure reliability so as to improve the use of situational awareness. WSNs can be deployed on demand, quickly, and in a highly responsive manner. Unmanned aerial vehicles (UAVs) serve as valuable data mules in smart cities, facilitating the collection of data from ground sensors. These data are then transmitted to a remote monitoring center for analysis, aiding informed decisionmaking. Furthermore, UAVs can act as relay nodes, fostering improved communication among ground sensors in the network.

4.3.3 Green Building Based on Solar Energy

In light of the increasing concerns about global warming and environmental degradation, the development of green buildings has become a significant priority.
Among pollution-free renewable energy sources, solar energy has gained recognition for its superiority and is now receiving greater attention in the construction
of green buildings. Solar energy offers effective ways to conserve non-renewable
energy, mitigate the pace of global climate warming, and reduce environmental
pollution. The adoption of renewable energy technologies has a tremendous impact
on the development of smart cities, leading to self-sustaining urban environments
with zero pollution. Implementing rooftop solar energy, which is a zero-pollution
energy source, empowers smart city households to meet their energy requirements
sustainably. By efficiently utilizing the available rooftop solar energy and implementing net metering facilities, not only can the burning of fossil fuels and pollution be reduced, but it also promotes resource-saving and environmentally friendly
policies.

4.3.4 Green, IoT, Cloud-Enabled Platforms

The advancement of communication and sensing technologies has led to the interconnectedness of various objects through the IoT. The scope of the IoT encompasses all aspects of smart cities. Continuous development in IoT technologies
has made its components smarter by employing adaptive communication networks,
processing capabilities, data analysis, and storage. A wide array of IoT devices,
including sensors, cameras, actuators, radio frequency identification (RFID) systems, drones, and mobile phones, collaborate and communicate to offer a diverse
range of applications in real-time monitoring, e-healthcare, smart transportation,
industrial digitalization and automation, and home automation. The integration of
IoT technology reduces power consumption in big data processing and transmission, leading to an improved quality of life in smart cities and contributing to a
greener and more sustainable world while maintaining a high quality of life [14,
15]. Presently, smart cities are becoming even smarter with the application of AI.
Examples of AI-based technologies include sensors integrated into smart transportation systems and cameras within smart monitoring systems. The IoT plays
a significant role in the enhancement of smart cities, impacting various aspects,
particularly in public transportation. The IoT is extensively utilized in smart
transportation systems to facilitate the easy tracking of vehicle movement and
efficient management services, thereby reducing traffic congestion. Additionally,
the deployment of different sensors at strategic locations contributes to enhanced
vehicle safety [16].

The green IoT plays a significant role in reducing carbon emissions and power
consumption. Green technologies can help promote sustainable development in a
smart city by minimizing the impact on the environment while still meeting the needs
of the community. By adopting sustainable solutions, smart cities can reduce their
carbon footprint, improve air quality, and promote a more sustainable future.

4.4 GREEN TECHNOLOGY IN THE SMART CITY: USE CASES

Green technology plays a critical role in developing sustainable and eco-friendly smart cities. Some use cases of green technology in smart cities are as follows.

4.4.1 RENEWABLE ENERGY SOURCES

Smart cities can use renewable energy sources such as hydro power, solar, and wind to generate electricity, reducing the dependence on non-renewable sources. Cities can use solar panels to power street lights, traffic signals, and buildings. Renewable energy plays a critical role in building sustainable and eco-friendly smart cities. By integrating renewable energy, smart cities can reduce their dependence on fossil fuels, decrease their carbon footprint, and promote sustainability. Renewable energy can also provide a reliable and cost-effective source of energy for cities. Here are some ways in which renewable energy can be integrated into smart cities [17, 18]:

- **Solar energy:** Smart cities can use solar energy to generate electricity for public lighting, traffic signals, and buildings. Solar panels can be installed on rooftops, parking lots, and other open spaces to capture solar energy. Smart grid systems can also help to manage the distribution of solar energy.
- **Wind energy:** Smart cities can also use wind turbines to generate electricity. Wind turbines can be installed on rooftops, onshore or offshore, to capture wind energy. Smart sensors can be used to monitor the performance of wind turbines and optimize their output.
- **Hydro power:** Smart cities can use hydro power to generate electricity. Hydro power can be generated by tapping into the energy of rivers, tides, and ocean currents. Smart sensors can be used to monitor the flow of water and optimize the output of hydro power systems.
- **Geothermal energy:** Smart cities have the opportunity to leverage geothermal energy for heating and cooling purposes in buildings. This sustainable energy source can be harnessed by drilling deep into the earth's crust to capture the natural heat energy it provides. To ensure optimal efficiency, smart sensors can be employed to monitor the temperature and regulate geothermal systems accordingly, thereby optimizing their performance.
- **Energy storage:** Smart cities can use energy storage systems to store excess renewable energy generated during peak production hours. Energy storage systems can be used to provide electricity during periods of low renewable energy production.
- **Microgrids:** Smart cities can effectively utilize microgrids as a means to manage the distribution of renewable energy. Microgrids provide a decentralized approach to controlling the flow of renewable energy from various distributed sources, such as rooftop solar panels and wind turbines. By implementing microgrids, smart cities can efficiently manage and optimize the utilization of renewable energy resources, ensuring a reliable and sustainable energy supply.

4.4.2 ENERGY-EFFICIENT BUILDINGS

Smart cities can promote the construction of energy-efficient buildings that use less energy for heating and cooling. They can use advanced technologies like green roofs, energy-efficient Heating, ventilation, and air conditioning (HVAC) systems, and smart lighting systems that can adjust to the presence or absence of people. Energy-efficient buildings are a crucial component of smart cities. They play an important role in promoting energy efficiency and minimizing energy consumption, carbon emissions, and operational costs. Energy-efficient buildings are vital to the success of smart cities. By promoting energy-efficient building design, automation, lighting, HVAC systems, building envelopes, and energy monitoring systems, smart cities can significantly reduce energy consumption, carbon emissions, and operational costs. Table 4.1 depicts various technologies in which energy-efficient buildings can form part of smart cities.

4.4.3 SMART WATER MANAGEMENT

Smart cities can use technologies to monitor water usage and detect leaks, reducing water wastage. They can also encourage the use of low-flow toilets and faucets,

---

TABLE 4.1
Various Technologies Enabling Energy Efficient Building for Smart Cities

Technology	Description
Building Design	Smart cities can promote the construction of energy-efficient buildings by encouraging the use of building designs that optimize energy efficiency. Buildings can be designed to capture natural light, reduce heat gain and loss, and promote natural ventilation.
Building Automation Systems	Smart cities can install building automation systems that optimize the use of energy. These systems can control lighting, heating, and cooling systems to reduce energy consumption.
Smart Lighting Systems	Smart lighting systems can be used in buildings to reduce energy consumption. These systems use occupancy sensors and daylight sensors to control lighting and reduce energy consumption.
HVAC Systems	Smart cities have the potential to employ advanced HVAC systems that are designed to optimize energy consumption. These innovative systems can incorporate technologies such as geothermal heat pumps, radiant heating and cooling, and demand-controlled ventilation. By utilizing these energy-efficient HVAC systems, smart cities can effectively reduce energy consumption while maintaining comfortable indoor environments.
Building Envelope	Smart cities have the opportunity to encourage the implementation of building envelopes that prioritize energy efficiency. The building envelope serves as a physical barrier between the building's interior and exterior environment. By adopting smart strategies, the envelope can be designed to minimize heat loss, minimize air leakage, and maximize solar heat gain. This approach ensures optimal energy performance and contributes to the overall sustainability of the building in a smart city context.
Energy Monitoring Systems	Smart cities can install energy monitoring systems that track energy consumption in buildings. These systems can identify energy waste and provide feedback to building occupants on energy usage.

rainwater harvesting, and greywater recycling systems. By implementing smart water management systems, smart cities can reduce water waste, conserve resources, and improve water quality. This can lead to cost savings, increased efficiency, and a more sustainable future. Smart water management plays a vital role in the development of sustainable smart cities. As urban areas experience rapid population growth, escalating water demand, and the impacts of climate change, effectively managing water resources becomes increasingly complex. Therefore, implementing intelligent solutions for water management is crucial in ensuring the long-term sustainability and resilience of smart cities. Smart water management deals with water resources efficientlyusing advanced technologies and data analytics. It includes monitoring and measuring water quality, detecting leaks and wastage, optimizing usage, and reducing consumption.

Smart water management systems use sensors and IoT devices to monitor supply and demand, detect leaks, and measure quality. This data are then analyzed to optimize usage and identify areas for improvement. In smart cities, water management can be integrated with other smart systems such as transportation, energy, and waste management. For example, the data collected from smart water management systems can be used to inform energy consumption decisions or to optimize waste management systems.

4.4.4 ELECTRIC VEHICLES

Smart cities can promote the use of electric vehicles (EVs) by setting up charging stations and providing incentives to users. This will reduce the carbon footprint of the transportation sector, which is a significant contributor to greenhouse gas emissions. EVs are gaining popularity in smart cities as they offer various environmental advantages, cost savings in operation, and the potential for seamless integration into smart transportation systems. The rise in EV adoption contributes to reducing carbon emissions, promoting cleaner air quality, and supporting sustainable mobility solutions in smart city environments. Additionally, EVs offer the potential for enhanced connectivity and interoperability, enabling them to be seamlessly integrated into the overall smart transportation infrastructure. Smart cities leverage the latest technology to improve the quality of life for residents, including better transportation systems that are both efficient and sustainable. EVs play an essential role in building sustainable and intelligent transportation systems in smart cities. By integrating EVs with the latest technology, cities can create cleaner, safer, and more efficient transportation systems that benefit both residents and the environment. Here are some ways in which EVs are being integrated into smart cities:

- **EV charging infrastructure:** Smart cities are installing EV charging stations at various locations such as public parking lots, shopping centers, and highways. This infrastructure helps to promote the use of EVs by providing drivers with convenient places to charge their vehicles.
- **EV car-sharing:** Smart cities are encouraging the use of EV car-sharing programs as a way to reduce traffic congestion and carbon emissions. These programs allow people to rent EVs for short periods, giving them the flexibility to use a car only when they need it.

- **Intelligent traffic management systems:** Smart cities are using technology such as real-time traffic data and smart traffic signals to optimize traffic flow, reduce congestion, and decrease travel times for EV drivers.
- **EV fleet management:** Many smart cities are adopting EVs as part of their municipal fleets, such as police cars and garbage trucks. This helps to reduce emissions and operating costs while showcasing the benefits of EVs to the public.

4.4.5 WASTE MANAGEMENT

Smart cities can use technology to manage waste efficiently, reducing the amount that goes to landfills. They can use sensor-enabled garbage bins that can alert authorities when they are full, reducing the need for frequent pickups. Additionally, smart cities can promote recycling and composting programs to reduce the amount of waste generated. Smart waste management plays a critical role in creating sustainable and livable cities. By leveraging technology and data analytics, cities can optimize waste management systems, reduce costs, and promote a more sustainable and circular economy. Waste management is an important aspect of creating a sustainable and livable city. Smart cities leverage technology to manage waste in a more efficient and sustainable way. Table 4.2 shows how smart cities are transforming waste management

TABLE 4.2
Various Ways in Which Smart Cities Are Transforming Waste Management

Procedure	Description
Smart waste bins	Smart cities are installing sensors in waste bins to monitor the waste levels and optimize the collection schedules. This technology reduces the number of unnecessary trips by collection trucks, saving fuel and reducing emissions.
Waste-to-energy	Smart cities are implementing waste-to-energy technologies, such as incineration, anaerobic digestion, and gasification, to convert waste into energy. This approach reduces landfill space and generates renewable energy, contributing to a more sustainable and circular economy.
Recycling	Smart cities are promoting recycling through the use of smart recycling bins that provide feedback and incentives to users. This technology encourages residents to recycle more effectively, reducing the amount of waste sent to landfills.
Waste reduction	Smart cities are implementing waste reduction strategies such as composting, food waste reduction programs, and plastic bag bans. These strategies help to reduce waste generation and promote a more sustainable and circular economy.
Data analytics	Smart cities are using data analytics to monitor waste generation patterns, optimize waste collection schedules, and identify areas for improvement. This technology helps to reduce waste management costs and improve the efficiency of waste management systems.

4.4.6 Urban Agriculture

Smart cities can promote urban agriculture by setting up community gardens, rooftop gardens, and vertical farms. These initiatives can provide fresh produce to city residents, reducing the carbon footprint of the food industry and promoting a healthy lifestyle. Urban agriculture is an emerging trend in smart cities, with the potential to transform the way we produce and consume food. Urban agriculture involves growing crops, raising livestock, and producing food in urban environments. Urban agriculture has the potential to transform the way we produce and consume food in smart cities. By promoting sustainable and innovative agricultural practices, cities can improve food security, promote biodiversity, and create more livable and sustainable communities. The integration of green technology plays a pivotal role in transforming smart cities into sustainable, efficient, and livable urban spaces. Embracing these technologies enables cities to significantly reduce their environmental footprint, enhance the well-being and quality of life for residents, and build a more resilient future. By leveraging green technology, smart cities can pave the way for a harmonious coexistence between human activities and the natural environment, fostering a sustainable and prosperous urban ecosystem. Here are some ways that smart cities are promoting urban agriculture:

- **Rooftop gardens:** Smart cities are actively promoting the establishment of rooftop gardens on both commercial and residential buildings. These gardens offer multiple benefits, including the provision of fresh produce, mitigating the urban heat island effect, and fostering biodiversity within the city. By integrating rooftop gardens into their urban landscape, smart cities embrace sustainable practices that enhance food security, mitigate climate change impacts, and create vibrant green spaces for residents to enjoy.
- **Vertical farming:** Mart cities are actively endorsing vertical farming as a means to optimize space utilization and enhance crop yields. Vertical farms utilize cutting-edge technologies like hydroponics and LED lighting to cultivate crops within controlled environments. This innovative approach not only maximizes space efficiency but also offers several environmental benefits. Vertical farming significantly reduces water consumption compared to traditional agriculture and minimizes the reliance on pesticides. By embracing vertical farming, smart cities prioritize sustainable and resource-efficient food production, contributing to food security and reducing the ecological footprint associated with conventional farming methods.
- **Community gardens:** Smart cities are supporting community gardens as a way to bring people together, promote social cohesion, and provide fresh produce to residents. Community gardens also help to beautify urban areas and promote biodiversity.
- **Urban farms:** Smart cities are promoting the development of urban farms, which can be located on vacant lots or in other underutilized spaces. Urban farms provide fresh produce, create jobs, and contribute to food security in the city.

- **Smart irrigation:** Smart cities are using technology such as sensors and smart irrigation systems to optimize water usage in urban agriculture. These technologies help to conserve water, reduce the cost of irrigation, and increase crop yields.

4.5 SOME MAJOR ISSUES FOR ENABLING THESMART CITY

Enabling a smart city using green engineering presents several challenges.A comprehensive approach is required that addresses the technical, financial, and regulatory challenges. By addressing these, cities can promote sustainable development, reduce their carbon footprint, and build a more resilient and livable community.

- **Limited resources:** One of the biggest challenges in enabling a smart city using green engineering is the availability of limited resources. To achieve sustainability, requires significant investments in renewable energy sources, green infrastructure, and technology. The cost of implementing these solutions can be high, and cities may not have the financial resources to invest in them.
- **Lack of public awareness and education:** Many people may not be aware of the importance of green engineering in building a smart city or may not understand how their actions impact the environment. Therefore, there is a need to educate the public about the benefits of green engineering, promote sustainable behavior, and encourage public participation in sustainability efforts.
- **Regulatory and policy frameworks:** There is a need for appropriate regulatory and policy frameworks that incentivize the adoption of green engineering solutions. These frameworks should provide incentives for the private sector and individuals to adopt sustainable practices and technologies.
- **Technical challenges:** The development of sustainable solutions using green engineering can be complex and requires technical expertise. Developing and implementing new technologies can also be challenging due to the lack of standardization and interoperability issues.
- **Maintenance and upkeep:** Green engineering solutions require regular maintenance and upkeep to ensure their continued operation and effectiveness. This can be costly and time-consuming, and cities need to develop strategies to ensure the long-term sustainability of these solutions.
- **Handling big data:** The generation of vast amounts of data from diverse sensors and simulations poses a significant challenge in terms of storage capacity for individual computers. Storing such a massive volume of data becomes a complex task. Additionally, managing and manipulating heterogeneous data in different file formats, including images, videos, texts, and audio, further adds to the challenges in data management. Finding effective solutions to handle and maintain such diverse and extensive datasets is a critical undertaking in the realm of data management.

- **Data security and privacy:** Security and privacy concerns are of utmost importance in smart cities. The raw data generated within these cities may contain confidential or sensitive information that requires protection against unauthorized access and usage. Smart cities are vulnerable to various types of cyberattacks, including denial of service, phishing, flooding, and man-in-the-middle attacks. The transmission of data through public networks in smart city applications poses the risk of unauthorized access by intruders. Uploading data to remote servers through unauthorized Wi-Fi connections further exposes the system to security threats. While distributed open-source big data platforms like Hadoop are widely used for data storage and analytics, they often lack adequate security measures to address these challenges effectively. Safeguarding the security and privacy of data in smart cities remains an ongoing concern that requires robust and comprehensive solutions.

- **Efficiency:** Different applications within smart cities may vary in complexity and require varying response times. When it comes to environmental sustainability, response time is generally less critical. However, utilizing open-source big data platforms like Apache Hadoop, Spark, HDFS, and MapReduce for data analysis and pattern identification across different domains introduces inevitable performance issues. Processing spatiotemporal data using these platforms without modification poses challenges. The development of an efficient spatiotemporal computing platform is still in its early stages, and optimizing big data computing platforms to effectively support and implement smart city applications remains a significant challenge. Further advancements are needed to address these performance concerns and provide robust solutions for smart city data processing and analysis.

4.6 CONCLUSION

This study has examined the utilization of cutting-edge technologies such as AI, smart sensors, and the IoT in the development of smart cities. The integration of AI/ML and the IoT has been particularly advantageous in smart home initiatives. To effectively transform a city into a smart one, it is crucial to optimize infrastructure utilization and promote interoperability and scalability of solutions. Addressing challenges such as rising carbon emissions and power consumption can be achieved through the application of green IoT practices. While smart cities enhance the quality of life, certain considerations must be given to security issues, effective management of big data, performance optimization, scalability, and interoperability when integrating the IoT and sensors. In summary, the implementation of green engineering principles in smart cities can contribute to sustainable development by minimizing environmental impact while meeting community needs. Embracing sustainable solutions enables smart cities to reduce their carbon footprint, enhance air quality, and foster a more sustainable future.

REFERENCES

1. Yigitcanlar, T.; Kamruzzaman, M.; Buys, L.; Ioppolo, G.; Sabatini-Marques, J.; da Costa, E.M.; Yun, J.J. Understanding 'smart cities': Intertwining development drivers with desired outcomes in a multidimensional framework. *Cities* 2018, *81*, 145–160.

2. Macke, J.; Casagrande, R.M.; Sarate, J.A.R.; Silva, K.A. Smart city and quality of life: Citizens' perception in a Brazilian case study. *J. Clean. Prod.* 2018, *182*, 717–726.

3. Chithaluru, P.; Turjman, F.; Kumar, M.; Stephan, T. I-AREOR: An energy-balanced clustering protocol for implementing green IoT in smart cities. *Sustainable Cities and Society* 2020, *61*.

4. Kumar, R.; Banga, H. K.; Kaur, H. Internet of Things-supported Smart City platform. Published under licence by IOP Publishing Ltd, 2020.

5. Hsu, Ching-Hsien, Nithin Melala Eshwarappa, Wen-Thong, Chang, Chunming Rong, Wei-Zhe Zhang, Jun Huang, Green communication approach for the smart city using renewable energy systems. 2022, *8*, 9528–9540.

6. Nguyen, N.T.; Leu, M.C.; Liu, X.F. Rtethernet: Real-time communication for anufacturingcyberphysicalsystems. *Trans. Emerg. Telecommun. Technol.* 2018, *29* (7).

7. Lloret, J., Sendra, S., González, P. L., & Parra, L. (2019). An IoT group-based protocol for smart city interconnection. In *Smart Cities: First Ibero-American Congress, ICSC-CITIES 2018*, Soria, Spain, September 26–27, 2018, Revised Selected Papers 1 (pp. 164–178). Springer International Publishing.

8. Gao, J.; Wang, H.; Shen, H. Machine learning based workload prediction in cloud computing. In *2020 29th International Conference on Computer Communications and Networks (ICCCN)*, IEEE, 2020, pp. 1–9.

9. Bani-Salameh, H.; Al-Qawaqneh, M.; Taamneh, S. Investigating the adoption of Big Data management in healthcare in Jordan. *Data* (2021), *6* (2), 16.

10. Al-Kiyumi, R.M.; Foh, C.H.; Vural, S.; Chatzimisios, P.; Tafazolli, R. A fuzzy logic-based routing algorithm for lifetime enhancement in heterogeneous wireless sensor networks. *IEEE Trans. Green Commun. Netw.* 2018, *2* (2), 517–532.

11. Neo, B.S.; Chen, G. *Dynamic Governance: Embedding Culture, Capabilities and Change in Singapore.* World Scientific Publishing, 2007, p. 3.

12. European Commission. A digital agenda for Europe: A Europe 2020 initiative (Smart Cities), Institute of Knowledge at Singapore Management University.

13. Bureau of Energy. *Energy Statistics Handbook 2018.* Taiwan Ministry of Economic Affairs, October, 2019, p. 87.

14. Allam, Z.; Dhunny, Z.A. On big data, artificial intelligence and smart cities. *Introduction to IOT*, 2019, *89*, 80–91.

15. Bibri, S.E. The IoT for smart sustainable cities of the future: An analytical framework for sensor-based big data applications for environmental sustainability. *Sustain. Cities Soc.* 2018, *38*, 230–253.

16. Misra, S.; Mukherjee, A.; Roy, A. *Introduction to IoT.* Cambridge University Press, 2021.

17. Glasmeier, K.A.; Nebiolo, M. Thinking about smart cities: The travels of a policy idea that promises a great deal, but so far has delivered modest results. *Sustainability* 2016, *8*(11).

18. Glasmeier, A.; Christopherson, S. Thinking about smart cities. *Cambridge J. Reg. Econ. Soc.* 2015, *8*, 3–12.

5 Sustainable Smart City Transformation

A Call for Data Governance Using a Data Mesh Approach

Mark Wall
Tata Consultancy Services, London, United Kingdom

5.1 INTRODUCTION

The objective of this chapter is to help contextualise relevant models and themes from Urban Planning, Economic Geography, Business Strategy, Management Strategy and Analytics, Telecommunications and Cybernetics for the growing body of research on green computing, sustainability and smart cities. The aim is to set the scene for robust data-driven collaboration between industry, academia and government by leveraging research at the intersection of geography, communications and technology from The University of Manchester. If we are to meet local and global sustainability objectives, and secure an inclusive and cost effective future for people, we cannot leave this future to a single stakeholder group or investor. In this context, I aim to bridge the gap between academia, government and urban planning with industry and business. Section 5.2 outlines the academic literature surrounding the economic and human geographic impact of communications technology in the context of smart cities. Section 5.3 draws on key observations from me, based on research sponsored by the University of Manchester, and then outlines a critical realist framework of analysis for sustainable smart cities. It is noted here that critical realist approaches have a strong affinity for enterprise approaches to consulting assessments in the firm and therefore could be applicable for the assessment of smart city sustainable outcomes at global and local levels.

5.2 THE CONTEXT OF SUSTAINABLE SMART CITIES: A PARADIGM CHALLENGE AND 'URBAN SPLINTERING'

The smart city in 2023 poses a paradigm challenge to understanding for research purposes. This is well documented in the academic literature on the emergent internet and 'cyberspace' since the 1990s [1]. There are many varied contexts and agencies

DOI: 10.1201/9781003388814-5

which are in play in the process of making a sustainable smart city a reality which has implications for the adoption of green computing. These agencies include the futurists and technology corporations, as well as the policy makers at the national and international-regional level, urban planners and local governments as well as local communities themselves, which are now superseded by the many social network communities. Fundamentally, cities are spaces for living in, for places of human inter-action and understanding and meaningful contextualisation to take place. A recent report by the OECD posed the question 'Do Smart Cities Benefit Everyone?' [2]. This chapter attempts to provide some answers through providing a critical realistic framework designed in 1999 before briefly highlighting ways to accelerate adoption of a data-driven approach to organising and operationalising sustainable smart city outcomes using a green computing approach. In this chapter the green computing considerations are from the perspective of data and analytics. What kind of data management and analytical capabilities are needed to limit the harmful impact on the environment, including the efficient, lean deployment of computational power to support sustainable outcomes?

The challenge is compounded by the backdrop of the convergent common issues facing many cities across the globe in 2023: access to resources, overcrowding, healthcare, escalating costs, climate change and urban heat island effects, and ever increasing demand for infrastructure. In January 2023 temperatures across Europe broke all historical records and London experienced temperatures of 3.5 °C above average. In this respect, London is on a trajectory to face similar issues that sub-tropical cities have faced for hundreds of years: flooding and high rainfall events. Solutions to these challenges like climate change are increasingly coordinated through agencies such as the World Economic Forum. Furthermore, today, the most advanced developments of cities are now also hidden from view, buried in the network of cables, the metro data centres now being installed into the central offices of telecommunications firms, and smart sensors collecting petabytes of data for further analysis on distributed cloud technology. An even more extreme example of 'invisible cities' is the development of the concept of 'metaverses'. This is a very different context of understanding from where we have come from as local societies and peoples:

"Where the early market economies grew out of the temporal and spatial regulari-ties of city life, today's are built on the logical or 'virtual' regularities of electronic communication, a new geography of nodes and hubs, processing and control centres. Networks of computers, cables and radio links govern where things go and who has access to what [3]".

Technology and communications advancements have steadily over the last 50 years, and now in an accelerated way in the last 10 years, created a vacuum in the research frameworks which planning agencies, businesses and policy makers have traditionally relied upon to make material decisions about locations and the urban environment. As a result, the new paradigm challenge in the 'real-time age' has three component parts identified as the challenge of invisibility; the challenge of concep-tualising space and time; and the challenge to urban policy [4].

The result of this lack of framework has created an 'urban splintering' over the last two decades in many cities across the globe [5, 6]. Enhanced regulation of the telecommunications providers who build their infrastructure into the dense urban fabric are aware of their role in the advancement of smart cities. Most major global telecommunications firms, including Verizon, Comcast (US), BT (UK), Orange, DTAG (Europe), Airtel (India), and MTN (Africa), have placed stakes in the future of smart cities, driven by the prospect of growing new revenue streams from the Internet of Things. The tele-management forum created a manifesto for smart city platforms, and has a dedicated initiative to help bring government, industry and business ecosystem investment together into a platform model [7].

Meanwhile, regulations which through historical inertia are concerned with fair access and creating a competitive business environment have also shifted towards ensuring the network infrastructure is safeguarded from growing threats of fraud and cybersecurity. The government of the United Kingdom recently enacted the Telecommunications Security Bill (November, 2021). Urban policies which are steadily shifting now on enhancing sustainability have lacked a common standardised platform in which to harness investment at local, regional and international levels in a coordinated task force which adheres to local sovereignty and more recent enhanced regulatory controls.

It is without a doubt that the timing is now right for communications services providers to host sustainable smart city infrastructure on distributed cloud infrastructure which can help make progress towards sustainable outcomes at local and global levels through trusted leadership and assistance in regulating local data sovereignty, whilst providing a new breed of latency aware applications which are built into the very fabric of wireless, fixed and satellite communications networks.

The hyperscalars like Amazon, Google and Microsoft have come under increasing pressure to adhere to local regulatory standards of data protection and prevent their own businesses extending the global technology domains of search and advertising into local places and communities [8]. Another example is Facebook's resale of geopolitically sensitive electorate data to firms such as Cambridge Analytica for political advertising purposes [9].

A new conceptual framework is needed to assess the viability of smart cities as living breathing human environments to help coordinate secure and privatised data-driven research across the globe and at local domain level. In terms of actual deployments, and driven by sustainable smart city outcomes, the telecommunications industry is currently best equipped to help fill this gap, in partnership with local start-ups and global technology providers, as well as academia and government, whilst helping to bridge the conceptual and regulatory gap.

5.2.1 THE CONCEPTUAL CHALLENGE OF THE GLOBAL–LOCAL CONTEXT

It is worth taking a moment to understand the empirical grounding of the paradigm challenge. Turning to the multi-faceted discipline of Human and Physical Geography, until the 1970s geographers approached cities and regions as though they developed within the external environment of an 'objective' time and an external 'Cartesian' space consisting of points that were positioned on a map using coordinates [10]. Until very recently "geography [...] treated space as the domain of the fixed, the undialectic, the immobile – a world of passivity and measurement rather than action and meaning" [11, p. 37].

Since the 1980s the interactions of people with places and communities and behavioural geography has emerged. Bruno Latour provides a refreshing approach, suggesting the need to study space and technology as a *process* in which new territorial formation may recombine in networks.

Most of the difficulties we have in understanding science and technology proceed from our belief that space and time exist independently as an unshakeable frame of reference inside which events and place occur. This belief makes it impossible to understand how different spaces and different times may be produced inside the networks built to mobilise, cumulate and recombine the world [12].

Smart cities are territorial formations, serving to localise the globally interconnected digital worlds, whilst simultaneously 'spatialising' the city locality and thus providing a means through which social networks in cities can recombine and optimise the consumption of space and time at various 'scales' or levels in networks, and so prepare for the limits to scalability of urban dwelling through the application of lean and efficient algorithms that are able to accurately model the human and physical processes at play in creating sustainable outcomes at local, city and global levels.

5.2.2 Changing Urban Policy: The Multi-Network Approach

This leads on to another major challenge for sustainable smart city organisation and conceptualisation, namely the challenge of understanding space as an overlay of multiple intersecting networks rather than a defined territory with set boundaries, which presents a philosophical shift. Many urban planners argue that fixed 'Cartesian' or 'Euclidean' space is no longer applicable to urban planning [4].

There is a stronger need to account for the real-time flows of capital, information services and power that silently flow through the city [13]. Thus Graham [4] also lists the contemporary city as the 'non-place urban realm' [14]; the 'informational city' [15]; the 'wired city' [16]; 'the telecity' [17]; 'the intelligent city' [18]; 'the virtual city' [19]; and similarly 'the virtual community' [20] or the 'electronic community' [21], which is more akin to the idea of an 'imagined community' [22].

Almost 25 years ago, Singapore embarked on a journey to create a 'cyber-city' that was actively exploring civic enhancement through the use of telematics to create greater urban connectivity, and to develop the local intellectual community: in the 1990s, "the government's hope is to transform the city-state into a networked, intelligent island, using information and communication technology to utilise its intellectual capital and sustain economic growth" [23, p. 117]. However, today the social, economic and political trends combine with emerging technologies to result in a social and ecological splintering of many communities and cities. In summary, this splintering refers to the dismantling of public networks like water [24], electricity, transport and telecommunications that were standard in many cities through privatisation, deregulation and globalisation. Singapore is bucking this trend, where a high level of government investment over the last 25 years has helped it reach global status as one of the top sustainable smart cities in the world for many consecutive years by a collective focus on the top priorities of health, mobility and supporting a rich diverse ecosystem of start-ups and mobile applications.

This is helped by the forward looking urban planning framework in place in Singapore which has a systemic approach with three tiers: a long-term plan over at least 50 years, a master plan for the medium term, and short-term plans, the first two of which are prepared by the Urban Redevelopment Authority and the last by multiple agencies. This is a topic we will revisit briefly below in the context of management cybernetics and the viable system model (VSM).

In many developed nations, like the USA and the UK, the challenge to urban planners concerns the historical inertia, and to develop new ways of thinking and acting that reflects the growing digitisation of urban life [4]. A future research topic is that those countries who have historically less structure and historical precedence concerning urban planning and investment have an opportunity to more rapidly shift to a network approach for sustainable smart cities using an updated form of digitised and networked urban planning.

It would benefit this chapter if we were to now consider a brief summary of the differences between traditional urban places and the urban digital spaces which sustainable smart cities as a platform grow into.

Table 5.1 shows electronically mediated space is causing the city to become more 'fluid', 'disembedded' and 'immaterial'. What are the methods we should turn to in studying the sustainable smart city with its intangible multi-networks and silent infiltration? Many approaches have emerged within the last decade and all have a common theme of political economy and critical theory. Significantly, it was those working in the traditions of political economy and critical theory, attuned as they were to the study of invisible relations, who produced many of the most comprehensive and perceptive meanings available on the geographical dimensions of the information society [25, 26].

Since this study in the late 1990s the effect of splintering urbanism has even caused a 'post-network' framework to emerge [27], pointing to the fact that no single network is capable of meeting all sustainable outcomes in the face of changing climatic conditions and 'climate urbanism' [27]. This emerged through a wave of urban optimism between 2011 and 2013 followed by a trough of doubt and concern between

TABLE 5.1
Urban Places Versus Urban Digital Spaces

	Traditional Urban Space	Digital City Space
Access	Streets, roads, buildings	Social networks, clouds, software
Scalability	Limited by time constraints, solved by minimising space constraints	Overcome space constraints by minimising time constraints, e.g. through communications
Locations	Defined by territory	Defined by network connections
Visibility	Actual Space – visible	Digital space – reproducible and invisible
Physics of space	Cartesian space – fixed coordinates can be mapped, objectively marking places. They are fixed in space, typically for years, decades, centuries.	Logical space – nodes in a logical time-space network can change dynamically in much shorter timescales. Motion or flux.

2013 and 2017. More recently, Broto and Westman have highlighted a renewed wave of pragmatism for addressing climate change within the context of cities [27, 28].

5.2.2.1 Beware of the New Age Futurologists

Futurology was previously widely criticised for the way in which it describes the effects of digitisation on the cities we live in without drawing from any empirical base at all. Kitchin [23] suggests there are three main reasons for the largely technophilic predictions emanating from authors [29–34].

Firstly, authors fall into the trap of confusing information with knowledge, and so far even with recent OpenAI innovations, such as ChatGPT, artificial intelligence has not yet in 2023 created new knowledge, but rather enables knowledge workers to enhance productivity. Secondly, smart city technologies are just another stage of development, layered on top of pre-existing network and computational technologies. One example of this is that many of the Internet of Things communications protocols leveraged early investments in 2G Edge-based networks, although many have since been phased out. It is not always certain that newer technologies will completely replace older networks and methods, as was recently highlighted by the cessation of the Tradelens blockchain program for distribution [35]. Thirdly, discussions tend to be focused on the impact of the global or local enterprise in the city and the benefits to be gained by the corporations within the confines of their own networked enterprises.

Recently however, futurology has made way for a new breed of management science which is increasingly assessing the interconnectedness and co-existence of global businesses within industry and location-specific contexts. This movement is creating a shift to the adoption of new ecosystem-based business models and data operating models. The shifts often intersect with new enabling technologies, such as AI, social media, analytics and blockchain.

5.3   KEY OBSERVATIONS, PROCESSES AND MODELS

5.3.1   Actor–Network Theory

Figure 5.1 shows the way in which global financial forces are directly related to the local community through the 'sustainable smart city', bypassing the national scale. In the diagram, the role of the smart city is to relate forces, for example new 'green' investment capital in the global network of corporations to the local smart city.

This is achieved by creating a valid representation of locality as a 'green' space in the smart city infrastructure, acceptable to the managers and policy makers, whose function is (in a democratically elected government) to represent the local community interests which are bound up in this complex network [1]. By producing valid representations of sustainable outcomes and incentives, and accurate and timely digital services, capital flows are actively localised within the new global–local 'glocal' network and help to reproduce the city as a place. In essence a dependent space, and space of engagement [36], is created for the local population by producing symbolism, metrics and key performance indicators relating the global sustainability context

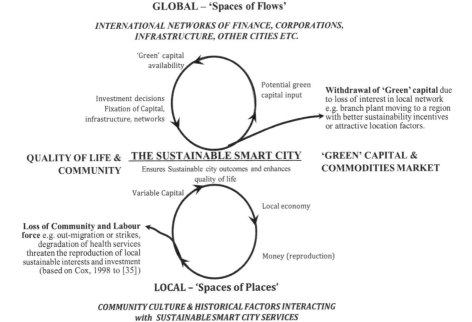

GLOBAL – 'Spaces of Flows'

INTERNATIONAL NETWORKS OF FINANCE, CORPORATIONS, INFRASTRUCTURE, OTHER CITIES ETC.

'Green' capital availability

Investment decisions
Fixation of Capital, infrastructure, networks

Potential green capital input

Withdrawal of 'Green' capital due to loss of interest in local network e.g. branch plant moving to a region with better sustainability incentives or attractive location factors.

QUALITY OF LIFE & COMMUNITY

THE SUSTAINABLE SMART CITY
Ensures Sustainable city outcomes and enhances quality of life

'GREEN' CAPITAL & COMMODITIES MARKET

Variable Capital

Local economy

Loss of Community and Labour force e.g. out-migration or strikes, degradation of health services threaten the reproduction of local sustainable interests and investment (based on Cox, 1998 to [35])

Money (reproduction)

LOCAL – 'Spaces of Places'

COMMUNITY CULTURE & HISTORICAL FACTORS INTERACTING with SUSTAINABLE SMART CITY SERVICES

FIGURE 5.1   Actor–network theory in relation to the global–local dualism in the sustainable smart city context [1].

to those regions' specific local culture and historical context in an interactive process (also known as a dialectic).

Simultaneously occurring global and local forces interact to produce a new form of regulation based on networks rather than dualisms like global/local, micro/macro. This implies that social spaces can be codified into a network and secondly can be contextualized anywhere. Figure 5.1 also shows how the new relations of production and consumption in the information economy produce the sustainable smart city. In this light, the new 'green' capital creates and operates 'a space of flows' which enable certain groups to form their own 'interpersonally networked subculture' and construct a space across the connecting lines of the space of flows [15] (Figure 5.2).

The creation of the space of flows is problematic because 'the more organisations depend on flows and networks, the less they are influenced by the social contexts associated with the place of their location' [15]. Thus the 'space of flows' begins to dominate the 'space of places' [15]. In other words, without the right regulation, the focus for investment can become less fixated in time and space. The task is to 'understand both the economic forces and socio-economic relationships within the world economy and of the unique features that represent local and historical variability' [37]. We have seen this problem manifest in many ways in highly developed nations, such as the decline of the so-called 'high street' in the UK and, more recently, the decline of the shopping mall in the USA caused by the rapid adoption of home-based e-commerce services made available through mobile devices and computers. At a higher super-regional scale, the political–economic phenomenon known as Brexit

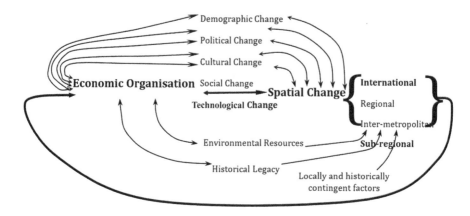

FIGURE 5.2   The interrelationships of economic organisation, spatial change and scale [1].

may also be regarded as a reaction against the spatial and economic changes induced by European policy makers within the 'space of flows' as a reassertion of the UK as a space of 'places' in its own sovereign right.

Contrasting the idea of flow space, others argue that 'permanence' is the key to our understanding of society [38]. In this scenario, change and instability are the norm, and humans attempt to create stability by creating 'permanences' or places from the 'flow of social and material life' [38]. Cities or places 'take on an appearance to a greater or lesser extent through collective "bounding activities"' [38, p. 310].

In this way, the sustainable smart city is an object created out of the need of local communities to harness the wider forces in the global space of flows of green capital and other flows across multiple and fragmented networks. It could also be a defensive reaction against the global economic shift that has switched its focus now to creating sustainable outcomes – and in so doing to enhance the stability of community in the face of the mobility of global 'green' capital or even to shield from 'black swan' events [39, 40].

The model depicted in Figure 5.1 shows the symbiotic interrelationship between historical–local factors in determining spatial change within the context of wider global economic restructuring. This was created within the embedded research of urban policy and human geography at the University of Manchester by me and depicts how a symbiosis between technology factors of smart cities interact with socio-economic forces and the community in real places. The model represents an interpretation at the time, that environmental resource and business investment factors operate at the local level, and so does historical legacy and other locally contingent factors such as culture.

5.3.2   NEO-MARSHALLIAN NODES

A key finding in research on virtual cities [1] was that 'new transnational urban hierarchies' are being created by agencies which operate transnationally, with new inter-urban relationships being forged on the basis of the functional role(s) individual cities are playing in supply chains, delivering digital products and services as well as physical goods.

For example, London emerged steadily as the financial capital in the European financial hierarchy. Whilst it is recognised that 'very different geographical outcomes' will emerge, some evidence suggests that new types of places are emerging, defined by their roles in global production '*filieres*' (a term describing the linkages developed between corporations during the production processes) [1].

Amin and Thrift [41, p. 571] suggest the 'neo-Marshallian nodes' model will replace the core-periphery model, referring to Alfred Marshall and his work on small firm districts in England. They argue that whilst the idea of a core region is less relevant in the world economy, centres are also needed as points 'at which knowledge structures, many of which carry considerable social barriers to entry, can be tapped into'. This new model of urban socio-economic development has also been triangulated more recently in the popular business press, as Steve Case, founder of AOL, observed that cities that were once marginalised are going through a reorganisation and reasserting themselves as 'entrepreneurial powerhouses' within the globally connected network of cities. New Orleans for example is emerging as a new centre for Ed-tech and asserting itself on the global stage as a key node in the education services value chain [42, p. 3].

In this process of connecting cities with industrial networks and global corporations [43] it seems plausible that sustainable smart cities will have the potential to make an impact on global level sustainability targets, whilst coordinating outcomes at the global level. For example, the issue of global climate change will not be resolved singularly by the action of any one city, but will require coordination and cooperation and collaboration across cities on different continents. Simultaneously, such global targets must eventually be implemented and contextualised locally if they are to be trusted, accepted and made operational in the city and local corporate-industrial complex processes. This would warrant further empirical research and policy assessment using an independent critical realist approach. The sustainable smart city, along with the local government structures and other corporate agencies, are operating in a reflexive or dialectic relationship; in other words, the city structures and interactions themselves are recreated by the very digital structures and infrastructure put in place. In the social sciences, this is described by structuration theory [44] and also in the field of biological sciences as autopoiesis, meaning the biological and cognitive processes through which organisms self-regulate, maintain and grow themselves [45, 46].

5.3.3 MATHEMATICS OF CITIES

A sub-field in mathematics is providing a body of research around the constraints of scalability within cities [47, 48]. In parallel, over the last 15 years, the field of topological data analysis has rapidly emerged in data science to assist the application of network science and topology analysis across very large datasets. This is creating a new emergent conceptual framework which applies information theory into the contextual domains created by spatiality and places [49]. This has the potential to be applied to the dynamics of spaces of engagement and information flows across smart city infrastructure [36].

From a green computing perspective, it is imperative that efficient algorithms are applied to the different mathematics of cities, from a physical modelling perspective (climate, topography) and an urban and human geographical analysis (measuring social connectivity, community interactions and flows of information). For example, within the 5G spectrum, mmWave is highly susceptible to physical aspects of the city and climate effects. This has implications for networks ('spaces of flows') relying on wireless communications in the context of climatic change, more severe weather events, and urban micro-climates [50].

Importantly, in the context of geography and urban planning, it also has potential to make the spaces of flows more visible for analysis and tracking. It can also help to make the shift in the concept of a city as a fixed objective structure into one which is focused on connectivity, scale and laws of limitation [6, 28].

No doubt, topological data analysis in the highly contextualised domains of smart cities as real places will provide rich research grounds. This will create an informed quantitative and qualitative approach to solving the sustainable development of cities through infrastructure investments and optimisation. In a geographical context, cities may be regarded as localised contexts, bounded by culture, language and historical factors. In other words, ontologies and sub-domains of highly contextualised knowledge broadcast through communications [51].

When combined with computation and telecommunications, smart cities have the potential to provide an automated, contextual entry point to domain insight within the lattices of highly connected global networks. This is creating an exciting intersection of geographical information theory with mathematics and management cybernetics, where vector-based models can be applied to the domains of pure theoretical space, modelled using topologies, and then meshed onto the physical topographical spaces of cities fixated in time-space.

This could lead to a more computational approach for business automation, such as that provided within Management Cybernetics, as well as academic disciplines like Geography itself, combining pure computational or model spaces with physical space. One example of this is using cosine similarity to create insights of similar patterns between users of a city space or network (topologies) to make recommendations of real physical venues and locations (topographies) in the city to individuals or groups.

5.3.4 Lessons from Management Cybernetics

In the absence of a policy framework, or proven reference architecture for a sustainable smart city, one field of study which offers a useful model of understanding is Management Cybernetics.

Stafford Beer from the 1950s to the 1990s combined the fields of Management Science with Cybernetics and Applied Systems Thinking. He held board-level positions in operations and cybernetics at United Steel and IPC (International Publishing Corporation, now known as Reed Elsevier) and many other large complex global organisations, including in government. Beer was schooled in the UK at Whitgift School in Surrey, and in later years held visiting professorships in cybernetics at many universities, including the University of Manchester Business School (1969–1993) [52].

5.3.4.1 The Underlying Complexity and 'Variety'

Management Cybernetics appears to have lost traction in the 1980s. A noteworthy point is that, just before this time, project Cybersyn was deploying IBM mainframe technology for the Chilean government, led by Beer, when it was overthrown in a coup led by the Chilean army. Despite many successes in systems implementation, this was seen as a high-profile failure for Management Cybernetics, although not directly caused by Beer.

More recent authors and urban policy researchers have attributed the waning of Cybernetics in the 1980s and 1990s to the inherent underlying complexity in the city's socio-political context [53]. Stafford Beer attributed this to the rise of the Internet from ARPANET around the same time. However, it is also noted it did not prevent the increasing maturity of city planning in the progressive deployment of technology and geographical information systems [53].

5.3.4.2 Trust in Systems: Autocracy vs Democracy

Another key observation worth mentioning attributes a lack of trust in a system which was regarded at the time as very technologically advanced and by outsiders as autocratic. Today, management hierarchies in established large corporate organisations are also focused more on remote shareholders pursuing common established financial metrics. City management structures and hierarchies are in theory at least somewhat more democratic and have a priority for citizens over shareholders and the pursuit of financial metrics. However, similar problems of internal competition for limited internal and externalised resources can also lead to challenges of coordination in smart cities between resource networks. Management Cybernetics could provide a solution as well as help to cut through costly bureaucracy through automation, and so enable enhanced qualities associated with agility, responsiveness and flexibility.

5.3.4.3 The Brain of the Firm

One of the core concepts in the 'brain of the firm' [52] is the viable system model which describes the fundamental cybernetics and communications systems that are required to sustain the ongoing viability of a firm's operations in the context of the 'variety' of the business within the operational processes, the management, the strategic objectives and external operating environment. As such, the model provides an informed framework in which to research the viability or sustainability of smart cities. Let's explore it in brief detail.

5.3.4.4 The Viable System Model (VSM)

For the purposes of this chapter, a brief high-level overview of the VSM is described, absent of any diagrams. The objective is to give enough understanding to seed sufficient insight to support a motion that a new approach to a domain-driven data operating model for sustainable smart cities can be supported by the existing body of operational research from Management Cybernetics. A refreshed approach, more focused on enabling democracy and a less autocratic style of processes, can be realised by adopting the practices of modern platform design as well as a more distributed data mesh and 'domain driven' system design. A key aspect of sustainable systems is that they need to be adaptable and flexible if they are to survive [54, pp. 89–139]. In a more volatile, uncertain, complex and ambiguous (VUCA) world, a VSM approach can support the continuous variability in both bottom-up processing and top-down strategic planning. It is proposed that VSM provides a cognitive intelligence system capable of providing anticipatory responses to the demands of sustainable smart city outcomes, for people, planet and profit. (For a detailed introduction to the role of VSM in making change happen, [54] is an excellent entry point.)

Systems 1, 2, 3, 4 and 5 of the VSM define different kinds of activities that occur in any organisation. They are recursive, meaning each system 1 primary activity is itself a VSM. In other words, a VSM can be composed of other VSMs.

System 1 is essentially the most operational, low-level system, which involves the different linear product processes, management information, management and the environmental variables which each process variation interacts with.

System 2 consists of the processes that harvest information in order to optimise between the system 1 linear processes. Beer noted that the management team which is looking after each discrete process needs a higher level of management in order to coordinate across the linear processes. System 2 allows the higher order system 3 to monitor system 1 activities. Beer describes the role of system 2 as helping to prevent oscillations in process throughput. This might be associated with Jevons Paradox, which in the recent literature describes the 'rebound effect' which the consumption of a resource exhibits once technology optimises its efficiency [52].

System 3 is there to control the internal activities of system 1 as well as the super-vision and regulation of system 2. The fourth subsystem is responsible for integrating internal and external factors aimed at balancing the current operational system with the demands of the broader context in future operational requirements. This is also referred to as the 'intelligence function' which is responsible for delivering adapta-tion to the changing external circumstances identified by system 5. It requires chan-nels to the environment and dynamic interactions with system 3. The intelligence of system 4 doesn't operate in a vacuum, but rather, stems from the ethos and character-istics of system 5.

System 5 balances short-term internal functioning with the longer-term internal and external contexts, including decisions to direct the entire organisation to ensure its sustainability. It defines the identity of the organisation and directs insight to the feedback for system 4 and subsequently system 3. It is the basis of the policy of the organisation and the reason for its existence, defining why it exists and operates.

In summary, the VSM can be applied not only to the firm, but to the smart city itself. It provides a rich model for considering the ongoing sustainability of cities by providing a system that incorporates the operating model and data systems needed to control the operations of the city.

5.3.5 CRITICAL REALIST RESEARCH METHODOLOGY FOR SMART CITIES

Future empirical research on sustainable smart cities should focus on the assess-ment, validation and optimisation of coordinated sustainable smart city outcomes. This will contribute to a growing corpus of knowledge that will assess the current state of applications, make recommendations and roadmaps to help validate policies, and provide ongoing optimisations aligned to technical process improvements at different scales.

Researchers in academia, government and corporations can consider the use of multiple research methods in order to triangulate the complex linear processes as well as the feedback systems in play in the creation of sustainable local city level and global level sustainable outcomes.

The use of multiple methods directly links the methodological opportunities presented by this fragmentation in urbanism and embraces the fact that many forces are in operation in making sustainable smart cities viable. It also represents a polyvocal approach to the research design. In this way 'the researcher does not necessarily privilege a particular way of looking at the social world' [8]. This was highlighted during seminars on 'Social Informatics' in Manchester [55].

Multiple methods may be understood as being the situation in which a number of complementary methods are employed to address different facets of a research question, or to address the same question from different perspectives [8].

The use of multiple methods may be particularly apt for the study of sustainable smart cities, which could have 'multiple realities'. Computer mediated communications technology systems exhibit a fair amount of interpretative flexibility. That is, they can mean different things to various groups of people and their use continues to be reinterpreted with the passing of time [56].

Furthermore, technology is rapidly evolving. The political reality of a sustainable smart city may differ from its economic and social realities. Another way of looking at the multiple realities would consider the difference between the original purpose for which they were developed and funded and the way in which they are appropriated by citizens, local governments, transient populations, and local and remote corporations. In the context of virtual communities, this process was highlighted during seminars on 'Social Informatics', headed by Rob Kling at Manchester Metropolitan University [57].

One key idea of social informatics research is that context matters in influencing the ways that people appropriate technology and its consequences for work, organisations and other social relationships. Here, social context is indexed by a particular incentive system for using technologies, or disincentives for using technologies, in particular ways [55, 57].

We can relate this back to Conway's Law in that the original funding and model for the smart city platform may well be an expression of the original organisation(s) which developed it. Through use and iterative agile development, the platforms can be appropriated by the different agencies, managers, administrators as well as everyday users.

5.3.5.1 Applying Critical Realism to Sustainable Smart Cities

The three guidelines in executing critical realist research in human geography are iterative abstraction, qualified grounded theory method and methodological triangulation [33, 58]. Furthermore, Harvey [14, 38, ch. 1] holds great emphasis on the need to study 'processes, flows, fluxes and relations over the analysis of elements, things, structures and organised systems' for the purposes of defining his principles of dialectics. He argues that this a deep ontological principle: 'things, structures, systems

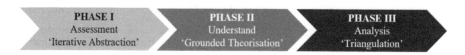

FIGURE 5.3   Phases of critical realist methodology.

do not exist outside of or prior to the processes, flows and relations that create, sustain or undermine them'. Thus the object of study through the research methodology can extend from the structure of the smart city as an object, to the processes of its sustainable production and its ongoing viability (Figure 5.3).

5.3.5.2 The Need for a Progressive Methodology

Why is a broad methodology needed for the empirical research on smart cities? It is needed to explore the underlying mechanisms of sustainable smart city development and the reproduction of geographical scale. In other words, the interactions occurring through the sustainable smart city networks and their processes at the local city level and global macro-level are complex. For greater understanding of the complexity in the creation of a theory and methodology see [1, 26, 59].

Sustainable smart cities should be viable when they are able to simultaneously serve the local city dwellers, improve the peripheral development through the knock-on effects of economic development, as well as act in coordination with research and overcome global-scale challenges such as climate effects. Above all, an approach which addresses human inclusivity and diversity is needed. The new cultural geography's concern with human experience and meaning requires subtle research techniques 'capable of exploring the realities of everyday lives as they are explained by the people that live them' [60, p. 458]. This, Pile argues, has driven the adoption of qualitative methods, which are 'empirically sensitive and politically radical' [60, p. 458]. In this sense, they are entirely suited to critical realism for exploring human agency; and the smart city managers, administrators and funding decisions are considered as real causes shaping the way in which a locality is represented in the global economy through the services, images and context it creates.

The analysis of virtual cities on the global–local nexus demonstrates how textual analysis can combine with quantitative statistical methods to study the transient phenomenon of virtual cities in depth from multiple perspectives, thus enabling an inclusive and polyvocal approach [1].

Importantly for smart cities research, the need to adopt the ontology and language of the various study domains was noted in the study. Research frameworks for sustainable smart cities have a variety of disciples and a body of research to build on, across the domains of Geography (both physical and human sub-disciplines), as well as Management Science, Cybernetics, Political Science, and the technical fields of data and information management.

5.3.6 A Data-Driven Approach for Making Smart Cities Sustainable

This section now briefly covers domain-driven approaches to technology elaboration and then discusses the potential of smart cities to capture opportunities in local development. City planners and city authorities can augment the data mesh approach in order to better integrate the spaces of flows with the city as permanent places.

Today, many researchers have focused on new models of distributed platform design and development to enhance agility. Operating models using more distributed team topologies can help streamline and navigate complexity in system development [61]. This could be a useful area of future research in the context of 'domain driven' platform design for sustainable smart cities. More complex team organisation models other than traditional hierarchies could help align the sustainable smart city to the variety that's inherent in designing and building for the complexity and so help build better trust.

In the light of new approaches to digital platform design and development, which are arguably more democratic and engaging and less autocratic than the centralised mainframe systems of the 1950s and 1960s, it is worth reconsidering the key concepts of Management Cybernetics, especially in the light of the observed processes and 'empirical laws' mentioned in this section.

'Data mesh' is a more recent approach to data management and offers a number of alternative options for data management from traditional centralised information governance models [62]. It does this through the process of creating 'data as a product', and enabling access to data using application programming interfaces (APIs).

5.3.6.1 A Data-Driven Operating Model for Smart Cities

A centralised and federated approach to data management is needed to enable smart cities to collaborate effectively across diverse groups, with increasing autonomy, and contribute to a centralised well-governed operating model. It is argued that using a data mesh approach to data management enables common intelligence and model features to be established as an ongoing asset, whilst enabling the local agility and flexibility needed to adapt a city's services to the ongoing population, infrastructure and network changes.

A data mesh enables a more decentralised model of data-driven governance for the sustainable smart city. In Management Cybernetics, the ongoing 'variety' and variability of city level processes are modelled and captured in the 'System 1' component of a VSM. Furthermore, ecosystem interactions and communications that operate at the strategic level in the spaces of flows (Figure 5.1) could be modelled within a VSM as system 5. This would model the processes through which the sustainable smart city is able to direct the resources within the city to ensure its long-term sustainability. Systems 2, 3 and 4 could be created in partnership with business, firms and government at national and local city levels to monitor, calibrate and communicate smart city projects towards inclusive, democratised and desired sustainability targets within the global context of sustainable development.

A distributed VSM could also be applied to systematically model the interactions of the city with firms. Data management within firms is better funded and more mature than in smart city deployment. Within global firms, data management maturity is progressing to a level where ecosystem level intelligence and composite service patterns can be managed and automated through digital platform businesses. Platform business models are capable of operating *within* sustainable smart cities but also collaborating *across* them. This enables the spaces of flows to integrate and

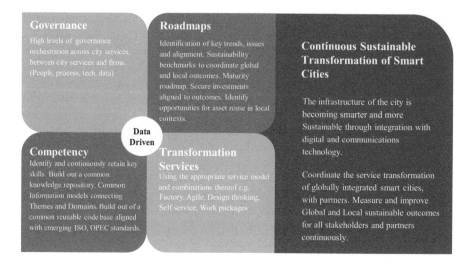

FIGURE 5.4   Data-driven operating model dimensions for sustainable smart cities.

contextualise into the city domain (spaces of places), the subdomains as well as diverse social perspectives from local historical and cultural aspects.

A data-driven operating model across global smart cities can be organised by augmenting existing investment streams from government at the national level, the local city level in the office of the town hall, as well as inward investment by local firms and foreign direct investment to create global and local level sustainability outcomes. A typical operating model derived from global businesses would need to cover the dimensions of governance, roadmap alignment and coordination, services as well as competency creation. By taking a data-driven approach, smart cities can adopt a composite service model enabling services to be composed from other services and so promoting the reuse of projects and services (Figure 5.4).

5.3.6.2   Current Challenges and the role of Common Standard Key Performance Indicators (KPIs) to Overcome Them

The OECD has highlighted a number of challenges to smart city initiatives in OECD countries including budget constraints, lack of supportive infrastructure, lack of human resources to analyse the data, and potential territorial divides. We can also add to this the potential conflict between local regulatory frameworks which provide directions on how local citizen data can be stored and processed versus the requirement to augment global resource and cloud technology platforms to help overcome technical challenges of access to skills and scalability.

Global delivery coordination across cities in digital transformation programmes and delivery is now a well proven and robust model, which accelerated during the lockdowns of the COVID-19 pandemic. Today digital delivery programmes are deploying automation through DevOps and the continuous integration and continuous deployment (CI/CD) of software, often on cloud platforms. The costs of sustainable smart city data and analytics can be optimised by harnessing global delivery resources, as well as adopting cloud native CI/CD development methods. The use of modern technology infrastructure practices and resources such as containers can enable distributed development and open source software development practices [63] which could implement emerging global standards and KPIs for sustainable smart cities.

5.3.6.3   Emerging Common Standards to Enable Digital Asset Reuse

Sustainable smart city standards and indicators have recently been updated in the ISO with prescribed data sources. This is covered in ISO 37120:2018 sustainable cities and communities. The indicators of smart city services and quality of life are set out in ISO 37122 and resilient city standards are prescribed in ISO 37123. This group of sustainability standards for smart cities would highly benefit from the creation of a blueprint for sustainable smart city digital services through a platform. Such a digital platform would have the ability to contextualise local services, whilst creating a standard metadata catalogue repository which could be adopted by the city, government and firms. This could create an innovation ecosystem which promotes collaboration and reuse across sustainable smart city initiatives through the adoption of common standards and APIs.

Adopting cloud models of development, even within highly localised 'sovereign' data clouds, can establish practices that enable better return on investment. By adopting common standards and APIs, software projects can be promoted and shared with other smart cities or reused in the future via container lifecycle management. This would support a close alignment between roadmaps with city budgets and the sometimes-varied political and financial contexts through which smart city initiatives are constrained to evolve.

5.3.6.4   Benefits of a Domain-Driven Approach

A domain-driven design approach for data platforms is distinct from a 'use case'-driven design approach. This focuses on modelling software and data to match a domain, and utilises input from the domain experts and stakeholders such as urban planners, policy makers, citizen groups and service providers. The structure and language of the platform would have a close resemblance to that of the city itself through a creative collaboration between technical and domain experts in order to refine a conceptual model that addresses city sustainability challenges.

Domain-driven architecture can provide benefits such as flexibility. Microsoft recommends it for complex domains where the model provides direct benefits in formulating a common elaboration of the domain [64], making it well suited for smart cities. Domain-driven design acknowledges multiple kinds of models that can be systems of abstraction that are used to create a common language. This is well aligned to the empirical research framework mentioned in this chapter, critical realism, which is recommended for adopting multiple methods to enable a 'polyvocal' approach. A data-mesh-enabled domain-driven platform can capture polyvocal data feeds from diverse groups and stakeholder domains across the city. In this way, a decentralised, more democratic management cybernetics model of the ongoing evolution of sustainable smart cities can embrace the requirements for viability as well as promoting qualities such as trust and democracy amongst diverse groups of citizens and other stakeholders.

5.3.6.5   Capturing Opportunities in Local Development

Interactions between the global firms and locality are historically problematic. As Peter Dicken explains, transnational companies are formed of 'Web's of Enterprise' within networks of relationships in the world economy [65]. Yet global firms are dependent on their location for a competitive edge within the global economy and depend on local resources for leadership within networks of 'internalised relations' and 'externalised relations': 'there is a basic tension between globalizing pressures on the one hand and localising pressures on the other, which makes any notion of a pure global integration [...] completely inappropriate'[65, p. 207].

Taking this argument further, the results of the research on virtual cities [1] uncover many location specific interactions between the local and global scales. In fact the dataset that was collected does expose evidence that 'nodalisation' or place-making between cities on the global scale is occurring.

The polarisation of attitudes between local negativity and wider scale optimism suggests that the virtual city could well be creating new 'imagined communities' at the scale of the global or macro-regional. Whilst it could be said that the virtual city is destroying local community, it could be enhancing a new distant sense of place on the global scale as is described in the chi-squared statistical analysis demonstrating the relationships between global and local perceptions of virtual communities [1, pp. 81–82] ('time for family and friends'; 'world' residual, positive = 5.7 at 0.003% chance of error and 'communicating with locals online', 'world' positive residual = 8.3 but with a high 1% chance of error) [1, p. 82]. Critically, this is, in my opinion, symptomatic of the type of 'glocal' mediation that occurs through new technologies for three reasons: it indicates a low level of local-to-local interaction; a high level of global-to-local interaction; and the possibility of bypassing the national scale as shown by the greater degree of UK-based negative attitude towards communicating with local people [1, p. 82]. Consequently a globally networked sense of place is created [1]. The implication for smart cities is that there is a high degree of propensity already occurring in digital communication channels for regional level governance and controls to be either enhanced or bypassed completely through the deployment of communications and media technologies into the city infrastructure. If this is a real possibility, it has implications for ties and joint outcomes between local and global communities, cities and nation states, and even between nation states and supra-regional bodies such as the EU.

As far back as 1999, the propensity for a highly positive sense of place broadcast from the city and communicated to other global locals was far more positively received by respondents across the globe, when compared to other cities in the UK. It not only highlights the competitive nature of cities and locations within the UK, it also suggests a future research paper could study the propensity for smart cities to collaborate globally to drive interrelated sustainability outcomes.

5.3.6.6   Increasing Connectivity between Local Organisations and Environmental Externalities

The smart city is a means to reduce social transaction costs by providing a space of permanence in the spaces of flows and by providing an endurable node within the dynamic "spaces of flows".

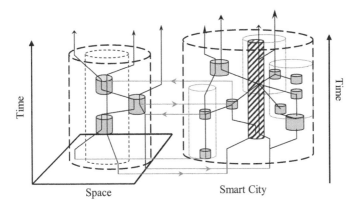

FIGURE 5.5   Representation of daily time-space paths in smart cities [1].

Figure 5.5 is extended from research on daily time-space paths [66] and shows representation of the increased number of transactions that can occur over time when the daily time-space paths interact and cross in a smart city infrastructure. The nodal coordinating role of the smart city over time is shown (striped cylinder). All transactions occurring in the smart city originate from existing 'domains' (grey cylinders) and 'stations' (dashed cylinders) in real space. The figure shows how the smart city (striped cylinder) is a potential media node to coordinate and maintain the locality as it is represented and interacted with through the internet and other digital and social media platforms.

Hägerstrand [67] produced a qualitative, humanistic study on the nature of people as agents, engaged in projects that take time through movement in space. He described individual 'life paths in time-space', represented in Figure 5.5 as the arrowed lines. They ranged from daily routines such as shopping to those on the scale of a lifetime such as migration. The only constraint to movement was the friction of distance. Social movements usually had 'coupling constraints' that dictated the need to have the time-space paths of at least two individuals to intersect to accomplish any social transaction. These transactions occurred within a geographical pattern of availability (being places where certain economic activities occur, e.g., shopping, indicated by the dotted cylinders) and domains (places for social interaction). Today, friction of distance and coupling restraints are less relevant due to the invention of the smartphone and increased bandwidth enabling real-time video calling. If we recall Jevons paradox, the removal of a constraint by innovation will typically increase the demand for that resource, in this example – social interactions. Figure 5.5 shows how the evolution of communications is enabling the number of social transactions accomplished to increase, and to reduce in cost. Importantly, all transactions in cyberspace must originate from and return to real space. Consider how the smart city aims to create scale by localising and rationalising social transactions within the context domain of a physical city. This has implications for sustainable smart cities because the communications and connectivity infrastructure of the city itself play a key role in optimising the social transaction costs, facilitating the delivery of efficient digital services, but also potentially bypassing national and regional levels of political governance over the city.

5.3.6.7   Modelling Connectivity versus the City as a Space of Permanence

Relative space can 'be understood as a relationship between objects which exists only because objects exist and relate to each other' [68, p. 13]. There are 'multiple geometries from which to choose, and the spatial frame depends critically on what is being relativized and by whom' [69, p. 272]. This has the potential to create conflicts in the smart city over local representations of place and space, depending on what is being relativised and by whom. Potentially, multiple urban realities could coexist in the same physical space, as digital spaces of flows and representations interact with the city and through the social structures within it [1].

Historically, one of the key focal issues with the community network movement is social inclusion [70], but if we are to become truly inclusive in the information society, if the transition is to benefit everyone, then the emphasis ought to shift to 'representation'. If the new politico-symbolic space is 'cyberspace', and the smart city is an expression of that movement, then those groups absent will become disempowered to a greater extent in the information society. As one virtual city manager described it: 'They want to do everything, but they're not going to do anything, they're not going to they're not doing everything, they can't do everything. It's like building a city isn't it?' 'We just want to help to bring the whole thing together all the communications together so the people of Leeds and everybody else has a wonderful service' [1, p. 10].

A further review of the new connected models of cultural-cognitive configurations of space and place as well as leading research in economic geography is offered in [71].

How this newly digitised economic space is modelled, processed, and visualised presents a new technical challenge. Where multiple representations of spaces connect with cultural-cognitive configurations of spaces and place, smart city digital services will require adequate infrastructure upon which to capture, represent and process the real-time event-based flows from every perspective.

How can this space of flows be modelled and captured for analysis by city planners? In information science, graph-based data modelling is distinct from entity relationship modelling. It can help capture the spaces of flows through a model that is composed of nodes, edges and properties. Nodes are instances of data that represent objects which are to be tracked. Edges represent the relationships between nodes, and the properties represent information associated with nodes. Edges also can store attributes that are captured over time that help to define the spaces of flows and the expression of relative space. In this way, a rich view of the relationships between objects within the smart city can be monitored and analysed and a correspondingly new insight derived. In the VSM, graph models could be used to capture the interrelationships within the processes of system 1, but also the connectivity between systems 1–5 (Table 5.2).

Classical entity relationship modelling, formalised by Codd in the 1970s with the 'Third Normal Form' or 3NF model of data, requires that a fixed entity or 'thing' is modelled and fixed in the logical and physical representation at design time. Relationships between objects are static and have to be defined once at design time. This is opposed to the graph model whereby relationships are rich, evolutionary, ephemeral and diverse. They can also be discovered by mining event data and also

TABLE 5.2

Identifying Differences and Distinctions When Modelling 'Spaces of Places' and 'Spaces of Flows' for Sustainable Data and Insights

	Spaces of Fixed Places	Spaces of Flows
Modelling approach	City as object	City as flows
Design focus	Entities	Relationships, edges
Geographical Information System (GIS) data	Raster and vector (as planar object; border of an object)	Vector as connection
Mathematics	Set theory, difference	Topology, connections
Algorithmic affinity	Classification, clustering of objects in a real bounded space	Cosine similarity; measurement of theoretical and 'model' distance between edges in a vector 'space'
Examples of what is modelled	Classical geographic space is measurable using physical geographical reference of space and time	Ability to measure classical fixed spatial dimensions as well as theoretical and fluid 'spaces' with high dimensionality, e.g., differences in citizen or consumer preferences or relationship to locations in space and time
Database	Relational	Graph database, vector database, NoSQL
community	Real community, neighbours in the city	Virtual community; communities of interests across city spaces, between cities and firms
Sustainable objectives	Local outcomes in the city	Globally coordinated outcomes aligned to key themes, e.g., climate change

inferred by machine learning algorithms such as cosine similarity. In contrast, the relativity in the 3NF model is fixed in representation and cannot have properties assigned to it. The majority of firms today all have their routines in system 1 running through this model of information, but recent analyst research suggests that the graph technology adoption is set to grow exponentially over the next five years [72].

5.3.6.8 Data Mesh Approach for Data Management of Sustainable Smart Cities

Research on data management approaches within cities remains relatively sparse. Due to time limitations, it was not possible to research data management deployments fully. A brief literature review uncovered studies from 2015 onward [53]. This could be an area of future research.

A data mesh approach within the context of smart cities would be a step change in the kind of data capabilities deployed today. Within the context of data management in the firm, it also represents a paradigm shift from centralised data integration and management for data reporting purposes, to a 'data as a product' approach. This introduces process and organisational changes that will see the Mayors Office or City Hall managing data as a tangible asset for operational uses. A data mesh attempts to link data producers in more agile and meaningful ways, and importantly reduce the overheads of IT management and provisioning. If implemented correctly and governed in the right way, this can drastically reduce time to market for new digital products and services by increasing the level of autonomy whilst promoting the reuse of design and development pipelines, and so reduce data engineering overheads. This is in addition to the benefits of adopting a domain-driven approach to data engineering.

The decentralisation strategy of a data mesh can also help sub-domains within cities accomplish outcomes aligned to the needs of different stakeholders and representations in the city more efficiently. Other enabling technologies coming to the fore include service mesh, containers, software defined networking in communications and cloud services providers, microservices, real-time data streams which can process flowing data as well as data at rest, parallel processing technologies, and sovereign data clouds.

5.4   CONCLUSIONS AND NEXT STEPS IN RESEARCH

This chapter has made a brief attempt to bring a critical realist framework to bear for the analysis of sustainable smart city outcomes and which is agnostic as to the location of the city itself and the stakeholder interests involved. It is hoped that in doing so, future researchers will be able to leverage and reuse the independent research as it was applied to virtual cities in the late 1990s for the rapidly emerging global network of smart cities to be aligned with the sustainability agenda. As such, it will be possible to adopt a truly polyvocal method to assess sustainable outcomes from multiple perspectives. In doing so, the sustainability for diverse groups within cities could be captured, assessed and communicated. Moreover, we can learn from the lessons of the past and so avoid a repeat of problems caused by organisations holding themselves accountable without an independent objective framework [73, 74]. A metrics framework for sustainable outcomes should not only be standardised to enable instantiation at different 'levels' across 'spaces of flows', but also consider the needs for a balance between authenticity and measurement across stakeholder interests. A review of existing research [75] and literature within the context of the methodologies proposed in this chapter along with recommendations would be useful.

This chapter has identified a number of future research topics that could prove effective at undergraduate and doctoral level, within and across geography, business, computer science, urban development, management and cybernetics.

A comparative study of smart city initiatives across regions of highly differentiated levels of economic development could lead to insights about overcoming historical inertia within aspects of urban planning and historical infrastructure investment. Conversely, cities in countries with less nationalised urban planning laws have an opportunity to transform more rapidly to embrace a network approach for sustainable smart city outcomes. This is of course assumed that adequate communications and technology infrastructure investment and resources are available. What the minimum viable platform requirements are could again be another topic of focus. This topic could be co-owned within the telecommunications firms in partnership with academic research councils and national and local governments as well as a startup ecosystem. In more developed markets, this could be a substantial market opportunity to leverage reuse of digital assets and infrastructure across larger scale urban deployments into adjacent markets. In other large global cities experiencing challenges with basic health levels of citizens, sanitation and poverty, a minimal viable smart city platform in partnership with local telecommunications providers could help host a sustainable smart city 'green computing' project. This would be deployed in cities and aid the efficient collection of datasets to help the projects sustain a data-driven approach, whilst conforming to local and national data control and privacy regulations.

Another area of research intersecting computer science with organisation design could study the potential for operating models that embrace distributed team topologies [61] for centralised and distributed governance. This could be a useful area of future research in the context of 'domain driven' platform design for sustainable smart cities as well as helping limited resource and capital to collaborate remotely on smart city development programmes and overcome the challenges [2]. Such a research framework would need to be sufficiently funded to support the ongoing monitoring and reporting and to run in partnerships with telecommunications and technology infrastructure providers who could help navigate highly complex areas of local data regulation and industry collaboration.

Figure 5.6 shows how a centralised and federated model of information and data governance could operate to manage the ongoing viability of the data models required to support sustainable outcomes for the smart city. The equivalent level of systems in Beer's viable system model would correspond as follows: system 5 is City Hall (in Figure 5.6), responsible for governing the ongoing evolution of the sustainable outcomes of the smart city. System 4 is the combined work of Central Data Governance and Smart City Sub Domain services, integrating internal and external factors aimed at balancing the current operational systems with the demands of the broader context in the future operational requirements. This is also referred to as the 'intelligence function'. The Centralised Standard Meta-Model is the equivalent of system 2 in the VSM. System 3 is the Central Process Management (Process 0) box, and system 1 processes are mapped across the federated domains of City domains 1, 2... n. System 2 allows the higher order system 3 to monitor system 1 activities. In the context of

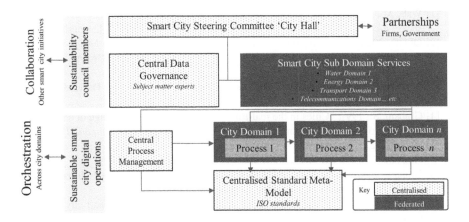

FIGURE 5.6 Centralised-federated model of information governance for the sustainable smart city.

sustainability, the role of system 3 is to make operational decisions that avoid the conflict of sustainable outcomes at different scales, and across different city domains or networks.

The assessment, validation and optimisation of coordinated sustainable smart city outcomes between cities is another identified research area. This will contribute to a growing corpus of knowledge on the neo-Marshallian node mode of urban development [71] in contrast to traditional core–periphery models in geography. Such a study could assess the current state of smart city applications, processes and data capabilities and make recommendations and roadmaps to help validate policies and provide ongoing optimisations aligned to improvements in outcomes at different scales as well as improving returns on investment. Studies could also leverage the VSM from management cybernetics to create a gap analysis in the feedback systems. This could lead to further recommendations about the ongoing viability of a sustainable smart city.

Leveraging emerging global standards in KPIs for sustainable smart cities, combined with best practices, could help overcome challenges associated with cost efficiencies in the design, development, roll out and maintenance of applications and digital services. Adopting cloud models of development, even within highly localised 'sovereign' data clouds, could establish practices that enable better returns on investment. By adopting common standards and APIs, software projects could be promoted and shared with other smart cities or reused in future by using container lifecycle management.

A fuller empirical survey and maturity model relating to active and current data management deployments for sustainable smart cities was not possible due to time constraints. This would be a fertile area to research using the critical realist framework and to that end, I welcome industry, city planners, city officers, advisors and consultants to collaborate through the research landing page [76].

REFERENCES

[1] Wall, M., 1999. *Virtual cities on the global-local nexus. Glocalising space in cyberspace: A critical realist approach.* University of Manchester Press, Manchester.

[2] OECD, 2020. *Measuring smart cities performance. Do smart cities benefit everyone? Scoping note*, 2nd OECD Roundtable on Smart Cities and Inclusive Growth. December 2020.

[3] Mulgan, G., 1998. *Connexity: How to live in a connected world.* Harvard Business School, Harvard.

[4] Graham, S., 1997. "Cities in the real-time age: The paradigm challenge of telecommunications to the conception and planning of urban space". *Environment and Planning A* 29, 105–127.

[5] Graham, S., 2001. *Splintering urbanism.* 1st Edn. Routledge, London.

[6] Graham, S., Marvin, S, 2022. "Splintering urbanism at 20 and the 'Infrastructural Turn'." *Journal of Urban Technology* 29 (1), 169–175.

[7] TMForum, 2023. Smart city forum. https://www.tmforum.org/smart-city-forum/10 (January 2023).

[8] BBC News, 2013. Google Faces Streetview wi-fi snooping action. Retrieved from https://www.bbc.co.uk/news/technology-24047235 (December 2022).

[9] BBC News, 2022. Meta settles Cambridge Analytica scandal case for $725m. Retrieved from https://www.bbc.co.uk/news/technology-64075067 (December 2022).

[10] Lefebvre, H., 1991. *La Production de l'Espace*. Anthropos, Paris; English translation, 1991 *The Production of Space*. Blackwell, Oxford.

[11] Soja, E., 1989. *Postmodern geographies: The reassertion of space in critical social theory*. Verso, London.

[12] Latour, B., 1987. *Science in action: How to follow scientists and engineers through society*. Open University Press, Milton Keynes.

[13] Thrift, N., 1996. "New urban eras and old technological fears: Reconfiguring the good-will of electronic things". *Urban Studies* 33 (8), 1463–1493.

[14] Webber, M., 1968. "The post city age". *Daedalus* 97, 1091–1110.

[15] Castells, M., 1989. *The information city: Information technology, economic restructuring and the urban-regional process*. Blackwell, Oxford.

[16] Dutton, W.H. (Ed.), 1996. *Information and communication technologies: Visions and realities*. Oxford University Press, Oxford.

[17] Fathy, T., 1991. *Telecity: Information Technology and its impact on city form*. Praeger, London.

[18] Laterasse, J., 1992. "The intelligent city: Utopia or tomorrows reality?". In F. Rowe & P. Feltz (Eds.), *Telecom, companies, territories*. Presses de L'ENPC, Paris.

[19] Martin, J., 1978 *The wired society*. Prentice-Hall, London.

[20] Rheingold, H., 1994. *The virtual community*. Addison-Wesley, Reading, MA.

[21] Poster, M., 1995. *The second media age*. Polity Press, Cambridge.

[22] Anderson, B., 1991. *Imagined communities: Reflections on the origins and spread of nationalism*. Verso, London.

[23] Kitchin, R., 1998. *Cyberspace: The world in the wires*. Wiley, Chichester.

[24] Dong, J., 2006. Splintering urbanism and sustainable urban water management. Research Article, Griffith University Press.

[25] Couclelis, H., 1996. "Editorial: The Death of Distance." *Environment and Planning B* 23, 387–389.

[26] Harvey, D., 1989. *The condition of postmodernity. An enquiry into the origins of cultural change*. Blackwell, Cambridge, MA.

[27] Castán Broto, V. and Westman, L.K., 2020. Ten years after Copenhagen: Reimagining climate change governance in urban areas. *Interdisciplinary Reviews: Climate Change*, *11*(4). https://doi.org/10.1002/wcc.643

[28] Vanesa Castán Broto, 2022. "Splintering urbanism and climate breakdown". *Journal of Urban Technology*, *29*(1), 87–93. DOI: 10.1080/10630732.2021.2001717

[29] Makridakis, S., 1995. "The forthcoming information revolution: Its impact on society and firms". *Futures* 27 (8), 799–821.

[30] Williams, R., 1998. "Beyond the dominant paradigm: Embracing the indigenous and the transcendental". *Futures* 30, 223–233.

[31] Stevenson, T., 1998. "Netweaving alternative futures: Information technocracy or communicative community?". *Futures* 30, 189–198.

[32] May, G.H., 1998. "New technology and the urban environment". *Futures* 30 (9), 887–899.

[33] Phillip, L.J., 1998. "Combining quantitative and qualitative approaches to social research in human geography - An impossible mixture?". *Environment and Planning A* 30, 261–276

[34] Tehranian, M., Ogden, M., 1998. "Uncertain futures: Changing paradigms and global communications". *Futures* 30, 199–210.

[35] Litan, A., 2022. *Maersk IBM TradeLens shut down after ASX cancellation*; *Ending an era of costly enterprise blockchain*. Gartner, New York.

[36] Cox, K.R., 1998. "Spaces of dependence, spaces of engagement and the politics of scale, or: Looking for local politics" *Political Geography* 17 (1), 1–23.

[37] Johnston, R.J., 1984. "The world is our oyster". *Transactions Institute of British Geographers* 9, 443–459.

[38] Harvey, D., 1996. *Justice, nature, and the geography of difference*. Blackwell, Cambridge, MA.

[39] Taleb, N., 2010. *The Black Swan: The impact of the highly improbable*. Penguin, London.

[40] Taleb, N., 2013. *Antifragile: Things that gain from disorder*. Penguin, London.

[41] Amin, A., Thrift, N., 1992. "Neo-Marshallian nodes in global networks." *International Journal of Urban and Regional Research* 16, 571–587.

[42] Case, S., 2016. *The third wave. An entrepreneurs vision of the future*. Simon & Schuster Paperbacks, New York.

[43] Amin, A.; N. Thrift, 2009. "Neo-Marshallian nodes in global networks". *International Journal of Urban and Regional Research, 16*, 571–587.

[44] Giddens, A., 1991. "Structuration theory: Past, present and future". In: C.G.A. Bryant & D. Jary (Eds.), *Giddens' theory of structuration: A critical appreciation* (pp. 201–221). Routledge, London.

[45] Varela, F.G., Maturana, H.R., Uribe, R., 1974. *Autopoiesis: The organization of living systems, its characterization and a model*. University of Chile, Santiago.

[46] Turpin, M., 2016. "Autopoiesis and structuration theory: A framework to investigate the contribution of a development project to a rural community". *Systems Research and Behavioural Science* 34 (6), 671–685.

[47] West, G., 2017. *Scale: The universal laws of growth, innovation, sustainability, and the pace of life in organisms, cities, economies and companies*. Penguin Random House, USA.

[48] West, G., 2018. "Opinion: The science of sustainable cities". In Issue 13 *Urban Solutions*, Jul. 2018.

[49] Feng, M., Porter, A., 2021. Spatial applications of topological data analysis: Cities, snowflakes, random structures, and spiders spinning under the influence. *Physical Review Research*, Revision 3.

[50] Love, D.J., Castellanos, M.R., Song, J., 2017. "Millimetre wave communications for 5G networks". In Wong, W.S., et al. (E.s), *Key technologies for 5G wireless systems*. Cambridge University Press, London.

[51] Hagerstrand, T., 1986. "Decentralization and radio broadcasting: On the possibility space of a communication technology". *European Journal of Communication* 1, 7–26.

[52] Beer, B., 1972. *Brain of the firm: A development in management cybernetics*. 2nd Edn. Wiley & Sons, London.

[53] Kitchin, R., 2017. "The realtimeness of smart cities, Maynooth University Press (IE)". *Tecnoscienza Italian Journal of Science and Technology Studies* 8 (2), 19–41.

[54] Hoverstadt, P., 2020. The viable system model. In Reynolds, M., Holwell, S. (Eds.), *Systems approaches to making change: A practical guide*. Springer, London.

[55] Kling, R., 1998. *Transcript of discussion paper social inclusion for social informatics*. Seminar Series at Manchester Metropolitan University, 12th August 1998.

[56] Paccagnella, 1997. "Getting the seats of your pants dirty: Strategies for ethnographic research on virtual communities". *Journal of Computer-Mediated Communication, 3*(1). https://doi.org/10.1111/j.1083-6101.1997.tb00065.x

[57] Kling, R., Lamb, R., 1996. *Bits of cities: Utopian visions and social power in place-based and electronic communities*. Indiana University.

[58] Yeung, H., 1997. "Critical realism and realist research in human geography: A method or a philosophy in search of a method". *Progress in Human Geography* 21 (1), 51–74.

[59] Graham, S., Marvin, S., 1996. *Telecommunications and the city: Electronic spaces, urban places*. Routledge, London.

[60] Pile, S., 1991. "Practising Interpretative geography". *Transactions of the Institute of British Geographers* 16, 458–469.

[61] Skelton, M., Pais, M., 2019. *Team topologies: Organizing business and technology teams for fast flow*. IT Revolution Press, Portland.

[62] Dehghani, Z., 2022. *Data mesh: Delivering data-driven value at scale*. O'Reilly, Media.

[63] Burns, B., Oppenheimer, D., 2015. *Design patterns for container-based distributed systems*. Google Whitepaper.

[64] Microsoft, 2009. *Microsoft application architecture guide*. http://msdn.microsoft.com/en-us/library/ee658117.aspx#DomainModelStyle

[65] Dicken, P., 1998. *Global shift [Transforming the world economy]*. 3rd Edn. PCP, London.

[66] Hagerstrand, T., 1970. "What about people in regional science?" *Papers of the Regional Science Association* 24, 6–21.

[67] Hägerstrand, T., 1975. "Space, time and human conditions". In A. Karlqvist, L. Lundvist, & F. Snickars (Eds.), *Dynamic allocation of urban space* (pp. 3–14). Saxon House, Farnborough, UK.

[68] Harvey, D., 1973. *Social justice and the city*. University of Georgia Press. http://www.jstor.org/stable/j.ctt46nm9v

[69] Harvey, D., 2006. *Spaces of global capitalism: Towards a theory of uneven geographical development*, Verso, London.

[70] IBM, 1997. *The net result: Social Inclusion in the Information Society*, IBM Press, London. Retrieved from http://www.local-level.org.uk/uploads/8/2/1/0/8210988/netresult.pdf on 21 November 2023.

[71] Suwala, L., 2021. "Concepts of space, refiguration of spaces, and comparative research: Perspectives from economic geography and regional economics". *Forum: Qualitative Social Research* 22 (3), Art. 11.

[72] Avidon, E., 2021. Gartner predicts exponential growth of graph technology. *TechTarget article*. Retrieved from https://www.techtarget.com/searchbusinessanalytics/news/2525 07769/Gartner-predicts-exponential-growth-of-graph-technology (12 Jan 2023).

[73] Hoskin, K., 1996. The 'awful idea of accountability': Inscribing people into the measurement of objects. In *Accountability: Power, Ethos and the Technologies of Managing*, R. Munro and J. Mouritsen (Eds). London, International Thomson Business Press.

[74] Muller, J. Z., 2018. *The tyranny of metrics*. Princeton University Press, Princeton.

[75] Roessing, C., Helfert, M., 2021. "A comparative analysis of smart cities frameworks based on data lifecycle requirements". In *Proceedings of the 10th international conference on smart cities and Green ICT systems - Volume 1: SMARTGREENS, ISBN 978-989-758-512-8* (pp. 212–219).

[76] Wall, M., 2023. Smartcities landing page for research. https://cognitivecity.substack.com/

6 Secure Surveillance in Smart Cities

A Comprehensive Framework

Ashish Kumar
XIM University, Bhubaneswar, India

6.1 INTRODUCTION

In today's rapidly advancing technological era, the concept of a "smart city" has gained significant traction as a way to make cities more efficient, sustainable, and safe. This is accomplished through the use of connected devices and the Internet of Things (IoT) [1–3]. A smart city uses technology and data to improve communication and collaboration among the city's agencies, to reduce its environmental impact, and to make better use of resources. In addition, it aims to improve the quality of life for its residents in a variety of ways, including improved transportation, better public services, increased sustainability, and better healthcare. One of the key components of a smart city is the use of surveillance technology [4, 5]. Public areas, traffic, environment, and civic infrastructure are just some of the many things that are tracked by surveillance cameras and sensors. Surveillance cameras [6–8] can be used to keep an eye on busy intersections, parks, and other public areas to reduce criminal activity and make the city a more secure place. The footage of cameras can be used to help in criminal and emergency investigations. Another use of surveillance cameras and sensors is to keep an eye on traffic, locate traffic jams, and fine-tune traffic lights. This has the potential to lessen traffic and facilitate travel. Furthermore, environmental monitoring can be performed by sensors and cameras to keep an eye on the state of the air, water, and ecosystems. This can help locate and fix the city's environmental issues, making it safer and more sustainable. In the event of a crisis, such as a natural disaster, a fire, or an accident, surveillance cameras and sensors in smart cities may help us to respond quickly. The speed, efficiency, and the response time of emergency services can be improved. Also, in the case of urban infrastructure development, bridges, highways, and public buildings can be watched and managed so that issues can be immediately isolated, and maintenance and repairs can be carried out more efficiently.

Despite the several uses of a surveillance system, it is crucial that such a system is secure and prevents unauthorized access to important data [9, 10]. The system should also be compliant with laws and regulations, and should preserve the privacy and

DOI: 10.1201/9781003388814-6

data protection rights of individuals. Personal data such as photos and position data acquired through surveillance cameras and sensors may be sensitive and should be safeguarded against unauthorized access, use, or disclosure. In addition, cameras and sensors are frequently linked to the internet, making them open to hacking and other cyber attacks. Unsecured surveillance systems can give attackers access to critical data as well as control over the cameras and sensors. Such systems can undermine confidence and raise privacy and security concerns. Also the systems that are prone to tampering, manipulation, and abuse may render surveillance data untrustworthy and undermine the system's effectiveness. Hence, secure surveillance systems should be designed to capture as little data as possible while protecting it with robust encryption and other security measures. Secure systems should be built to prevent manipulation and abuse while also ensuring the integrity of the data collected. Finally, for a smart city to function properly, it must establish and retain community trust. By offering openness, accountability, and adherence to laws and regulations, secure surveillance technologies can assist in creating and sustaining community confidence.

To ensure the sustainability and eco-friendliness of surveillance systems in smart cities, it is crucial to embrace green computing principles. By utilizing energy-efficient technologies and optimizing resource usage, these systems can effectively reduce their carbon footprint. Smart cameras and sensors are designed to consume minimal power while still delivering reliable surveillance capabilities. Furthermore, by integrating renewable energy sources like solar panels or wind turbines, the surveillance system can operate on clean energy, thereby contributing to a greener environment. The adoption of green computing practices enables smart city surveillance systems to enhance public safety while minimizing their environmental impact.

This study investigates the challenges associated with implementing effective security measures, as well as identifying prospective research opportunities in this area. It also aims to provide a comprehensive framework for a surveillance system in smart cities.

6.1.1 CONTRIBUTION

In this chapter, a comprehensive framework for a surveillance system for smart cities and the challenges of integrating multiple security mechanisms within it will be discussed. It also includes a literature survey of existing surveillance systems for different applications. In addition, this chapter also highlights the limitations of existing systems and identifies potential research opportunities.

6.1.2 CHAPTER ORGANIZATION

Section 6.2 examines the framework of the smart city surveillance system. Section 6.3 describes the adoption of green computing practices in surveillance systems. In Section 6.4, a literature survey of existing surveillance systems is presented, and in Section 6.5 the issues related to the development of security procedures for a smart city surveillance system are examined. Section 6.6 discusses the prospects and applications of secure surveillance systems. Section 6.7 finally concludes the chapter.

6.2 A FRAMEWORK FOR SMART CITY SURVEILLANCE SYSTEMS

A smart city surveillance system is a network of sensors, cameras, and other devices that are used to monitor and manage various aspects of a city, including public safety, traffic management, and environmental monitoring. The goal is to improve the efficiency and effectiveness of city services and to enhance the overall quality of life for citizens. The framework for a smart city surveillance system is illustrated in Figure 6.1.

The main components of a smart city surveillance system are:

1. **Sensors and devices**: These are the physical components of the system, including cameras, sensors, and other devices that collect data about the city.
2. **Data storage and processing**: Data collected by the sensors and devices is stored in a central database and processed using advanced analytics software to extract meaningful insights and make decisions.
3. **Control center**: The control center is the hub of the system, where data is analyzed and decisions are made. It is typically staffed by trained professionals who monitor the system and respond to events as needed.

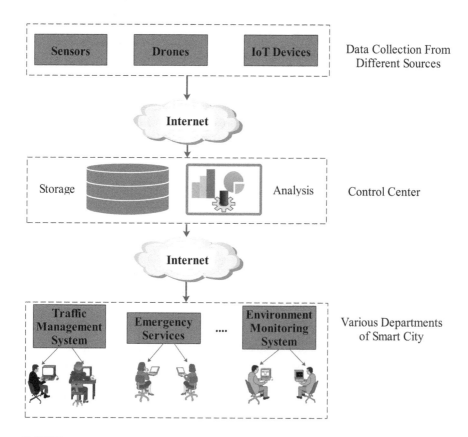

FIGURE 6.1 Framework for a smart city surveillance system.

4. **Communication and connectivity**: The smart city surveillance system must be able to communicate and exchange data with other systems and devices, such as traffic management systems and emergency services. This is typically done through a network of communication infrastructure, including fiber optic cables, wireless networks, and satellite communications.
5. **User interface**: The user interface is the way in which the system is accessed and controlled by city officials, emergency responders, and other authorized users. This can be through a web-based interface, a mobile app, or other means.

6.3 ADOPTION OF GREEN COMPUTING PRACTICES IN SURVEILLANCE SYSTEMS

The concept of green computing recognizes the significant energy consumption, electronic waste generation, and carbon emissions associated with information technology (IT) infrastructure. It aims to mitigate these environmental issues through various strategies and technologies. Adoption of green computing is crucial in the secure surveillance of smart cities for several reasons. Firstly, smart cities deploy a vast number of surveillance cameras and sensors to ensure public safety and monitor various aspects of city life. However, these systems consume significant amounts of energy. By adopting green computing practices, such as using energy-efficient hardware and optimizing power usage, the energy consumption of surveillance systems can be minimized. This is particularly important for smart cities that strive to reduce their carbon footprint and promote sustainability. Secondly, smart cities are characterized by their scale and complexity, requiring an extensive surveillance infrastructure. Green computing practices, including cloud-based solutions and virtualization, offer scalable and cost-effective deployment of surveillance systems. Cloud-based platforms can handle the increasing data load and provide storage and processing resources as needed. This eliminates the need for excessive hardware investment and reduces maintenance costs, making surveillance operations more efficient and economically viable.

Reliable and resilient operation is another critical aspect of secure surveillance in smart cities. Continuous and uninterrupted operation of surveillance systems is essential. By adopting green computing practices, such as energy-efficient hardware design and redundant system architectures, surveillance systems can operate reliably with minimal downtime. This ensures that the surveillance infrastructure remains operational even during power disruptions or other adverse conditions, maintaining public safety and security. Furthermore, green computing practices align with the environmental goals of smart cities. By employing energy-efficient components, optimizing resource utilization, and implementing responsible disposal practices, smart cities can minimize electronic waste generation and reduce the use of non-renewable resources. Green computing contributes to environmental conservation efforts and demonstrates the commitment of smart cities to sustainability. Data security and privacy are paramount in secure surveillance. Green computing practices encompass these aspects by implementing secure data transmission protocols, robust access controls, and encryption techniques. Green computing ensures that security measures are effectively implemented while minimizing the impact on system performance and energy consumption, safeguarding the sensitive information captured by surveillance systems. Lastly, green computing adoption in secure surveillance contributes to the positive perception and acceptance of smart city initiatives. When citizens witness environmentally responsible practices being adopted, such as energy-efficient surveillance systems, it enhances their trust in the technology and their willingness to participate in smart city initiatives. Green computing not only provides tangible benefits in terms of energy efficiency and cost-effectiveness but also fosters a favorable image of smart cities as environmentally conscious and technologically advanced urban environments. Implementing green computing in surveillance systems for smart cities involves various techniques and strategies. Here are some key techniques required for implementing green computing in such systems:

1. **Energy-efficient hardware**: Utilize energy-efficient components for surveillance cameras, servers, storage devices, and networking equipment. Look for hardware that meets energy efficiency certifications, such as Energy Star.
2. **Power management**: Implement power management techniques to optimize energy usage. This includes configuring devices to enter low-power modes during idle periods, scheduling shutdowns or sleep modes when not in use, and employing dynamic frequency scaling to adjust power consumption based on workload.
3. **Video compression**: Apply video compression techniques, such as H.264 or H.265, to reduce the size of video files. Compressed video files require less storage space and result in lower energy consumption for storage devices.
4. **Cloud-based solutions**: Consider utilizing cloud-based surveillance solutions. Cloud platforms provide scalable resources, allowing efficient allocation of computing power and storage based on demand. This reduces the need for on-site hardware, leading to energy and cost savings.
5. **Virtualization**: Implement virtualization techniques to consolidate physical servers into fewer machines. Virtualization optimizes resource utilization and reduces power consumption by running multiple virtual machines on a single physical server.
6. **Intelligent video analytics**: Employ artificial intelligence and machine learning algorithms for intelligent video analytics. This enables the system to detect specific events or objects of interest, reducing the need for continuous video recording and minimizing resource consumption.
7. **Data center efficiency**: If deploying on-premises data centers, optimize their efficiency by using efficient cooling systems, managing airflow, and employing energy-efficient server racks and power distribution units.
8. **Renewable energy sources**: Consider powering surveillance systems with renewable energy sources, such as solar or wind power. Install solar panels or wind turbines to generate clean energy to offset the power consumption of surveillance infrastructure.
9. **Responsible disposal**: Implement proper e-waste management and disposal practices for outdated or faulty surveillance equipment. Ensure that electronic waste is recycled through certified recycling facilities to minimize environmental impact.

10. **System monitoring and optimization**: Continuously monitor and optimize the performance of surveillance systems to identify areas where energy efficiency can be improved. This includes monitoring power usage, analyzing resource utilization, and fine-tuning system configurations for optimal efficiency.

11. **Stakeholder education and awareness**: Raise awareness among system operators, administrators, and end-users about green computing practices and their importance. Educate stakeholders about energy-saving techniques, responsible data management, and the environmental benefits of green computing.

By implementing these techniques, surveillance systems in smart cities can become more energy-efficient, cost-effective, and environmentally friendly, while ensuring the security and safety of the city's inhabitants.

6.4   REVIEW OF SECURE SURVEILLANCE SYSTEMS

Surveillance systems are used in a variety of settings to monitor and manage various aspects of daily life, including public safety, traffic management, and environmental monitoring. These systems rely on sensors, cameras, and other devices to collect and analyze data, and can provide a range of benefits, including improved efficiency and effectiveness of services and enhanced quality of life for citizens. However, the use of surveillance systems also raises concerns about privacy and security.

A number of studies have examined the use of surveillance systems in various contexts, as well as the challenges of ensuring the security and privacy of these systems. Many studies have examined the use of surveillance systems in the transportation sector, specifically in the context of intelligent transportation systems (ITSs). Recently, Abbas et al. [11] proposed the use of blockchain and IoT for secure transportation systems [12]. Prior to that, Lei et al. [13] presented a dynamic key management using blockchain for ITSs. Later, Zhang and Wang [14] designed a smart signal control system using blockchain to reduce the involvement of humans in signal management systems. Thereafter, Ahmed et al. [15] and Badr et al. [16] designed architectures for smart parking systems. The architecture proposed by Ahmed et al. solves the problem of centralization [17] by using a consortium blockchain that includes different parking service providers of a city in one stage. In the above cited works, the integration of data from various sources, including cameras, sensors, and GPS systems, has not been considered. However, integrating data from multiple sources can provide significant benefits in terms of improved traffic flow and safety. The above-mentioned works also neglected to address security issues such as the possibility of data breaches and unauthorized access to sensitive information.

Khan et al. [18] explored the use of surveillance systems in public spaces, including the use of CCTV cameras for crime prevention and the monitoring of public behavior. This study proposed the use of a blockchain platform for the verification and authenticity check of data for smart cities. The authors found that these systems can be effective in reducing crime and improving public safety, but also noted that the use of surveillance in public spaces raises concerns about privacy and the potential for abuse of personal data. Nikooghadam et al. [19] designed a lightweight authentication protocol for an Internet of Drones for the surveillance of the smart city. They claimed that their protocol is secure and resists all known attacks.

Dagher et al. [20] examined the use of surveillance systems in the healthcare sector, specifically in the context of electronic health records (EHRs). They aimed at providing secure and efficient access to health records using a blockchain-based framework. The authors found that the use of EHRs can provide significant benefits in terms of improved patient care and efficiency, but also identified a number of security challenges, including the risk of data breaches and the need to ensure the confidentiality and privacy of patient data.

Table 6.1 shows a brief summary of the literature related to surveillance systems. The literature suggests that systems can provide significant benefits in a variety of contexts, but also highlights the need to carefully consider the security and privacy implications of these systems. Ensuring the security and privacy of surveillance systems requires the implementation of strong technical and organizational measures to prevent data breaches and unauthorized access to sensitive information. Notably, none of the existing studies have explored the integration of green computing practices into surveillance systems, which presents an intriguing avenue for future research.

TABLE 6.1
Summary of Literature Related to Surveillance Systems

Reference	Year	Application	Techniques Used	Limitations
Abbas et al. [11]	2021	Secure transportation system	Blockchain, IoT	Possibility of data breaches and unauthorized access
Lei et al. [13]	2017	Dynamic key management for ITSs	Blockchain	Lack of integration of data from multiple sources
Zhang and Wang [14]	2019	Smart signal control system	Blockchain	Possibility of data breaches and unauthorized access, lack of data from multiple sources
Ahmed et al. [15]	2019	Smart parking system	Blockchain	Lack of privacy
Badr et al. [16]	2020	Smart parking system	Blockchain	Lack of data from multiple sources, possibility of unauthorized access, no secure storage
Khan et al. [18]	2020	Surveillance system for crime prevention	Blockchain	Lack of data from multiple sources, lack of privacy
Nikooghadam et al. [19]	2021	Smart city surveillance	Internet of drones	Lack of data from multiple sources, lack of privacy, no secure storage, lack of data processing
Dagher et al. [20]	2018	Health care surveillance	Blockchain	Lack of privacy and anonymity

6.5 CHALLENGES IN SECURITY TECHNIQUE IMPLEMENTATION

There are several challenges to consider when implementing security mechanisms for a smart city surveillance system. Some of these challenges include:

1. **Data privacy and security**: One of the main concerns with smart city surveillance systems is the potential for abuse of personal data. It is important to ensure that the system is designed with strong privacy and security measures in place to protect the data of citizens. This may include measures such as encryption of data, secure storage of data, and strict access controls to prevent unauthorized access to sensitive information.
2. **Robust integration with other systems**: A smart city surveillance system must be able to communicate and exchange data with other systems and devices, such as traffic management systems and emergency services. This requires careful planning and coordination to ensure seamless integration and to prevent security breaches.
3. **Device security**: The sensors and devices that make up a smart city surveillance system are vulnerable to tampering and hacking. It is important to implement strong security measures at the device level to prevent unauthorized access or manipulation of the data being collected.
4. **Network security**: The smart city surveillance system relies on a network of communication infrastructure to transmit data from the sensors and devices to the control center. This network must be secured to prevent unauthorized access or tampering with the data being transmitted.
5. **User access and management**: The smart city surveillance system must be accessed and controlled by city officials, emergency responders, and other authorized users. It is important to implement strict access controls and user management processes to prevent unauthorized access or misuse of the system.
6. **Scalability**: A smart city surveillance system must be able to scale as the city grows and changes. This requires careful planning and a flexible design that can adapt to changing needs.
7. **Maintenance and upkeep**: A smart city surveillance system requires ongoing maintenance and upkeep to ensure that it is functioning properly and producing accurate and reliable data. This can be a significant ongoing cost for cities.
8. **User adoption and acceptance**: It is important for a smart city surveillance system to be user friendly and easy to use for city officials and other authorized people. If the system is difficult to use or not well accepted, it may not be used to its full potential.

Overall, implementing security mechanisms for a smart city surveillance system requires careful planning and consideration of a wide range of technical and organizational factors. It is important to strike a balance between the benefits of the system and the potential risks it may present.

6.6 RESEARCH OPPORTUNITIES AND APPLICATIONS

Developing a secure surveillance system that can effectively address security challenges is an ongoing challenge for researchers and security practitioners. The emergence of new security threats, such as cyber attacks and terrorism, highlights the need for more advanced and secure surveillance systems. Here, research opportunities in developing surveillance systems that can address security challenges will now be discussed:

1. **Cyber security**: As surveillance systems become more connected and integrated, they become more vulnerable to cyber attacks. Cyber security is a significant challenge in developing secure surveillance systems. There is an opportunity for research in developing more secure and robust cyber security mechanisms that can protect surveillance systems from cyber attacks.

2. **Integration of artificial intelligence (AI) and machine learning (ML)**: The integration of AI and ML technologies in surveillance systems can significantly improve their performance. AI and ML can be used for real-time data analysis and decision-making, and for improving the accuracy and reliability of the system. However, there are challenges in developing AI and ML algorithms that can effectively address security challenges, so there is an opportunity for research here.

3. **Privacy preservation**: Privacy preservation is another significant challenge in developing secure surveillance systems. The use of such systems can raise concerns about privacy and civil liberties. There is an opportunity for research in developing more advanced techniques for obscuring faces and other identifiable features of individuals. These techniques should balance the need for surveillance with the protection of privacy.

4. **Communication protocols**: Robust integration requires seamless communication and data exchange between different systems and devices. Therefore, research can be conducted to develop communication protocols that are secure, reliable, and efficient. This will help ensure that smart city surveillance systems can seamlessly communicate with other systems and devices, without any security breaches or communication errors.

5. **Interoperability**: Interoperability is critical for the seamless integration between different systems and devices. Research can be conducted to develop standards and protocols that enable different systems and devices to work together. This will help ensure that smart city surveillance systems can integrate with other systems and devices, regardless of their manufacturer or technology.

6. **Integration of green computing approach**: Integrating green computing principles into the development of surveillance systems presents exciting research opportunities, particularly in addressing security challenges. By combining energy-efficient technologies with robust security measures, researchers can explore innovative approaches to enhancing the overall sustainability and resilience of these systems. For instance, exploring the potential of edge computing, where data processing occurs closer to the source, could reduce the need for extensive data transmission and minimize

security risks associated with data transfer. Additionally, integrating AI and ML algorithms can enable real-time threat detection and response while optimizing energy consumption. Furthermore, investigating novel encryption techniques and privacy-preserving protocols could ensure the confidentiality of surveillance data without compromising system performance. By delving into these research avenues, the integration of green computing principles could lead to the development of surveillance systems that not only prioritize sustainability but also effectively address the evolving security challenges of our modern cities.

By addressing these research opportunities, it is possible to develop more accurate, reliable, and secure surveillance systems that could help improve public safety and security. The use of these systems could significantly improve public safety and security, and help prevent and deter criminal activities and security threats. As technology continues to advance, the applications of the systems are expected to expand, making them even more valuable tools for maintaining public safety and security. Some of the applications are:

1. **Public safety and crime prevention**: Secure surveillance systems can be used to monitor public areas and prevent crime. The system can be used to detect potential threats and alert security personnel in real time. The use of these systems can significantly improve the effectiveness of law enforcement agencies in maintaining public safety and security.
2. **Traffic management**: Secure surveillance systems can be used to monitor traffic flow and identify potential traffic incidents. This would help in improving the efficiency of traffic management and reducing congestion.
3. **Border security**: Secure surveillance systems can be used to monitor border areas and detect potential threats. This would help in improving border security and preventing illegal immigration.
4. **Retail security**: Retail security is another important application of secure surveillance systems. Retail stores can use these systems to prevent theft, detect shoplifting, and monitor employee behavior. The use of such systems in retail security could help reduce losses due to theft, improve employee productivity, and enhance customer satisfaction.
5. **Critical infrastructure protection**: Critical infrastructure, such as power plants, airports, and water treatment facilities, are vulnerable to security threats. Secure surveillance systems could be used to monitor critical infrastructure and detect potential security threats. Their use in critical infrastructure protection could help prevent terrorist attacks, improve emergency response, and enhance overall security.

6.7   CONCLUSION

The implementation of a secure and green surveillance system is a crucial component of a smart city. It would help in reducing incidents, improving response time, and making the city more efficient, sustainable, and safe. However, it is important to ensure that the system is secure, compliant with laws and regulations, and respects the privacy and data protection rights of individuals. To address these challenges, the smart city surveillance system should have a robust framework that integrates various security mechanisms such as encryption, access control, and tamper detection. Additionally, the system should be designed to minimize the collection of sensitive data, ensure its integrity, and maintain the trust of the public by offering transparency, accountability, and adherence to laws and regulations.

REFERENCES

[1] Madakam S., Lake V., Lake V., and Lake V., 2015. Internet of Things (IoT): A literature review. *Journal of Computer and Communications*, vol. 3(05), pp. 164.

[2] Farooq M.U., Waseem M., Mazhar S., Khairi A., and Kamal T., 2015. A review on internet of things (IoT). *International Journal of Computer Applications*, vol. 113(1), pp. 1–7.

[3] Gokhale P., Bhat O., and Bhat S., 2018. Introduction to IOT. *International Advanced Research Journal in Science, Engineering and Technology*, vol. 5(1), pp. 41–45.

[4] Gandy Jr OH., 1989. The surveillance society: Information technology and bureaucratic social control. *Journal of Communication*, vol. 39(3), pp. 61–76.

[5] Cayford M., and Pieters W., 2018. The effectiveness of surveillance technology: What intelligence officials are saying. *The Information Society*, vol. 34(2), pp. 88–103.

[6] La Vigne N.G., Lowry S.S., Markman J.A., and Dwyer A.M., 2011. *Evaluating the Use of Public Surveillance Cameras for Crime Control and Prevention*. Washington, DC: US Department of Justice, Office of Community Oriented Policing Services. Urban Institute, Justice Policy Center, pp. 1–52.

[7] Alexandrie G., 2017. Surveillance cameras and crime: A review of randomized and natural experiments. *Journal of Scandinavian Studies in Criminology and Crime Prevention*, vol. 18(2), pp. 210–222.

[8] Piza E.L., Welsh B.C., Farrington D.P., and Thomas A.L., 2019. CCTV surveillance for crime prevention: A 40-year systematic review with meta-analysis. *Criminology & Public Policy*, vol. 18(1), pp. 135–159.

[9] Kroener I., and Neyland D., 2012. New technologies, security and surveillance. In *Routledge Handbook of Surveillance Studies*, pp. 141–148.

[10] Kalbo N., Mirsky Y., Shabtai A., and Elovici Y., 2020. The security of ip-based video surveillance systems. *Sensors*, vol. 20(17), 4806.

[11] Abbas K., Tawalbeh L.A., Rafiq A., Muthanna A., Elgendy I.A., and Abd El-Latif, A.A., 2021. Convergence of blockchain and IoT for secure transportation systems in smart cities. *Security and Communication Networks*, pp. 1–13.

[12] Ganin A.A., Mersky A.C., Jin A.S., Kitsak M., Keisler J.M., and Linkov I., 2019. Resilience in intelligent transportation systems (ITS). *Transportation Research Part C: Emerging Technologies*, vol. 100, pp. 318–329.

[13] Lei A., Cruickshank H., Cao Y., Asuquo P., Ogah C.P.A., and Sun Z., 2017. Blockchain-based dynamic key management for heterogeneous intelligent transportation systems. *IEEE Internet of Things Journal*, vol. 4(6), pp. 1832–1843.

[14] Zhang X., and Wang D., 2019. Adaptive traffic signal control mechanism for intelligent transportation based on a consortium blockchain. *IEEE Access*, vol. 7, pp. 97281–97295.

[15] Ahmed S., Rahman M.S., Rahaman M.S., et al., 2019. A blockchain-based architecture for integrated smart parking systems. In *Proceedings of the IEEE International Conference on Pervasive Computing and Communications Workshops (PerCom Workshops)*, pp. 177–182, Kyoto, Japan.

[16] Badr M.M., Amiri W.A., Fouda M.M., Mahmoud M.M.E.A., Aljohani A.J., and Alasmary W., 2020. Smart parking system with privacy preservation and reputation management using blockchain. *IEEE Access*, vol. 8, pp. 150823–150843.

[17] Esposito C., Ficco M., and Gupta B.B., 2021. Blockchain-based authentication and authorization for smart city applications. *Information Processing & Management*, vol. 58(2), p. 102468.

[18] Khan P.W., Byun Y.-C., and Park N., 2020. A data verification system for CCTV surveillance cameras using blockchain technology in smart cities. *Electronics*, vol. 9, p. 484. https://doi.org/10.3390/electronics9030484

[19] Nikooghadam M., Amintoosi H., Islam S.H., and Moghadam M.F., 2021. A provably secure and lightweight authentication scheme for Internet of Drones for smart city surveillance. *Journal of Systems Architecture*, vol. 115, p. 101955.

[20] Dagher G.G., Mohler J., Milojkovic M., and Marella P.B., 2018. Ancile: Privacy-preserving framework for access control and interoperability of electronic health records using blockchain technology. *Sustainable Cities and Society*, vol. 39, pp. 283–297.

7 The Ecosystem of Smart City Surveillance Using Key Elements

Ramesh Ram Naik and Sunil Gautam
Nirma University, Ahmedabad, India

Rohit Pachlor
MIT ADT University, Pune, India

Sanjay Patel
Nirma University, Ahmedabad, India

7.1 INTRODUCTION

In today's world, surveillance cameras are ubiquitous. They can be found on street corners, in shops and businesses, and even in private homes. With the rise of smart cities, this trend is only set to continue. Have you ever noticed the countless surveillance cameras positioned around your city? They may seem invasive, but they play a crucial role in creating a safer environment for all. In fact, CCTV security systems are becoming an essential component of smart cities worldwide. From reducing crime rates to monitoring traffic flow, these cameras offer numerous benefits that can't be ignored. However, as with any technology, there's also a downside to consider. Join us as we explore both the advantages and disadvantages of using surveillance in urban areas and learn how you can ensure its effectiveness in keeping citizens secure!

But why are there so many surveillance cameras? The answer lies in their importance when it comes to security. CCTV security systems allow authorities to monitor public spaces for criminal activity or potential threats. At the same time, these cameras provide a deterrent effect that can prevent crimes before they happen. When people know that they're being watched, they're less likely to engage in illegal activities.

However, the downside is that constant surveillance can also lead to privacy concerns. It's important for individuals' rights and freedoms not to be infringed upon by overly invasive monitoring practices. To make sure your surveillance system strikes a balance between security needs and personal privacy rights, it's crucial to work with experienced professionals who understand both sides of the equation.

While we may not always like being watched by Big Brother-style technology everywhere we go, there is no denying that CCTV security systems have become an essential component of modern-day living, ensuring safety at all times. Now we will discuss the importance of CCTC security systems.

DOI: 10.1201/9781003388814-7

Under surveillance. Have you ever felt like someone was watching you? Well, in a smart city ecosystem that's exactly what's happening? Surveillance cameras are everywhere and they're constantly monitoring the daily activities of citizens.

Whether it's for security purposes or to gather data on pedestrian traffic, CCTV security systems have become an integral part of modern cities. As we move towards a more connected world, the implementation of these systems will only continue to increase.

While some may argue that this constant surveillance infringes on our privacy rights, others argue that it provides necessary protection from potential threats. It is up to each individual to decide where they stand on the issue.

Regardless of personal opinions, one thing remains clear: CCTV security systems play a crucial role in maintaining safety and order within urban environments. It is important for cities to invest in high-quality surveillance technology and ensure proper management and regulation to effectively utilize these tools while respecting citizens' privacy rights.

Surveillance system importance. Surveillance has become an integral part of modern society, and for good reason. It allows us to monitor activity in public spaces and deter criminal behaviour, making our communities safer. CCTV security systems have proven to be a valuable tool in achieving this goal. One of the primary benefits of surveillance is crime prevention. The mere presence of cameras can act as a deterrent to criminal activity. Criminals are less likely to commit crimes when they know that they're being watched and recorded, which ultimately makes our streets safer.

Additionally, surveillance footage can be used as evidence in court cases. This not only aids law enforcement agencies in catching criminals but also helps ensure that justice is served by providing concrete evidence. Moreover, surveillance systems can help protect businesses from theft or vandalism. With the ability to monitor their premises around the clock, business owners enjoy greater peace of mind knowing that any suspicious activity will be detected immediately. It's clear that surveillance plays an important role in keeping our cities safe and secure. By investing in CCTV security systems and other monitoring technologies we can continue to improve public safety while ensuring privacy concerns are addressed appropriately.

Due to their important role in enhancing urban life and their positive economic effects, smart cities have gained increased attention in the scientific literature and international. The term "smart city" refers to a notion that encompasses more than simply implementing technology in urban areas [1]. It can also be used to describe an intelligent, instrumented, and linked city [2]. Intelligent systems use modelling services, visualization, optimization, and other approaches to aid in making sound operational decisions, as opposed to instrumented systems, which can integrate and capture data through the use of sensors. Detecting violence, intelligent parking, intelligent street lighting, intelligent traffic management, and monitoring traffic jams are a few examples. "Interconnected" describes the incorporation of data into a computing platform in order to promote information sharing among multiple city entities. A city is deemed a "smart city" technologically if it has a significant application of information and communications technology throughout its core services and infrastructure. This technology has an impact on artificial intelligence services that answer

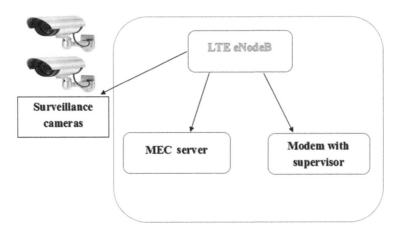

FIGURE 7.1   The system architecture in general terms.

as intelligently as thinking robots [3]. The entire design of a smart city surveillance system that employs an Internet of Things (IoT)-based video monitoring system will now be described. As shown in Figure 7.1, CCTV cameras placed at sensitive or predetermined spots throughout a city capture videos. Each video is saved with the time, date, and location.

By providing the information necessary for an immediate or future inspection, the monitoring of these locations enables us to spot people and strange elements in the environment.

Figure 7.1 shows the architecture in general terms. Based on prior preparation by security experts, the cameras are set up in key city places. Each camera is wired to a wireless base station for 4G transmission in charge of gathering the camera images that are broadcast and sending them to the server of MEC. MEC makes it possible to manage real-time photos, save records, and use facial recognition.

As all signals pass through them, the centralisation of wireless base stations could result in a bottleneck. Peak data volumes may have a negative impact on the obtainability and reliability of the scheme. The goal of obtainability is to minimise the time that system services are prone to failure while maintaining maximum availability. Reliability is the capacity of the system to operate correctly throughout period t.

The server at MEC is carefully examined in terms of software layers because it is responsible for a lot of processing. The MEC server's layering system is shown in Figure 7.1b and includes the following components: containers, a docker daemon, storage, hardware, and an operating system. The elements in the bottom layer affect those in the top layer. Records need to be kept in the storage component. Live images are managed by the docker daemon and container software components.

7.1.1   New Developments in CCTV Surveillance Technology

Video surveillance systems based on cutting-edge technology have come a long way since they were first introduced for prison security in the middle of the 20th century.

Their role in the public arena, privacy, and protection is radically expanding daily, revolutionising both public safety and people's daily lives.

1. **Local CCTV security system**: Today's CCTV cameras use PanTiltZoom technology, high-resolution recording, and a variety of lenses, including night vision. CCTV cameras record analogue video signals, which are then transferred to DVRs for storage. The operator can also synchronise motion-based video events with audio analysis thanks to its additional capabilities. CCTV cameras with high definition have replaced those with standard definition in recent years, and 2016 was another banner year for HD CCTV4.

2. **A network-based internet protocol (IP) surveillance system**: IP cameras are recommended for large installation locations with a well-established high bandwidth network:
   - The move from analogue to IP-based solutions is being spurred by a variety of benefits offered by digitisation and network capabilities over CCTV, such as real-time monitoring, remote access and management from anywhere, and digital zooming for crisper, higher-quality photos.
   - Lower operational and infrastructure costs incurred during installation.
   - Excellent integration potential with other systems, such as access control and building management systems.
   - Greater ease of distribution, immediate distribution, and faster data transfer through email and other communication systems with automatic alerts.
   - Simultaneous recording and playback with advanced search capabilities.
   - Video analytics extend beyond motion detection to determine humidity, temperature, colour, and noise.
   - Easy to install and transport to another site by employing existing wires while connecting directly to the network.

3. **System for IP surveillance based in the cloud**: The positioning of the wireless access points between the network and cameras is the sole distinction between the IP wireless security camera system design and the standard IP camera system design. With up to 328 feet of Ethernet cable and 1.5 miles of distance from the local area network, cameras can be installed thanks to this.

4. **Synthetic surveillance system**: The most cost-effective and efficient approach to update an existing, substantial CCTV infrastructure is with a hybrid solution that uses a video server to convert analogue CCTV cameras to IP cameras. Any type of camera can be connected into an IP surveillance system using a small stand-alone video server that can convert analogue signals to digital format and provide the analogue cameras' IP addresses.

5. **Systems for mobile video surveillance**: Another rapidly expanding market category for video surveillance is mobile video surveillance. Both analogue and IP systems can be used for mobile surveillance. This market consists of devices for installing surveillance in trains and automobiles. The rise in crimes on trains, buses, and school buses, which result in millions of dollars in losses, is what is driving the market for mobile video surveillance.

7.2 LITERATURE REVIEW

On congested streets, traffic monitoring is a difficult task, according to Patel Parin and Gayatri Pandi [4–8]. Manual, expensive, time-consuming techniques of traffic monitoring also include human administrators. Due to limited accessibility, massive stockpiling and video stream analysis are not feasible. Despite this, it is currently possible to conduct reconnaissance on video feeds created by traffic checking, as well as object detection and following, traffic pattern analysis, number plate identification, and other tasks. Standard database systems are unable to store and analyse the big data video streams due to their enormous size. Exam results are stored in the hive information stockroom, which is built on hadoop.

Most traffic volumes and speeds, and individual vehicle speeds, are included in this data, which provides an overview of the facts, questions, and research. We suggest using sophisticated video to handle street traffic research in order to avoid the requirements of path control and sensor exhibitions. Best-in-class object recognition algorithms were introduced by Shaoqing Ren et al. [9–12] which rely on area proposition computations to infer object areas. The operating time of these discovery systems has decreased thanks to innovations. Article limits and objectness scores are simultaneously predicted by an RPN, a fully convolutional structure, at each location [13].

The suggested approaches account for information affiliation by taking into account target appearances as well as their ephemeral adjacencies andcomparing identifications with small transitory holes to those with large transient holes. This is in contrast to typical positioning failures. With the help of open standards, IBM's S3 system may be quickly customised to meet the needs of various applications.

Existing individual recognisable evidence benchmarks and procedures, according to Tong Xiao et al. [14], mostly focus on coordinating reduced passerby photos across queries and rivals.

7.2.1 THE MARKET FOR VIDEO SURVEILLANCE IN INDIA, 2017–2023

India's video surveillance industry has expanded significantly over the past few years as a result of growing security concerns and steady infrastructure building throughout the nation. The installation of surveillance systems in both public and private sector facilities has become necessary due to the ongoing threat of terrorist strikes in the wake of the 2008 Mumbai attack and attacks on the Indian Parliament. These surveillance systems are being used more frequently in commercial buildings to secure their infrastructure, data, and compound security.

The analogue-based surveillance systems have historically generated the majority of market profits, and IP-based surveillance systems are expected to proliferate over the next several years as a result of rising awareness and falling costs.

In 2016, the northern area brought in the most money, with the largest contributors being the states of Delhi NCR, Haryana, Punjab, Uttaranchal, and Uttar Pradesh. Along with schools, airports, train stations, and power plants, hybrid video surveillance solutions are commonly used in these places.

The market for video surveillance is covered in-depth in the study by types, components, and verticals. Security concerns have grown in recent years as a result of an

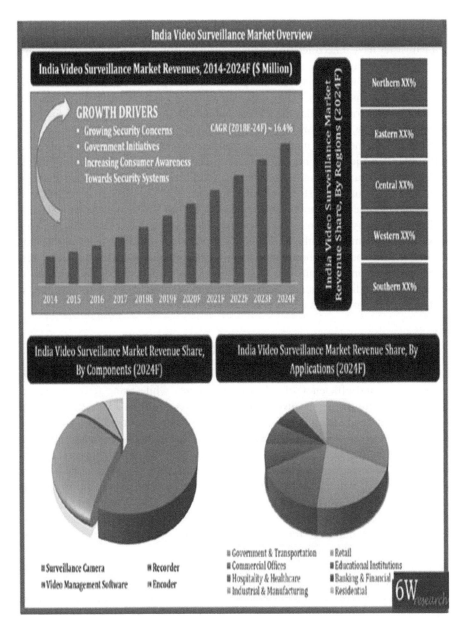

FIGURE 7.2   Indian video surveillance market overview [28].

increase in theft, terrorism, domestic violence, educational changes, and the market for video surveillance, as well as other factors. The government is stepping up its efforts in relation to smart city initiatives and city surveillance. Increasing disposable income fuels the demand for a more secure system, driving an increase in video surveillance (Figure 7.2) [15].

7.2.2 THE SURVEILLANCE AND SECURITY MARKET IN INDIA

All industries have a need for surveillance systems. Both analogue and IP-based systems offer advantages since they enable customers to select the most practical system for their needs. Due to their widespread use in power plants, airports, train stations, and educational institutions, hybrid video surveillance solutions also present potential.

The demand for surveillance systems and solutions is being driven by the infrastructure, banking, and financial industries.

Even though the Indian market for surveillance systems is expanding, it still faces difficulties in project sourcing and execution. The industry has a number of difficulties, including a lack of components, hefty import levies, poor supply chain management, and inadequate infrastructure. American exporters should be ready to deal with opaque and frequently unpredictable regulatory and tariff systems while doing so.

For US businesses to enter the Indian market, it is crucial to find reputable partners who are familiar with the country's market and with its legal procedures. Exporters from the USA should investigate market tactics, as there are diverse geographical opportunities, standards, languages, and cultural distinctions. As a face-to-face society, India frequently requires in-person meetings before commercial partnerships or agreements are formally established [16].

7.3 PROBLEMS WITH SURVEILLANCE IN SMART CITIES

Information and communication technology is used in smart cities to increase the sustainability and efficiency of urban environments while using fewer resources. Smart cities from the perspective of surveillance keep tabs on residents using strategically positioned sensors throughout the urban environment, which gather information on a variety of aspects of urban life [17]. Governments and other municipal authorities communicate, gather, and analyse data from these for transportation and environmental management [18, 19], traffic control, [20, 21] waste reduction, and [22] energy use. Better urban planning is made possible by this and for governments to adapt their services to the local population [23–25]. Many cities have used this system [26] and others, such as London, Barcelona, and Stockholm. Effective law enforcement, improved transit options, and modernised infrastructure systems, including the delivery of local government services via e-government platforms, have all benefited from the practical implementations of smart city technology [27].

Concerns have been raised that these systems could transform governments deploy data [28]. The vertical information exchange methods between citizens and the government have drawn criticism due to privacy concerns, which undercut the idea of urban anonymity [29].

7.4 EVALUATION METRICS

Sensitivity analysis evaluates the local influence of the data from an input on the data from an output with the aim of discovering the computer system's weak areas. From then, it aims to implement a variety of tactics in an effort to improve these systems under diverse circumstances.

7.5 EQUIPMENT FOR SURVEILLANCE SYSTEMS

For surveillance, two different types of cameras are used: analogue and digital IP cameras.

7.5.1 ANALOGUE CAMERAS

In order to manage sites with several cameras, more DVRs might be needed. Analogue cameras have more design alternatives, so you might be able to locate the ideal camera for your requirements at a lesser price than a digital one. Additionally, the network bandwidth that digital cameras consume won't be lost.

7.5.2 DIGITAL CAMERAS

IP cameras have a greater resolution than traditional cameras, producing crisper images, though, as we've already mentioned, they require more storage and more transmitting bandwidth. APoE switch, which has ports for many cameras, is used to link cameras to network video recorders (NVRs). While fewer connections are needed as a result, your network bandwidth may be hampered. The location of cameras in relation to the NVR is unrestricted, and footage can be viewed remotely via wireless network access. Using digital cameras provides a variety of extra functions, including digital zoom, smartphone notifications, motion-activated auto-recording, one-touch connections to authorities, and object identification. You might want to think about a hybrid video recorder capable of accommodating both kinds of cameras if you're making the switch from analogue to digital.

7.6 REQUIREMENTS

It's critical to evaluate your security requirements and spending capacity before choosing the cameras, recorders, and storage to utilise. Your system selections will be influenced by them. As you examine your needs, some factors to think about are:

1. **Amount of cameras**: You can determine how many cameras you'll need once you've decided which sections of your office you want to monitor. Remember that the field of view is reduced on analogue cameras. According to Customer First, an outsourced IT service, organisations require three to four analogue cameras for every IP camera. So, if you decide to use analogue, you'll probably need more cameras.
2. **Outdoor versus indoor cameras**: In order to meet the demands of outdoor monitoring, exterior cameras might need to cost more and/or have more features.
3. **Video calibre**: Which type of video resolution are you looking for? High quality is recommended since it improves the photos' integrity and may make it easier to identify people or pieces of evidence in the event of a crime.
4. **Storage**: The amount of storage you will need depends on the quality of your video, the number of cameras that are recording, and how long you intend to retain your video. A wide range of businesses offer online storage calculators.

5. **Camera design**: There are three types of cameras: bullet, dome, and pan-tilt-zoom. Another choice is to use a camera that can record audio or enable two-way audio communication. Consider selecting a camera with infrared LEDs if you need to record in low light.
6. **Compatibility**: Ensure that your recording system is compatible with the cameras you intend to use.

7.7   MASSIVE MONITORING

Widespread surveillance is intrinsically tied to the concept of smart cities. Continuous data flows from sensors, cameras, and tracking apps are essential to the benefits of smart city technology. Urban anonymity is lessened by mass surveillance using big data. But, because of the volume of data involved, this amount of privacy, according to proponents of smart cities, is comparable to that seen in small towns. A blockchain-based video surveillance system was developed in. The suggested system's major goal is to increase the number of authorised individuals who can access it in the case of an incident. Block Sees successful results make it possible to develop new distributed citywide surveillance systems.

7.8   THE BENEFITS OF SURVEILLANCE

Surveillance is an essential component of any smart city ecosystem. It has its benefits, including the prevention of crime and increased public safety. With CCTV security systems in place, it becomes easier to monitor public spaces and deter criminal activity.

One of the significant benefits of surveillance is that it helps create a safer environment for everyone. CCTV cameras can identify potential dangers before they occur, making it easier for law enforcement officials to respond quickly and prevent harm from happening. By enhancing public safety in this way, people feel more secure while going about their daily activities.

Another advantage of surveillance is that it provides valuable evidence when a crime occurs. The footage captured by these cameras can be used as evidence in court cases to prove guilt or innocence beyond reasonable doubt. This technology helps expedite investigations and simplify legal proceedings.

Moreover, surveillance also plays an important role in traffic management and emergency response situations. In the case of a road accident or emergency situation such as a fire outbreak or natural disaster, the authorities can use real-time monitoring information gathered from CCTV cameras to respond quickly.

The benefits provided by surveillance outweigh the negatives associated with privacy concerns if implemented effectively with adequate measures taken to ensure data protection and limited access control over sensitive personal data collected through video feeds obtained via these systems.

7.9   THE DOWNSIDE OF SURVEILLANCE

While surveillance is an essential component of smart city ecosystems, it also has its downsides. The constant monitoring and tracking of citizens can make them feel uncomfortable and unsafe. People should have the right to protect their personal information and activities.

Surveillance technology can also be misused by authorities or hackers if proper security measures are not taken. The misuse could lead to wrongful arrests, false accusations, or even blackmailing individuals with sensitive data obtained from surveillance footage.

Moreover, reliance on surveillance technology may cause people to become complacent about taking personal responsibility for their safety since they believe that cameras will always be watching over them. It might discourage people from reporting crime incidents as they assume that surveillance systems would have capture dany criminal activity in realtime.

The cost associated with implementing a CCTV security system can also be quite significant depending on factors such as location size and the complexity required; this expense eventually falls back on taxpayers' shoulders.

Therefore, while we recognise the importance of having a robust surveillance system in smart cities, policymakers must address these downsides to ensure transparency, accountability, and privacy protection for everyone involved in creating secure urban spaces.

7.10   HOW TO MAKE SURE YOUR SURVEILLANCE IS EFFECTIVE

Installing surveillance cameras in your smart city is one thing, but making sure they are effective is another. Here are some tips to ensure that your CCTV security system is efficient and useful. Firstly, consider the placement of your cameras. The ideal location would be high up and out of reach of potential vandals or criminals. Make sure you cover all areas of interest such as entrances, exits, high traffic areas, and parking lots.

Secondly, invest in quality equipment for better image resolution and storage capacity. Poor-quality images make it difficult to identify people or vehicles which defeats the purpose of having a CCTV security system.

Thirdly, ensure that there's adequate lighting in the surveillance area especially during night time because good lighting helps improve image clarity.

Fourthly, employ professionals to monitor footage 24/7 so that any suspicious activity can be detected on time before things get out of hand.

Conduct regular maintenance checks on the equipment to prevent malfunctions or camera failures which could leave gaps in coverage. By taking these steps you can have an effective surveillance system that enhances safety measures within your smart city ecosystem.

7.11   CONCLUSION

Surveillance is an essential part of a smart city ecosystem. By utilizing CCTV security systems intelligently, cities can become safer and more efficient. Surveillance technology has come a long way in recent years, allowing for better monitoring and analysis of data. However, it's important to balance the benefits with privacy concerns and ensure that surveillance is used ethically. Cities around the world are investing in smart technologies to improve their citizens' quality of life. With continued advancements in surveillance technology, we can expect even greater improvements in public safety and urban planning. As long as we remain vigilant about how these systems are implemented and used, we can create smarter cities that benefit us all.

The most undervalued but crucial component of surveillance systems is security. The processing of several video streams from diverse cameras scattered throughout the city presents a number of hazards and issues, which supports this. This study has shown a focus on citizen safety as well as crime prevention, which is an essential element needed for smart city security applications. We presume that there are a number of problems with the current security-related video analysis systems. In order to create intelligent camera surveillance systems that might possibly send out alarms automatically in the event of a disaster, such as invasion detection or fire detection, convolution neural networks were used. The results show that the suggested IVS might attain incredibly low false alarm rates. The research also looked at strategies to improve IoT security by resolving security issues with wireless sensor networks.

REFERENCES

1. Rho, et al. (2012). Advanced issues in artificial intelligence and pattern recognition for intelligent surveillance system in smart home environment. *Engineering Applications of Artificial Intelligence*, 25(7), 1299–1300.
2. Albino, et al. (2015). Smart cities: Definitions, dimensions, performance, and initiatives. *Journal of Urban Technology*, 22(1), 3–21.
3. Barve, et al. (2019). Stratum: A serverless framework for lifecycle management of machine learning based data analytics tasks. arXiv preprint arXiv:1904.01727.
4. Basu, et al. (2013, February). Wireless sensor network based smart home: Sensor selection, deployment and monitoring. In *2013 IEEE Sensors Applications Symposium Proceedings* (pp. 49–54). IEEE.
5. Baxter, S. (2015). Modest gains in first six months of Santa Cruz's predictive police program. *Santa Cruz Sentinel*. 2012.
6. Berg, N., Predicting crime, LAPD-style. *The Guardian*. Retrieved 2023-01-10.
7. Copus, C., et al. A focus on innovation: Cities and local political leadership. Retrieved 2023-01-12.
8. Gallo, et al. (2018, June). BlockSee: Blockchain for IoT video surveillance in smart cities. In *2018 IEEE International Conference on Environment and Electrical Engineering and 2018 IEEE Industrial and Commercial Power Systems Europe (EEEIC/I&CPS Europe)* (pp. 1–6). IEEE.
9. Hardy, Q. (2014). How urban anonymity disappears when all data is tracked. *New York Times Bits Blog*.
10. Harrison, C., Eckman, B., Hamilton, R., Hartswick, P., Kalagnanam, J., Paraszczak, J., & Williams, P. (2010). Foundations for smarter cities. *IBM Journal of Research and Development*, 54(4), 1–16.
11. Kim, et al. (2019, February). Autonomous network traffic control system based on intelligent edge computing. In *2019 21st International Conference on Advanced Communication Technology (ICACT)* (pp. 164–167). IEEE.
12. Klett, et al. (2014). Smart cities of the future: Creating tomorrow's education toward effective skills and career development today. *Knowledge Management & E-Learning: An International Journal*, 6(4), 344–355.

13. Lawler, L. (2019). Hardware, heartware, or nightmare: Smart-City technology and the concomitant erosion of privacy. *Journal of Comparative Urban Law and Policy*, 3(1), 207.

14. Chen, et al. (2013). Smart homecare surveillance system: Behavior identification based on state-transition support vector machines and sound directivity pattern analysis. *IEEE Transactions on Systems, Man, and Cybernetics: Systems*, 43(6), 1279–1289.

15. https://www.6wresearch.com/industry-report/india-video-surveillance-market-2017-2023-vsc-forecast-by-camera-specification-outdoor-indoor-verticals-regions-competitive-landscape#

16. https://www.trade.gov/market-intelligence/indias-surveillance-and-security-market

17. Liu, et al. (2019). When machine learning meets big data: A wireless communication perspective. *IEEE Vehicular Technology Magazine*, 15(1), 63–72.

18. Liu, et al. (2020). Traffic flow combination forecasting method based on improved LSTM and ARIMA. *International Journal of Embedded Systems*, 12(1), 22–30.

19. Liu, et al. (2019). A distributed node deployment algorithm for underwater wireless sensor networks based on virtual forces. *Journal of Systems Architecture*, 97, 9–19.

20. Siano,et al. (2016, June). Iot-based smart cities: A survey. In *2016 IEEE 16th International Conference on Environment and Electrical Engineering (EEEIC)* (pp. 1–6). IEEE.

21. Meaney, R. (2013). *Don't Even Think about it*. The O'Brien Press.

22. Min-Allah, et al. (2020). Smart campus – A sketch. *Sustainable Cities and Society*, 59, 102231.

23. Mora, L., et al. (2017). How to become a smart city: Learning from Amsterdam. *Smart and Sustainable Planning for Cities and Regions: Results of SSPCR2015*, 1, 251–266.

24. Nikouei, et al. (2018, October). Smart surveillance as an edge network service: From harr-cascade, svm to a lightweight cnn. In *2018 IEEE 4th International Conference on Collaboration and Internet Computing (CIC)* (pp. 256–265). IEEE.

25. Pardo, et al. (2011, June). Conceptualizing smart city with dimensions of technology, people, and institutions. In *Proceedings of the 12th Annual International Digital Government Research Conference: Digital Government Innovation in Challenging Times* (pp. 282–291).

26. Paskaleva, K. (2013). E-governance as an enabler of the smart city. In *Smart Cities* (pp. 45–63). Routledge.

27. Sharma, et al. (2019). Big data analysis in cloud and machine learning. In *Big Data Processing Using Spark in Cloud* (pp. 51–85).

28. Wang, et al. (2014). Smart cities of the future: Creating tomorrow's education toward effective skills and career development today. *Knowledge Management & E-Learning: An International Journal*, 6(4), 344–355.

29. Xu, et al. (2018, May). Real-time human objects tracking for smart surveillance at the edge. In *2018 IEEE International Conference on Communications (ICC)* (pp. 1–6). IEEE.

8 Security Enhancement and Applications in IIoT for Smart City Services

Vikash Kumar and Santosh Kumar Das
Sarala Birla University, Ranchi, India

8.1 INTRODUCTION

Industrial Internet of Things (IIoT) and its application are rapidly used in our daily life cycle. It main aims to make the system smart and efficient for smart city services [1]. The revolution of industry 4.0 and its related strategy is based on some computation and communication process [2]. There are several revolutions such as industry 1.0 to industry 4.0. Industry 1.0 is based on steam engine, industry 2.0 is based on electricity, industry 3.0 is based on automation, computer and some logic system, industry 4.0 is an advance computing system. It helps to manage several processing and networking system that helps to manage performance [3]. The concept of IIoT is derive from IoT which is design for the purpose of industry 4.0. The IIoT basically design for smart factory to manage different conflicting strategies of industry 4.0. It consists of both industrial internet and factory internet for communication purpose. The basic understanding of IIoT shown in Figure 8.1.

The basic components of industry 4.0 are smart or intelligent system that uses smart television, smart phone and smart home appliance system. The fusion of industry 4.0 along with Internet of Things (IoT) makes an efficient IIoT System that plays an important role in the current scenario [4]. The business role of IIoT are application and customer provider, platform provider, network provider and device provider. IIoT enabled systems are builds of sensors, actuator, other computational machine mainly for industrial use. These smart industries required interoperability and communication network to improve its throughput [5].

COVID-19 pandemic has caused large no of human death specially in city. This pandemic has enforced individual and system to think about our policies, priority and interest. This pandemic enforced us to build intelligent system such as smart health management system, smart, transportation system, smart education system, smart manufacturing system and by combining all these smart systems to build smart city [6]. IIoT enabled smart city generates data exponentially from different sector such as medical, finance, administration and utility sector through sensor, actuator and other smart devices [7].

Smart city is growing exponentially and generates huge amount of data. These data generally stored in cloud due to storage and processing constraint of IIoT systems. To provide security and privacy, encryption and decryption mainly applied on cloud server

DOI: 10.1201/9781003388814-8

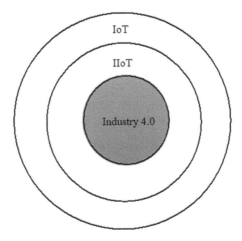

FIGURE 8.1   Basic concept of IIoT.

and required large no of keys to secure sensitive information. Blockchain based key management and storage management scheme are ideal choice in distributed environment of IIoT systems [8]. Traditional data security mechanism are not much suitable for IIoT environment and cause many cases of data leakage and privacy compromise.

Different proprietary based communication and operational devices and software are used in smart city. These technologies are not effective in cross platform scenarios to provide secure communication [9]. IPV6 based routing protocol is an ideal solution for effective data communication in resource constraint, low power and lossy network [10]. The existing authentication schemes are not much suitable for multifactor authentication in untrusted cloud server with huge and exponentially growing no smart devices. For proper data gathering and propagation of big data for IIoT systems, Ethereum merge mechanism can be used [11]. It control data propagation and improve data processing speed through classification and distributed integrity authentication of smart devices.

8.1.1   CONTRIBUTION

This chapter shows review of requirement of IIoT based on smart systems after covid 19 pandemic for smart city application. It discuss the application of IIoT systems such as smart manufacturing, smart transportation, smart supply chain management, smart healthcare, smart government and smart utility services. It discusses the challenges of IIoT systems such as heterogeneity, big data management, trust management, e-waste management, limited resources management and security requirement. This chapter develops and discusses the methodologies for security enhancement of IIoT based systems for smart city applications based on fusion of some techniques such as wireless sensor network or WSN, cloud computing and IoT. This chapter builds framework of secure cloud computing system of IIoT and shows relationship among cloud computing, IoT, IIoT, WSN. Finally, it shows collaboration of methodology of cyber physical system.

8.1.2 Organisation of Chapter

The rest of the chapter is divided into some sections as Section 8.2 highlight some related works as literature review. Section 8.3 discusses applications of IIoT enabled smart systems for Smart City applications. Section 8.4 discusses challenges of IIoT system for smart city applications. Section 8.5 discusses security and privacy requirement of IIoT enabled smart system. Section 8.6 discusses methodologies for security enhancement and its related application for IIoT. Section 8.7 conclude the chapter.

8.2 LITERATURE REVIEW

There are several works have been proposed in the last few years. The works are based on smart and intelligent system to resolve several issues with the context of IoT or any other purposes. Some of the works are discusses as Lombardi et al. [12] designed a new standard grading system for interstitial pneumonia based on quantitative ultrasonographic data of COVID-19 participants. Healthy volunteers, adult patients hospitalized to the emergency department with symptoms that may be connected to pneumonia, and clinical cases from internet databases were all analysed. On the basis of the recognized imaging indications, the algorithm and an expert operator (who was kept blind to the algorithm findings) awarded a Pneumonia score from 0 to 4 to each zone, and the patient Lung Staging was calculated as the highest observed score. Mukati et al. [13] designed a method using Internet of Things (IoT)-enabled technologies, healthcare assistance to COVID-19 patients was provided. Finally, prospective basic IoT applications for the medical field were recognised in the COVID-19 Pandemic, along with a brief description. During the COVID-19 pandemic, IoT makes significant gains in the medical profession, improving facilities and information systems. Better chronic illness management, medical emergencies, enhanced patient care, fitness, blood pressure management, health inspections, measures and control, cardio frequency inspection, and audiological assistance are all possible with IoT. It has the ability to monitor COVID-19 patients in a consistent and dependable manner, as well as boost the medical sector's personalisation. This technology will enhance patient health in the future, allowing for better treatment, and it will be used in the event of a COVID-19 pandemic.

Peter et al. [14] illustrated several requirement challenges and opportunities of IIoT based on business application. It helps to give new direction to the user for economies system. Each of the direction is based on industry 4.0. It helps in emerging economies and employ several factors that are associated with it. The application and opportunities help to manufacturing several driving systems that are used in real-life application. The empirical study and qualitative research are done that are describe in this article. It helps in prospect and expert analysis to develop some areas with IIoT. Das and Dey [15] proposed a book for constraint decision making system that helps to model several applications. The main aim of this work to highlight several approaches and methodologies that are used in constraint decision making system. It easily helps in several modern application and area. Most of the application and area are fusion of different intelligent techniques such as artificial intelligence, machine learning, deep learning, data science, etc. Each of the technique is used in IoT, IIoT, and other innovating areas.

Ding et al. [16] designed a feasibility study for multi-mode analysis in the area of IIoT. The work is based on medical analysis for internet of things that helps in data transmission. Data transmission is based on medical data system that helps to transmit data in the technology. The work is the fusion of both IoT and IoMT. The combination of both helps to stable robust system. It helps to protect the system from different anomaly and attack. It helps to diagnosis and resource allocation system based on intelligent data transmission system. Convolution neural network or CNN is used to manage and analysis of several stability of the system. Golchha et al. [17] designed a method for cyber attach system in IIoT that helps in voting system analysis. The work is based on learning approach methodologies that helps to employ voting analysis system in IIoT system. The work is design for smart technology analysis for interconnection analysis and its management. The work consists of multi-level security analysis system. It helps to detect several intrusion detections in the environment of IIoT. There are several techniques are used in this modelling such as Random Forest or RF, Histogram Gradient Boosting or HGB or Machine Learning or ML.

Jin [18] designed an efficient and robust signcryption system for IIoT that helps to manage heterogeneous environment of the network. The environment is the based-on extension of IoT that widely apply in industry system. The work is used several sensors that helps to allow certificateless cryptography analysis. It helps to model several cryptographies in the mode of random oracle model. There are several keywords are used in this system such as IIoT, heterogeneous system and modelling, certificateless cryptosystem, signcryption, etc. Finally, it helps to improve security and decease overhead of the system. S. K. Das [19] proposed a work for smart application and design system that helps to model several applications in the areas of smart innovation and its application. The work consists of several challenges, issues with the context of services that are used in the real-life application. It also helps to use some solution that are describe in these areas. Smart application increases frequently for managing the requirement of the user and customer. Its corporates several new and innovative techniques to make the system easy. Khalil et al. [20] proposed a method for localization system in IIoT system and its modelling. It helps in activation non-cooperative anchor analysis that helps in energy efficient management. The work mainly proposed for ultra-reliable and low-latency system for cooperative several operations. The work is adaptive to manage several operations such as error adaptive, sequential and distributed. It helps in decision process for anchor selection and bottleneck analysis in non-cooperative environment. It helps in collaborative and distant monitoring for wide area application management.

Sasikumar et al. [21] proposed a method for trust management in IIoT for empowering system. The work is based on twin empowered system and its management. It helps in rapid development for managing several revolutions. The digital system is applied in limited resources for secure communication system. There are several sensor and data are used in application management. It helps in proof of authority and random generation analysis. It helps in industry revolution 5.0 to make secure system. Its main purpose to innovate digitalized automation with fusion of some techniques such as edge computing, smart factoring and distributed production system.

Muruganandam et al. [22] process operational control method for IIoT with the fusion of artificial intelligence, sensor and machine learning methods. It helps to make intelligent system based on smart factories. It also helps to make the system easy and effective for production system. This production helps in monitoring, controlling and operation management. The work is based on operation constrained process control (OCPC) to prevent several errors. It helps to manage several smart machine analyses for effective operation. It helps to manage several control error analysis and training analysis system.

Das et al. [23] designed a work for non-linear communication in peer-to-peer network. The work is based on in hybrid mode application for non-linear formulation. It helps to model application in hybrid mode with fusion of some techniques such as game theory, non-linear geometric programming, intuitionistic fuzzy logic. The purpose of this work is to innovate and model the application in any networks that are used in industry and other area. It helps to resolve several issues based on requirement of the user and customer. Kilani et al. [24] designed a method for socialization system of smart communicative analysis. It helps to manage several objects of industrial analysis. The work is based on IIoT system to manage an interactive communication for smart analysis. It helps to exchange connection and information among the object to create several communicated objects. All the object are divided into two parts as homogeneous and heterogeneous in nature. The work easily helps in object communication to exchange message and adaptive system. The objects are autonomously, dynamically and community that are inspired with socio-inspired technique.

8.3   APPLICATION OF IIoT IN SMART CITY

The main objective of building smart city is to provide modern infrastructure, healthy and sustainable environment, quality of life and smart solutions of day to day life problems of citizen. Figure 8.2 contains applications of IIoT systems for smart city. We illustrate the different emerging IIoT based smart systems those make a city smart collectively.

Smart Manufacturing: Smart manufacturing plays a crucial role in smart cities. The purpose of smart manufacturing is to maximize production throughput, improve safety and quality, to make process easier for human, reduce energy consumption and pollution.

Smart Transportation: Smart transportation plays a vital role for building smart city. Many problems such as road accidents, traffic, parking issue and toll collection can be resolved by making IIoT based smart transportation system. By adopting Smart traffic management system, smart parking system, smart toll management system, live road update broadcasting systems and self-driving vehicle can make a city smart.

Smart Supply Chain Management: The rapid population growth in metropolitan cities required smart supply chain management systems of goods and services. IIoT based supply chain management system provides real time tracking of good and services supply, reduce the time of supply and make system transparent and easy to use.

Smart Healthcare: Smart healthcare is an integrated and significant part of a smart city. Basically it improve the reach of medical facility to the people via IIoT based real time healthcare monitoring system and wearable smart medical devices.

Smart Government: Smart government is one of the core characteristics of a smart city. Smart government means easy access of information, institutions and activities to the people by the smart government systems which can be build smart collaboration of information technologies, communication technologies and operation technologies. Smart government makes the government process digital, efficient and transparent.

Smart Utilities: It is not possible a city smart without making utilities services smart such as water, electricity and gas supply. Smart utilities system reduces the overconsumption of resources, improve safety, improve bill settlement by the use of smart digital meter and smart grid management system.

8.4   IIoT RESEARCH ISSUE FOR SMART CITY APPLICATIONS

IIoT enabled systems relies on multiple technology such as information technology, communication technology and operational technology. Also these devices are dynamic, distributed, resource constrained, different proprietary software and hardware, working in exponentially growing environment creates many research challenge. Vulnerability of individual technology plus vulnerability created due to collaborations of these technology creates many research issue such as heterogeneity, scalability, big data management, Denial of Service, trust issue, security issue, networking issue, communication issue, electronic-waste management and limited resource issue. Figure 8.2 shows major research challenge for IIoT system for smart city applications.

Heterogeneity: Heterogeneity is one of the main characteristics of IIoT systems because all the information technology, communication technology and operational technology hardware and software are made by different vendors and have different standard. To build a IIoT system for smart city applications in this heterogenous environment is one for the main research challenge.

Big Data Management: Since no of devices such as sensors, actuator, data processing and data storage devices grow exponentially in smart city and these devices produce huge amount of data. This big data has characteristics such as volume, variety and velocity. To build a intelligent system, which perform analysis, process the big data and produce insightful information for the improvement of smart city functionality, is a major research challenge.

Trust Management: Smart city contains numerous smart devices connected with huge intelligent network. The vulnerability of single device or network can affect entire system. The safety, reliability and resilience of entire system depends on trust of one system to another one. To build an intelligent trust management system, for IIoT based Smart city applications, is a major research challenge.

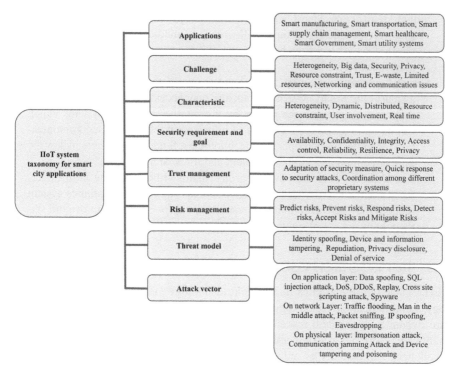

FIGURE 8.2   IIoT system taxonomy for smart city applications.

Security: Security is one of the major challenge for adopting IIoT enabled intelligent system for smart city applications. As computing power of devices increased exponentially, it open the door for many cyber-attacks. Security requirement can be characterised as confidentiality, integrity, availability and access control. To achieve these security features in resource constrained smart devices is big challenge. Designing light weight security protocol, suitable for heterogeneous, resource constrained, dynamic and distributed intelligent IIoT enabled system, is an open challenge.

E-waste Management system: As the no of electronics devices grow exponentially and every device has a limited life period. These devices become waste after a period. These huge e -waste become a challenge for environment of a smart city. To build a environment friendly e- waste management system, which can control, recycle and manage the huge e- waste, is a big research issue in the near future.

Resource Management: Resources such as water, gas, electricity and land are limited. Use of resource with proper planning saves these resources for future and save environment. To build an intelligent resource management system which can stop wastage of resources and promote green energy source for sustainable environment of smart city.

8.5   SECURITY AND PRIVACY REQUIREMENTS IN ADOPTING IIoT BASED SYSTEMS FOR SMART CITY APPLICATIONS

A smart city refers to modern technology enabled, self-sustainable, environmental friendly, safe and secure city that uses IIoT enabled smart systems which made of smart collaboration of modern information and communication technology with operational technology to solve modern city problem such as traffic, transportation, growing population, safety, security, health facility, environment change and limited resources. These smart systems are heterogenous, resource constrained, dynamic, distributed and direct user involvement which make them vulnerable to security attack. Traditional security mechanism is not much suitable due to its divers nature and resource contained. Figure 8.2 shows security and privacy taxonomy of IIoT enabled smart system for smart city applications. Remainder of this section mainly focus on security requirement of IIoT enabled smart systems for smart city applications.

Confidentiality: Smart city stores huge amount of confidential data such as citizen personal information, objects location and object sensitive information. Confidentiality refers to ensure all the communication of data and storage of data should be in encrypted form.

Authentication: Authentication process in smart city ensure that only authorized person and system can access the legitimate resources and services. Light weight authentication protocols are needed since traditional authentication mechanism is not much suitable when huge no of heterogenous and resources constrained smart devices communication to each other.

Availability: Availability in smart city means to make easily available system and services when needed. To ensure the availability of resources and services, IIoT systems for smart city applications should be made robust, attack resistant, temper proof and secure against DoS and DDoS attack. Availability ensure the data availability and backup even after the cyber-attack.

Integrity: Integrity ensure the identification of any modification both on smart devices and data shared among different smart devices, databased and users. Traditional integrity control protocol such as MD5, SHA512 are not much suitable for resource constrained IIoT based smart devices used for building smart city.

Access Control: With huge number of smart devices present in smart city, access control of devices, person and databases must be ensured through designing light weight access control mechanisms. IIoT enabled access control protocol are needed to ensure access permission in heterogenous, dynamic, resources constrained, different proprietary software and hardware system.

Reliability: Smart city contains different smart system such as smart transport system, smart government system, smart healthcare system and smart utility system. These systems generate huge amount of data through distributed sensors. All these information not reliable due to malicious attack. Reliability ensure a smart device to execute it operations without failure and provide reliable information to the citizen under predefine situation and time period.

Resilience: Smart city refer to combination of huge number of heterogenous, resources constrained, mobile and distributed IIoT enabled smart system working together to build smart city. Resilience ensure that all heterogenous smart devices work together in the challenging environment and ensure the system should be secure, robust, adjustable, scalable and available.

Intrusion Detection System: Smart city required an efficient intrusion detection system to monitor all the network devices, sensors and data storage servers. Even after all the security measure smart city is prone to security attack. A strong intrusion detection system monitor application layer devices such as application software, interfaces and databases, monitor network layer devices such as router, gateways and smart network devices, also monitor physical layer devices such as sensors, actuators and smart process control devices for suspicious activities and inform when needed.

8.6 METHODOLOGIES FOR SECURITY ENHANCEMENT AND APPLICATIONS

The methodologies of security enhancement and application are based on fusion of some techniques such as wireless sensor network or WSN, cloud computing, IoT. The combination of all the techniques helps to make intelligent and efficient system of IIoT Cloud computing stores the data with the help of internet for the purpose of secure and permanent data storage system. The concept of cloud computing helps in several applications and areas for handling several requirements of the users. It helps to make smart city by providing several application and services to the users. There are several basic components of smart city such as smart building, smart education, smart energy, smart governance, smart healthcare, smart security, smart town planning, smart transportation or mobility, smart water and sanitation, etc. These services are managed with the helps several components of cloud data centre such as core optical, core router, aggregation router, ToR switch, rack of servers, etc. It helps to make the system efficient and modelling that serve the facility to the customer based on conflicting demands.

Figure 8.3. shows the framework of security system with green computing for smart city. Light weight security protocols are suitable for IIoT systems used in smart city. These light weight protocols are energy efficient, storage efficient and resource constrained, are suitable for self-sustainability and environmental friendly. The purpose of green computing is green use, green design and green dispose. All these requirement of green computing can be achieved by proper use of secure IIoT based systems in different sector of a smart city. Light weight security protocols combined with IIoT and cloud makes the city smart as well as self-sustainability, saves cost, protect environment, reduce waste, provide healthy environment, enhance the sustainability of a city.

Figure 8.4 shows framework of secure cloud computing system of IIoT. It consists of several functionalities such as encryption data system, internet, control instruction system, query analysis and processing system. The combination of all

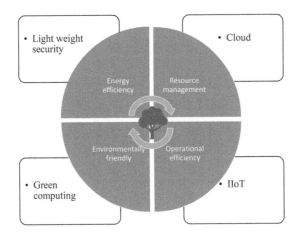

FIGURE 8.3   Frame work of security system with green computing for smart city.

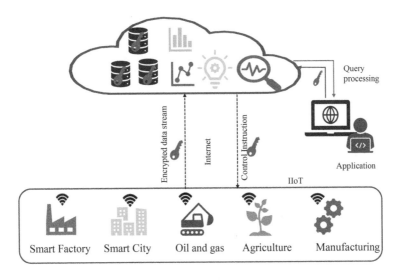

FIGURE 8.4   Framework of secure cloud computing system of IIoT.

information and components helps to model some components such as smart factory, smart city, oil and gas, agriculture, and manufacturing system. Figure 8.5 indicates the relation among cloud computing, IoT, IIoT, WSN based on application analysis and its usage.

There are several operations is carried out using internet. Due to availability of low-cost chips, internet, user compatible devices, market of Internet of Things (IOT) is booming day by day. According to precedence research, Industrial IOT (IIoT) market is exponentially increasing and expected to reach near 1742.8 US dollars, from current market share of 392.85 US dollars. Currently manufacturing, energy and oil sector hold almost 75 percent of market share and agriculture, healthcare accounts

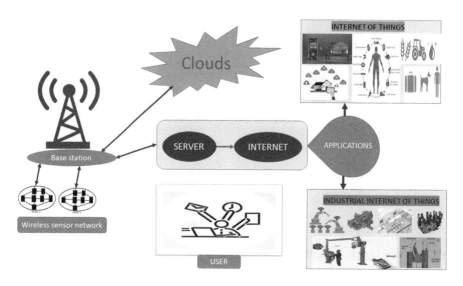

FIGURE 8.5   Relation among cloud computing, IoT, IIoT, WSN.

for less than 15 % market share. With advancement in device technology, increasing consciousness about health monitoring, research on tackling global food shortage using smart agriculture, it is expected that need of IOT will further exponentially increase in these areas. Also, almost 65% of consumer market of IOT is currently with 15% of world's population residing in Europe and North America. With increasing concentration of factories in Asia specific and increasing aspiring consumer market, it is expected that IOT will grow much faster than expected. These areas have huge market and potential for growth of IOT. In order to fulfil the needs of aspiring class at remote locations, researchers need to enhance scalability, efficiency, connectivity, flexibility, fast processing, cost efficiency and accuracy.

There comes the need of cloud which not only drastically reduces capital cost of expensive machinery, which in turn enhances remote computing but also helps in data integration, business continuity, data processing, security, communication, data training and scalability. New applications and start-ups in field of agriculture, infrastructure, medicine and machinery is scaling and research is going on to create new opportunities, applications, software and service for internet and IOT based next industrial revolution. This will depend on IOT, internet, data, cloud and wireless Sensor Network (WSN). WSN, which is combination of sensors and nodes are generally deployed in remote areas for environment monitoring, surveillance, automation, smart grid, assault detection, weather data collection and others. Integration of cloud, IOT and WSN will definitely help in creating an user friendly smart system. For smooth functioning of IOT and WSN integrated system, researchers need to focus on security, energy, application server, processing, coordination, transmitting of data and reducing human interaction. Overcoming challenges and better research will improve and completely change demand of IOT and IIoT. With such integrations, proposition of smart cities, smart agriculture, smart construction, smart industry, smart Grid, smart manufacturing, smart home, efficient & stable health monitoring

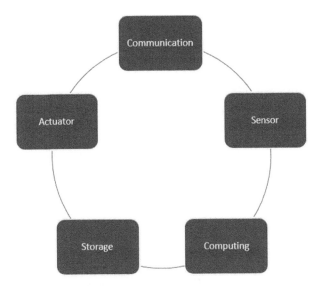

FIGURE 8.6   The methodology of cyber physical system.

and others will soon turn into reality. The demand of IoT and IIoT also consists of the application of future of factory or FOF that uses CPS system which is known as cyber physical system.

The concepts of future of factory of FOF is design for IIoT to process data and information for future use or future demand. Its efficiency is based on industrial sector to enhance the production. The production system interacts autonomously to improve workplace and enable predictive process. The cyber physical system of CPS enhances the operation of IIoT based on three methods cybernetics, mechatronics, design and process embedded system. The methodology of CPS shown in Figure 8.6 that contains five basic components such as computing, communication, sensor, actuator and storage. Sensor is used to sense the information and protect the system from different attacks. Computing system is used to compute different types of errors for communication system. communication is used to communicate one device to another device and resolve several attacks. Some of the attacks are man-in-middle attack, resource blocking attack, packet modification etc. Storage system is used to store data and information as well as data remanence. Actuator is used for authentication purpose and failure analysis that helps to manage deadline hit and miss analysis.

The combination of above-mentioned methodologies helps to resolve several issues such as open-source code issue, attack, vulnerabilities, poor testing, limited security integration, unpatched vulnerabilities, weak password analysis, overwhelming data volume, etc. Hence, it helps to provides several robust features such as domain naming system and filtering, authentication based on multi-functionalities, patch management, password management, security analysis and gateway management, encryption and decryption management, inventory management for device and broader application, device scan management, software update, etc.

8.7   CONCLUSIONS

The chapter highlighted several information based on different prospective of IoT and IIoT. It helps to reader and researcher about different paradigms that are incorporates in IIoT. It also helps to model several application and research innovation that helps to derive new device and application. The chapter also illustrates that how innovation of industry 4.0 requires the application of IIoT with fusion of new techniques and technologies. The application consists of different intelligent tools and methods for resolving several issues of the society. Each of the application consist of artificial intelligence tool that provides some agent-based facility and application in the industry 4.0. The working principles of IoT and industry 4.0 together used in IIoT to make the system efficient and dynamic.

REFERENCES

[1] Das, S. K., Dao, T. P., & Perumal, T. (Eds.). (2021). *Nature-inspired computing for smart application design.* Springer Singapore.

[2] Wang, W., Huang, H., Yin, Z., Gadekallu, T. R., Alazab, M., & Su, C. (2022). Smart contract token-based privacy-preserving access control system for industrial Internet of Things. *Digital Communications and Networks.*

[3] Zhao, Y., Yang, J., Bao, Y., & Song, H. (2021). Trustworthy authorization method for security in Industrial Internet of Things. *Ad Hoc Networks*, 121, 102607.

[4] Das, S. K., & Kumar, V. (2023). IoT security enhancement system: A review based on fusion of edge computing and blockchain. *Constraint Decision-Making Systems in Engineering*, 204–218.

[5] Panchal, A. C., Khadse, V. M., and P. N. Mahalle,. "Security issues in IIoT: A comprehensive survey of attacks on IIoT and its countermeasures," *2018 IEEE Global Conference on Wireless Computing and Networking (GCWCN)*, Lonavala, India, 2018, pp. 124–130, doi: 10.1109/GCWCN.2018.8668630

[6] Umair, M.; Cheema, M.A.; Cheema, O.; Li, H.; Lu, H. Impact of COVID-19 on IoT adoption in healthcare, smart homes, smart buildings, smart cities, transportation and Industrial IoT. *Sensors* 2021, 21, 3838. https://doi.org/10.3390/s21113838

[7] Jorge Duque (2023), The IoT to smart cities - A design science research approach. *Procedia Computer Science*, Volume 219, 2023, Pages 279–285, ISSN 1877-0509, https://doi.org/10.1016/j.procs.2023.01.291. (https://www.sciencedirect.com/science/article/pii/S1877050923002995)

[8] Liang Tan, Keping Yu, Caixia Yang, and Ali Kashif Bashir. 2021. A blockchain-based Shamir's threshold cryptography for data protection in industrial internet of things of smart city. In *Proceedings of the 1st Workshop on Artificial Intelligence and Blockchain Technologies for Smart Cities with 6G (6G-ABS '21)*. Association for Computing Machinery, New York, NY, USA, 13–18. https://doi.org/10.1145/3477084.3484951

[9] L. Fang, H. Zhang, M. Li, C. Ge, L. Liu and Z. Liu, "A secure and fine-grained scheme for data security in Industrial IoT platforms for Smart City," in *IEEE Internet of Things Journal*, Volume 7, no. 9, pp. 7982–7990, Sept. 2020, doi: 10.1109/JIOT.2020.2996664

[10] Kashif Naseer Qureshi, Shahid Saeed Rana, Awais Ahmed, Gwanggil Jeon, "A novel and secure attacks detection framework for smart cities industrial internet of things", *Sustainable Cities and Society*, Volume 61, 2020, 102343, ISSN 2210-6707.)

[11] Ravi Sharma, Balázs Villányi () A sustainable ethereum merge-based Big-Data gathering and dissemination in IIoT system, *Alexandria Engineering Journal*, Volume 69, Pages 109–119, ISSN 1110-0168, https://doi.org/10.1016/j.aej.2023.01.055., (https://www.sciencedirect.com/science/article/pii/S1110016823000820)

[12] Lombardi, F. A., Franchini, R., Morello, R., Casciaro, E., Ianniello, S., Serra, M., ... & Casciaro, S. (2021). A new standard scoring for interstitial pneumonia based on quantitative analysis of ultrasonographic data: a study on COVID-19 patients. *Respiratory Medicine*, 106644, https://doi.org/10.1016/j.rmed.2021.106644

[13] Mukati, N., Namdev, N., Dilip, R., Hemalatha, N., Dhiman, V., & Sahu, B. (2021). Healthcare assistance to COVID-19 patient using Internet of Things (IoT) enabled technologies. *Materials Today: Proceedings*, https://doi.org/10.1016/j.matpr.2021.07.379

[14] Peter, O., Pradhan, A., & Mbohwa, C. (2023). Industrial Internet of Things (IIoT): opportunities, challenges, and requirements in manufacturing businesses in emerging economies. *Procedia Computer Science*, 217, 856–865.

[15] Das, S. K., & Dey, N. (2023). *"Constraint Decision-Making Systems in Engineering"*, IGI Global.

[16] Ding, X., Zhang, Y., Li, J., Mao, B., Guo, Y., & Li, G. (2023). A feasibility study of multimode intelligent fusion medical data transmission technology of Industrial Internet of Things combined with medical Internet of Things. *Internet of Things*, 100689, https://doi.org/10.1016/j.iot.2023.100689

[17] Golchha, R., Joshi, A., & Gupta, G. P. (2023). Voting-based ensemble learning approach for cyber attacks detection in Industrial Internet of Things. *Procedia Computer Science*, 218, 1752–1759.

[18] Jin, C., Li, C., Qin, W., Chen, X., & Chen, G. (2022, December). *A secure and efficient heterogeneous signcryption scheme for IIoT*. In *Frontiers in Cyber Security: 5th International Conference, FCS 2022*, Kumasi, Ghana, December 13–15, 2022, Proceedings (pp. 3–17). Springer Nature Singapore, Singapore.

[19] Das, S. K. (2021). Smart design and its applications: Challenges and techniques. *Nature-Inspired Computing for Smart Application Design*, 1–6.

[20] Khalil, R. A., Saeed, N., Almutiry, M., & Alenezi, A. H. (2023). Energy-efficient anchor activation protocol for non-cooperative localization of Industrial Internet of Things (IIoT). *ICT Express*, https://doi.org/10.1016/j.icte.2023.01.004

[21] Sasikumar, A., Vairavasundaram, S., Kotecha, K., Indragandhi, V., Ravi, L., Selvachandran, G., & Abraham, A. (2023). Blockchain-based trust mechanism for digital twin empowered industrial Internet of Things. *Future Generation Computer Systems*, 141, 16–27.

[22] Muruganandam, S., Anas, A. S., Mohd, A. A. P., Manikanthan, S. V., & Padmapriya, T. (2023). Sensors and machine learning and AI operation-constrained process control method for sensor-aided industrial internet of things and smart factories. *Measurement: Sensors*, 100668, https://doi.org/10.1016/j.measen.2023.100668

[23] Das, S. K., Dey, N., Crespo, R. G., & Herrera-Viedma, E. (2023). A non-linear multi-objective technique for hybrid peer-to-peer communication. *Information Sciences*.

[24] Kilani, R., Zouinkhi, A., Bajic, E., & Abdelkrim, M. N. (2022). Socialization of smart communicative objects in industrial Internet of Things. *IFAC-PapersOnLine*, 55(10), 1924–1929.

9 Radio Frequency Identification and Authentication-Based Intelligent Parking Management System Using the IoT and Mobile Applications
An Implementation Point of View

Keyurkumar Patel
Rashtriya Raksha University, Gandhinagar, India

Manivel Kandasamy, Raju Shanmugam, Tejas Uttare, Niomi Samani, Mohammed Aqeel, and Sonia Joshi
Karnawati University, Gandhinagar, India

9.1 INTRODUCTION

The Internet of Things (IoT) refers to a collection of different kinds of daily gadgets and devices used in various sectors that are widening the aspect of the internet. These devices can share and receive data with other objects enabled due to their internet connectivity [1]. The IoT is basically a network of devices that can exchange data and information with other devices over the Internet. In the IoT, all objects or things are fully equipped with sensors, software, and machine learning techniques. The use of such objects drastically reduces human intervention and gives a hassle-free experience to its users. IoT devices collect data and send it to a central data server (Figure 9.1).

DOI: 10.1201/9781003388814-9

APPLICATIONS OF IOT

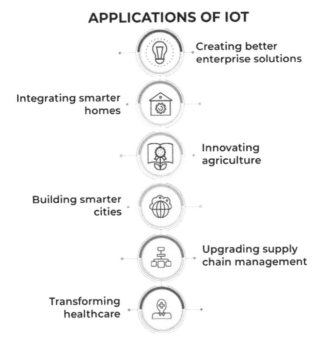

FIGURE 9.1   Applications of the IoT [2].

The sensor embedded in smart devices allows them to collect and access various types of information at any time [3]. In a short period of time, the collected data can be analyzed and then put into action or shared with other devices. This reduces the effort and time required for people to do the same work manually. The development of sensors and smart software in devices has expanded their use across a wide range of industries. These devices play an important role not only in homes but also in various industries such as manufacturing, healthcare, and agriculture. The most important use of the IoT is included in the structuring and development of smart cities [4–6]. A network of sensors in various capacities is distributed throughout the city for multiple things such as traffic management, waste management, streamlining streetlights, saving water, monitoring energy expenditure, and creating smart buildings [7].

Considering the current scenario of the parking system, it is observed that the method is highly ineffective. Intelligent parking solutions based on the IoT are being developed and implemented to address the growing demand for parking spaces in urban areas. These solutions leverage IoT technologies such as sensors, cameras, and data analytics to improve the efficiency and effectiveness of parking management systems. The basic principle of an intelligent parking solution based on the IoT is to use sensors to detect the presence or absence of vehicles in parking spaces. This information is then transmitted to a central system, which can analyze it to provide real-time information to drivers about available parking spaces. Some of the key benefits of these solutions include reducing traffic congestion, improving parking space utilization, and enhancing the overall parking experience for drivers [8, 9].

One example of an intelligent parking solution based on the IoT is the use of sensors embedded in parking spots to detect the presence of vehicles. These sensors can be connected to a central system that can monitor the status of each parking space in real time. This information can be used to create a digital map of available parking spaces, which can be displayed to drivers via mobile apps or other digital platforms.

Another approach is to use cameras and machine learning algorithms to analyze video feeds from parking lots and streets to identify available parking spaces. This approach can be more flexible and cost-effective than installing sensors, as it can be applied to existing infrastructure without significant modifications [10].

In addition to improving parking space utilization, intelligent parking solutions based on the IoT can also help reduce traffic congestion by providing real-time information about available parking spaces. This can help drivers avoid circling around looking for a parking spot, which can contribute to traffic jams and air pollution [11].

Overall, intelligent parking solutions based on the IoT have the potential to revolutionize the way we manage parking in urban areas. As technology continues to evolve and become more widespread, we can expect to see even more innovative solutions that further enhance the parking experience for drivers and reduce the environmental impact of transportation.

Further, an intelligent parking management system that utilizes radio frequency identification (RFID) technology and user preferences can greatly improve the efficiency and convenience of parking for drivers. RFID technology allows for contactless communication between a tag or label and a reader, making it an ideal solution for parking management [12].

The basic principle of this system is to equip vehicles with RFID tags or stickers, which are registered in a database along with the user's preferences such as preferred parking locations, payment methods, and other relevant information. As the vehicle enters a parking facility or area, an RFID reader detects the tag and automatically identifies the vehicle and user preferences.

Based on the user preferences and availability of parking spaces, the system can direct the driver to an available parking spot that best meets their needs. For example, if the driver prefers a parking spot close to the entrance, the system can direct them to the nearest available spot that meets this criterion. Similarly, if the user prefers a covered parking spot, the system can direct them to a suitable spot in a covered area [13, 14].

The system can also provide real-time information to drivers about the availability of parking spaces, which can be accessed through a mobile app or another digital platform. This can help drivers avoid wasting time looking for parking spaces and reduce traffic congestion.

At present there is still no management system for the large parking lots although all the technology is available, so people park their cars randomly. Once a car is stuck in a crowded area, it leads to traffic congestion and excess time to remove that car. Moreover, this entire chapter is based on green computing concepts to enhance the overall efficiency of smart parking management systems.

Green computing can play a significant role in enhancing the sustainability and efficiency of IoT-based smart parking systems. Here are some ways in which green computing can help:

- **Energy-efficient sensors:** Smart parking systems rely on sensors to detect the availability of parking spaces. By using energy-efficient sensors, the power consumption of the system can be reduced, thus making it more environmentally friendly.
- **Cloud computing:** Smart parking systems generate large amounts of data that need to be processed and analyzed in real time. By leveraging cloud computing, the system can avoid the need for on-premises data centers, which consume significant amounts of energy.
- **Low-power wireless communication:** IoT-based smart parking systems rely on wireless communication technologies to transmit data between the sensors, the gateway, and the cloud. By using low-power wireless communication technologies like LoraWAN or NB-IoT, the energy consumption of the system can be minimized.
- **Optimization algorithms:** Smart parking systems can use optimization algorithms to minimize the time and energy required to find a parking spot. These algorithms can take into account factors such as distance, availability, and energy consumption, thus reducing the overall environmental impact of the system.
- **Renewable energy sources:** By using renewable energy sources like solar or wind power to power the smart parking system, the reliance on traditional power sources can be reduced, thus making the system more sustainable.

Overall, by adopting green computing practices, IoT-based smart parking systems can reduce their carbon footprint and enhance their overall efficiency and sustainability.

The above-mentioned various IoT applications and their potential and significant contribution towards energy optimization and cloud usage motivate us to solve the complex parking management system concept at the university and then at the city level.

9.2   EXISTING METHODS AND RELATED WORK

The conventional method is to park a car according to the convenience of the driver. This process is labor intensive and time-consuming and needs a lot of manpower to guide all the vehicles all the time to park in the right place. However, the automated process of displaying parking slot availability to users and RFID-based authentication will restrict unauthorized entry as well. Hence, in this chapter an overall eco-system is developed using the integration of the IoT, different sensors and Arduino boards, and security enhancement to cater to the needs of the client who wants to access a secure parking guidance system with technological aspects. However, here a preliminary study of the development of an RFID-based intelligent parking system prototype is proposed. Later on, we will develop a full fledged smart parking management system with all the technological aspects by a fusion of sensors and other technological devices.

9.3 METHODOLOGY

The proposed automatic parking system tracks and regulates parking spaces to make the vehicle parking process more convenient, comfortable, and less time-consuming (Figures 9.2–9.4). The components used in the smart parking system are:

1. **Arduino UNO:** This is built around the microcontroller ATmega328P. Shields, additional circuits, and pins for digital and analog I/O make up the board. The Arduino UNO contains a USB port, a power jack, six analog

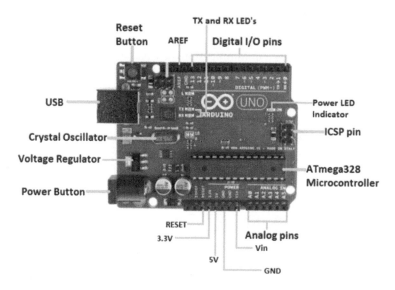

FIGURE 9.2   Arduino UNO [15].

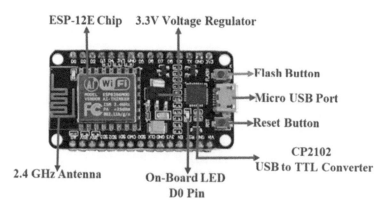

FIGURE 9.3   Node MCU [16].

FIGURE 9.4   Ultrasonic sensors [19].

pins, fourteen digital pins, and an in-circuit serial programming header. It is an integrated development environment. Both online and offline platforms can use it.

2. **NodeMCU:** This is an ESP8266-based open-source platform that enables the connection of objects and the transmission of data over the Wi-Fi protocol. The ESP-12E module is included with the NodeMCU ESP8266 development board. This module includes a Tensilica Xtensa 32-bit LX106 RISC microprocessor-equipped ESP8266 chip. This microprocessor supports a real-time operating system (RTOS) and has a clock frequency that can be adjusted between 80 and 160 MHz. For storing data and programs, the NodeMCU contains 4 MB of flash memory and 128 KB of RAM. Its strong processing power makes it perfect for IoT projects thanks to built-in Wi-Fi/Bluetooth and Deep Sleep functionality. A micro-USB port and VIN pin (an external power supply pin) can both deliver power to NodeMCU. Interfaces such as UART, SPI, and I2C are supported.

3. **Ultrasonic sensors:** A device that measures the distance to an object using ultrasonic waves is called an ultrasonic sensor. A transducer is used by the sensor to transmit and receive ultrasonic pulses, which communicate information about an object's vicinity. By reflecting off the borders, high-frequency sound waves produce different echo patterns [17–19].

The working principle of the ultrasonic module is very simple and less complicated. It emits a 40 kHz ultrasonic pulse that travels through the air and bounces back off the sensor if there is an obstacle or object. Calculating the transit time and sound speed will yield the distance. Ultrasonic sensors can detect things regardless of their color, surface, or composition [20].

RFID and authentication-based smart parking systems are designed to use the IoT to automate the process of parking vehicles in a lot or garage. These systems use RFID technology to identify and authenticate vehicles as they enter the parking area. Once a vehicle has been authenticated, the system can automatically assign it a parking space and direct the driver to the appropriate location.

There are several benefits to implementing an RFID and authentication-based smart parking system using the IoT. One of the main benefits is the improved efficiency of the parking process. By automating the assignment of parking spaces, the system can reduce the time it takes for a vehicle to find a spot and park. This can help to reduce congestion and improve the overall flow of traffic in the parking area.

Another benefit of RFID and authentication-based smart parking systems is the increased security they can provide. By using RFID technology to identify and authenticate vehicles, the system can prevent unauthorized vehicles from entering the parking area. This can help to reduce the risk of vandalism or theft, and can also help to improve overall safety in the parking area.

Overall, an RFID and authentication-based smart parking system using the IoT can be an effective way to improve the efficiency, security, and overall experience of parking vehicles in a lot or garage.

Moreover, this system largely depends on various sensors which generate a large volume of data. Hence, effective algorithms need to be designed for a data management system to improve the overall performance of the system. Further, effective solutions need to be developed in case of sensor failures due to various reasons such as adversarial weather conditions, power outages, or physical damage. In this integrated system, the system collects and stores sensitive information, such as location data, which must be protected from unauthorized access.

In order to make a convenient car parking experience, the proposed system will comprise eight major parts:

1. Authentication for the registered users;
2. Automatic parking management system without human intervention;
3. Automatic gate barriers for entry and exit;
4. Display of total availability in different parking zones, right at the entrance;
5. Information on online parking slot availability on smartphones (application programming interface (API) integration);
6. Easily accessible system;
7. Data security protocol implementation to restrict unauthorized access;
8. Development of continuous monitoring and updating of parking slot availability in case of sensor failures, power outages, etc. (Figure 9.5).

During the implementation of this approach, all university members will be verified using car number plate detection and then be allowed to cross the barricade. This process is done to ensure no unauthorized driver enters the parking space. Then users can access a web-based software platform that will enable them to locate the closest parking spot available for their vehicles. The inducted microcontroller (Arduino based) continuously reads the sensor values inserted in the parking area and updates this information online over the IoT server (Figures 9.6 and 9.7).

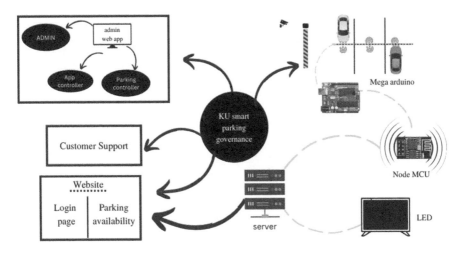

FIGURE 9.5 Architecture of smart parking governance system.

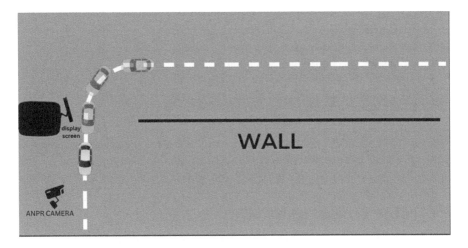

FIGURE 9.6 Entrance and verification of car using number plate.

9.4 OBJECTIVES

The objectives of an RFID and IoT-based smart parking guidance system are similar to those of an IoT-based system but with the added benefits of RFID technology. These objectives include:

- Providing real-time information on available parking spots through the use of RFID sensors and IoT technology.
- Improving the accuracy and reliability of parking data by using RFID tags to identify and track vehicles.

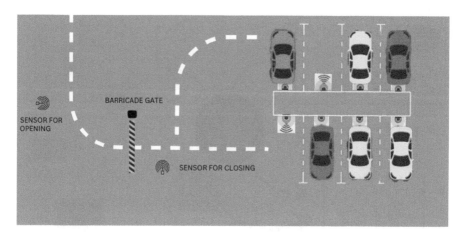

FIGURE 9.7   Barricade opening/closing and car parking.

- Enhancing the convenience and ease of use of the system for both drivers and parking operators through the use of RFID-enabled parking access and payment systems.
- Improving security and reducing fraud through the use of RFID-enabled parking access control and validation.
- Improving the scalability and flexibility of the system by using RFID and IoT technology to enable remote monitoring and management of parking resources.

A smart parking solution's primary goal is to accurately detect the presence or absence of a car in a specific parking place and transmit this information into a system for analysis so that parking attendants can access it.

9.5   PROTOTYPE IMPLEMENTATION

1. Ultrasonic sensors with Arduino Uno: Ultrasonic sensors are used in vehicle detection. This connection indicates that slots are available in the parking area. With the help of Arduino Uno, if any vehicle comes in front of the ultrasonic sensors, it detects them and the LED lights up, which shows that the particular slot is full. All the ultrasonic sensors are individually connected with Arduino Uno and with different ports. Here we use one ultrasonic sensor for one vehicle slot, as we have many slots in the parking area.

2. Stepper motor with motor driver integration with Arduino Uno and IR sensor for auto-functioning of the barrier: Here, first of all, the stepper motor with motor driver and IR sensors are integrated with Arduino Uno for I/O functions. The stepper motor is connected to a gate for opening and closing it, whenever the vehicle is detected. To integrate this entire hardware with the software, the appropriate code is written on Arduino Uno. To perform this task an object will come close to the IR sensor, which promptly detects

the object and sends the signal to the motor driver through Arduino Uno. Further, using motor driver signals the stepper motor will rotate 90 degrees so the gate attached to it will remain open until the vehicle has passed. As long as the IR sensor detects the object it will not rotate back to its initial position. After a few seconds, once the vehicle passes through the gate, it will automatically rotate to its initial position which will close the gate (Figures 9.8–9.10).

FIGURE 9.8   Ultrasonic sensors with LED connected with Arduino Uno top view.

FIGURE 9.9   Ultrasonic sensors with LED connected with Arduino Uno front view.

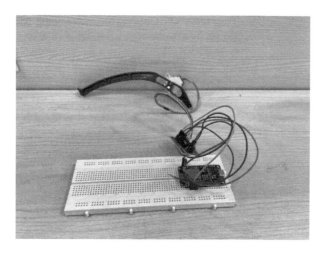

FIGURE 9.10   Stepper motor with a motor driver.

9.6 CONCLUSION

An RFID and authentication-based intelligent parking management system using the IoT and mobile applications offers a convenient and efficient way to manage parking slots. The system uses RFID technology to authenticate vehicles and guide drivers to available parking spaces. The implementation of the system involves integrating IoT devices and sensors to collect and process data and using a web-based platform to display real-time information and provide guidance. Overall, the system can improve the parking experience and reduce congestion, making it a valuable addition to smart cities.

REFERENCES

[1] Madakam S., Ramaswamy R., and Tripathi S. IT Applications Group Internet of Things (IoT): A Literature Review, National Institute of Industrial Engineering (NITIE), Vihar Lake, Mumbai, India.

[2] Singhvi R., Lohar R., Kumar A., Sharma R., Sharma L., and Saraswat R. IoT Based Smart Waste Management System: India perspective, *2019 4th Inter National Conference on Internet of Things: Smart Innovation and Usages (IoT-IU)*, Ghaziabad, India, 2019, pp. 1–6. https://doi.org/10.1109/IoT-SIU.2019.8777698

[3] Patil M., Chetepawad K., Shahu A., Swami S. *IOT-based smart car parking system*, IJARIIE, 2020.

[4] Mishra N., Singhal P., and Kundu S. Application of IoT Products in Smart Cities of India, 2020 *9th International Conference System Modeling and Advancement in Re search Trends (SMART)*, Moradabad, India, 2020, pp. 155–157.

[5] Khalil U., Uddin M, Malik O. and Hussain, S. A Blockchain Footprint for Authentication of IoT-Enabled Smart Devices in Smart Cities: State-of-the-Art Advancements, Challenges and Future Research Directions, in *IEEE Access*, vol. 10, pp. 76805–76823, 2022, https://doi.org/10.1109/ACCESS.2022.3189998

[6] Cheng B., Solmaz G., Cirillo F., Kovacs E., Terasawa K., and Kitazawa A. Fog Flow: Easy Programming of IoT Services Over Cloud and Edges for Smart Cities, in *IEEE Internet of Things Journal*, vol. 5, no. 2, pp. 696–707, April 2018.

[7] Rajaram M., Kougianos E., Mohanty S., and Sundaravadivel P. A wireless sensor network simulation framework for structural health monitoring in smart city ies, 2016 IEEE 6th International Conference on Consumer Electronics – Berlin (IC CE-Berlin), Berlin, Germany, 2016, pp. 78–82, https://doi.org/10.1109/ICCE- Berlin.2016.7684722

[8] Yin Y. and Dalin J. 2013. Research and Application on Intelligent Parking Solution Based on Internet of Things. In *Proceedings of the 2013 5th International Conference on Intelligent Human-Machine Systems and Cybernetics - Volume 02 (IHMSC '13)*. IEEE Computer Society, USA, 101–105. https://doi.org/10.1109/IHMSC.2013.171

[9] Shimi A., Ebrahimi Dishabi M.R., and Azgomi M.A. An Intelligent Parking Management System using RFID Technology based on User Preferences. Preprints.org 2020, 202011 0738.

[10] Tsiropoulou E., Baras J.S., Papavassiliou S., and Sinha S. (2017). RFID-based smart parking management system. *Cyber-Physical Systems*, *3*, 22–41.

[11] Mazlan M.S., Rahmi I., Hamid A., and Kamaludin H. 2018. Radio frequency identification (RFID) based car parking system. *JOIV: International Journal on Informatics Visualization*, 2(4–2), 318–322.

[12] Xiao X. and Zou Q. The architecture of the RFID-based intelligent parking system. *Advances in Intelligent Systems and Interactive Applications: Proceedings of the 2nd International Conference on Intelligent and Interactive Systems and Applications (IISA 2017)*. Springer International Publishing, 2018.

[13] Salah B. (2020). Design, simulation, and performance-evaluation-based validation of a novel RFID-based automatic parking system. *Simulation*, *96*, 487–497.

[14] Pradhananga P., Rezazadeh J., Salahuddin A., Shrestha A. (2022). RFID Based Remote Monitoring Techniques for IoT Smart Parking System. In: Daimi, K., Al Sadoon, A. (eds) *Proceedings of the ICR'22 International Conference on Innovations in Computing Research ICR 2022*. Advances in Intelligent Systems and Computing, vol 1431. Springer, Cham. https://doi.org/10.1007/978-3-031-14054-9_38

[15] Agrawal T., and Qadeer M. A. Tracing Path with Arduino Uno using GPS and GPRS/GSM. *2018 International Conference on Computing, Power, and Communication Technologies (GUCON)*, Greater Noida, India, 2018, pp. 1203–1208, https://doi.org/10.1109/GUCON.2018.8674953

[16] Parvati S. V., Thenmozhi K., Praveenkumar P., Sathish S., and Amirtharajan R. IoT Accelerated Wi-Fi Bot controlled via Node MCU. 2018 International Conference on Computer Communication and Informatics (ICCCI), Coimbatore, India, 2018, pp. 1–3, https://doi.org/10.1109/ICCCI.2018.8441215

[17] Brashdi Z., Hussain S., Yosof K., Hussain S., and Singh A. IoT-based Health Monitoring System for Critical Patients and Communication through Think Speak Cloud Platform, *2018 7th International Conference on Reliability, Infocom Technologies and Optimization (Trends and Future Directions) (ICRITO)*, Noida, India, 2018, pp. 652–658, https://doi.org/10.1109/ICRITO.2018.8748751

[18] Jayaysingh R., David J., Joel Morris Raaj M., Daniel D. and BlessyTelagathoti D. IoT Based Patient Monitoring System Using NodeMCU, *2020 5th International Conference on Devices, Circuits, and Systems (ICDCS)*, Coimbatore, India, 2020, pp. 240–243, https://doi.org/10.1109/ICDCS48716.2020.243588

[19] Dimitrov A. and Minchev D. Ultrasonic sensor explorer, *2016 19th International Symposium on Electrical Apparatus and Technologies (SIELA)*, Bourgas, Bulgaria, 2016, pp. 1–5, https://doi.org/10.1109/SIELA.2016.7542987

[20] Vukonić L. and Tomić M. Ultrasonic Sensors in IoT Applications," *2022 45th Jubilee International Convention on Information, Communication, and Electronic Technology (MIPRO)*, Opatija, Croatia, 2022, pp. 415–420.

10 Bike Sharing Systems
A Green IT Application for Smart Cities

Kaushal Nileshbhai Maniyar
and Jai Prakash Verma
Nirma University, Ahmedabad, India

Neha Sharma
TCS Pune, Pune, India

10.1 INTRODUCTION

The role of the Internet of Things (IoT) in smart cities is to improve the efficiency, sustainability, and quality of life for citizens by using connected devices and systems to gather and analyse data. The IoT can be used in various domains, such as transportation, energy management, waste management, public safety, health care, and more [1]. By providing real-time information, the IoT can help optimize resource utilization, reduce waste and emissions, improve traffic flow, enhance public safety, and support informed decision-making. The goal of the IoT in smart cities is to create an interconnected, data-driven ecosystem that can respond to the needs and challenges of the city and its residents in real-time. The way we interact with the world around us might be completely transformed by the IoT. Since there are billions of little devices connected to the internet, they may work together to carry out a variety of tasks, from monitoring our houses and automobiles to keeping track of our health and fitness. However, this also poses a huge problem in terms of energy use and emissions. The IoT's excessive growth in communication volume has the potential to raise energy use and carbon dioxide emissions from the devices themselves, which currently contribute more than 2.5% of the world's harmful emissions. In the majority of innovative IoT applications, smart battery-powered gadgets are anticipated to operate effectively for months or even years while working with other items to reliably do their assigned tasks.

The need for a green future is becoming more and more obvious as people across the world become aware of how technology affects the environment. Researchers are working very hard to reduce the environmental effects of these difficulties, having realized the need to reduce energy usage and carbon emissions from IoT devices. The Green Internet of Things (G-IoT) idea is being introduced to help with energy reduction and environmental preservation. The G-IoT intends to advance the creation of energy-efficient IoT hardware and software, including technologies that lower the

DOI: 10.1201/9781003388814-10

energy needed for communication as well as hardware and software that is energy-efficiently designed. Researchers are striving to address these issues by creating intelligent, energy-efficient gadgets and pushing the idea of the G-IoT in order to overcome the concerns that the IoT raises in terms of energy consumption and emissions. By doing this, the IoT can minimize its environmental effect while continuing to expand and adapt responsibly.

In addition to examining how the G-IoT can be accomplished, we will seek to give a thorough overview of the problems related to the environmental impact of using IoT technologies. We focus on the creation of a bike sharing system (BSS) and explore various IoT applications in the context of smart cities. We first provide a succinct history of BSS creation and application, with a focus on the Indian BSS landscape in order to contextualize this debate. We then provide an overview of the most important BSS use cases and a study of the difficulties in implementing it. We also provide a literature survey of past studies on BSS in addition to these qualitative observations. An experimental data analysis of a pertinent dataset is conducted in addition to the literature survey with the goal of identifying significant patterns and trends in the use of BSS. We then apply a variety of machine learning methods to the dataset to assess how well they forecast BSS usage trends. The findings of these studies are thoroughly explored, with a focus on their consequences for the upcoming development and application of BSS and other IoT-enabled technologies in the context of smart cities. Overall, the chapter significantly adds to the current conversation about how IoT technologies affect the environment and provides crucial information for the creation of sustainable urban mobility solutions.

10.2   THE ENVIRONMENTAL IMPACTS OF THE IoT

- The use of energy: The consumption of energy in the IoT can be attributed to various factors. Devices in the IoT need energy to perform tasks such as sensing, processing, computing, monitoring, and communication. Additionally, the data centres that support IoT applications also consume significant amounts of energy. Research has demonstrated that the majority of energy consumption occurs during communication and data transmission between devices. It has been noted that wireless connections, particularly 3G and 4G networks, consume significantly more energy than wired connections. For instance, 3G networks consume 15 times more energy than wired networks, while 4G networks consume 23 times more energy. Despite the high energy consumption, wireless access networks are crucial for the functioning of the IoT and cannot be disregarded.
- Elevated levels of carbon released into the atmosphere: The growth of the IoT and intelligent devices has resulted in a significant rise in energy consumption, leading to increased carbon emissions. The carbon footprint of the information and communication technology (ICT) industry has gone up significantly, reaching 30 million tons in 2015 from 6 million tons in 2012. This exacerbation of carbon emissions contributes to the problem of climate change. One of the environmental concerns is finding ways to address the carbon emissions caused by the ICT industry.

- E-waste: The increasing production and usage of ICT devices has led to a rise in electronic waste. In 2018, 49.8 million tons of electronic waste were generated worldwide, a significant increase from the 33.8 million tons produced in 2010, as reported by Statista.
- Resource depletion: The production of IoT devices requires the extraction and use of natural resources, such as minerals, which can lead to resource depletion and the degradation of ecosystems.

10.3   WAYS TO ACHIEVE THE G-IoT

The G-IoT refers to the design and implementation of IoT systems with the aim of minimizing their environmental impact. One of the key ways to achieve the G-IoT is to design devices that consume as little energy as possible. This can be done by using low-power microprocessors, optimizing power management systems, and using energy-efficient communication protocols.

- Sustainable materials: Using environmentally friendly materials in the production of IoT devices can help reduce their environmental impact. For example, using recycled materials or biodegradable materials can reduce the amount of waste generated by the production of these devices.
- Proper disposal of electronic waste: When IoT devices reach the end of their lifespan, it is important to dispose of them in a responsible manner to reduce the environmental impact of electronic waste. This can be done through electronic waste recycling programmes, which can reduce the amount of waste that goes to landfills and prevent the release of toxic chemicals into the environment.
- Energy-efficient data centres: The data generated by IoT devices is often stored and processed in data centres, which can consume a significant amount of energy. Implementing energy-efficient practices in data centres, such as using renewable energy sources and improving cooling systems, can help reduce the energy consumption of these centres.
- Sustainable deployment: The deployment of IoT systems should consider the environmental impact of devices, including the materials used in their construction and the energy used in their operation. By considering the full lifecycle of the devices, it is possible to minimize their environmental impact and promote sustainable deployment practices. By adopting these practices, it is possible to achieve the G-IoT and minimize the environmental impact of IoT systems.

10.4   APPLICATIONS OF THE G-IoT IN SMART CITIES

The G-IoT is the integration of IoT technologies with green initiatives to reduce the environmental impact of IoT systems and devices. In a smart city, some examples of G-IoT applications are [2]:

- Smart energy management systems: Use IoT devices to monitor and control energy usage in buildings, reducing energy waste and promoting the use of renewable energy sources.

- Smart waste management: Use IoT sensors and cameras to optimize waste collection routes, reducing emissions from waste trucks.
- Intelligent transportation systems: Use IoT technologies to optimize traffic flow, reduce emissions from vehicles, and promote the use of sustainable transportation modes such as electric vehicles. One example is the BSS, which we will discuss below.
- Intelligent building management: Use IoT sensors and controls to monitor and optimize heating, cooling, and lighting systems so as to reduce energy consumption and emissions.
- Smart agriculture: Use IoT sensors to monitor soil moisture and crop health, which reduces water usage and promotes sustainable agriculture practices.

10.5 A BIKE SHARING SYSTEM

A BSS is a network of bicycles available for shared use for individuals on a short-term basis. Users can rent a bike from a designated station and return it to another station when they are finished. The system typically uses a membership or subscription model, with users paying a fee to access the bikes. Bike sharing is a convenient, eco-friendly, and cost-effective transportation option for short trips, making it a popular choice for urban commuters, tourists, and students.

10.5.1 HISTORY OF THE BIKE SHARING SYSTEM

The BSS can be categorized into three generations [3, 4]:

- First generation: In this generation bikes are freely available to the public. One uses a bike then leaves it for next user. There is no theft protection provided in this generation.
- Second generation: At particular spots in the city, bikes may be borrowed and returned in exchange for a coin deposit. A fine is imposed when the bike is not returned. However, thefts still occur due to users adopting anonymity.
- Third generation: A number of technological advancements were made to the third generation of BSSs to make them smarter, including electronically locking bike racks or locks, telecommunications systems, smart cards and fobs, mobile phone access, and on-board computers.

The following is a brief history of the BSS, how it got started and evolved:

- 1965: The idea of the first BSS was proposed by Luud Schimmelpennink in association with the group Provo [5–7]. It was started in Amsterdam, the Netherlands. They painted 50 bikes in a white colour and made them available for the public to use freely [8]. This idea was named the "White Bicycle Plan". The majority of the bikes were stolen within a month, while the remaining ones were discovered in neighbouring canals [9]. This system comes under the first generation.

- 1995: In this year, coin based unlocking and refundable deposit based rental was introduced in Copenhagen, Denmark. The system was called "City Bikes". Riders could rent bicycles for an indefinite amount of time in a designated "city bike zone" by placing a refundable deposit at one of 100 unique locking bike stations [10]. The Copenhagen police severely enforced more than US$150 fines for failing to return a bicycle or cycling outside the bike sharing zone. This system comes under the second generation.
- 2005: In Rennes, France, the first cutting-edge, technologically driven, bike-sharing programme was introduced. Smart cards were used in the "Vélo à la Carte" system, which allowed users to hire and return bikes at automated kiosks [3]. This system comes under the third generation.
- 2013: In Mysuru, Karnataka, the country's first bike-sharing programme was launched in this year. The "Trin-Trin" public bicycle sharing (PBS) system was introduced with the intention of encouraging environmentally friendly and sustainable transportation in the city.
- 2018: Dockless bike-sharing systems became popular in many cities around the world, including in the United States, Europe, and Asia. Dockless bike share means bikes don't need a special station to park at, which can be expensive and limit how many bikes a city can have. Instead, with dock-less systems, bikes can be parked in a certain area at a bike rack or on the sidewalk.

10.5.2 A SURVEY OF THE INDIAN BIKE SHARING SYSTEM

The first BSS was introduced in India in 2013. After that many cities adopted BSSs:

• Mysuru – Trin-Trin	• Bhubaneswar – Yaana	• Vadodara – MYBYK
• Delhi – SmartBike	• Kolkata – Yulu	• Ahmedabad – MYBYK, Yulu
• Mumbai – MIBICI	• Chennai – Zoomcar PEDL	• Hyderabad – Pedal
• Bhopal – MYBYK	• Kochi – MYBYK	• Chandigarh – Yulu
• Jaipur – Namma Jaipur	• Nagpur – MYBYK	• Lucknow – Yulu
• Pune – PEDL, Yulu		

Trin-Trin, SmartBike, MIBICI, MYBYK, PEDL, Yulu, and Yaana are the main service providers of the BSS in India. Brief information about every service provider is given below.

- Trin-Trin: A PBS called Trin-Trin was launched in Mysuru, Karnataka, in 2016. The Mysuru City Corporation created the system with the intention of encouraging environmentally friendly and sustainable transportation in the city. Users of the Trin-Trin system can hire bicycles using a smartphone app or at one of the authorized Trin-Trin stations spread out across the city. By setting up an account on the mobile app or by physically going to a Trin-Trin station, users can sign up for the Trin-Trin system. Once logged in, customers may use a smart card or the mobile app to unlock a bicycle from a station. Users pay a small cost for the duration of the trip, and they may return the bike at any Trin-Trin station that is close to where they are going. In Mysuru, the Trin-Trin system has been effective in encouraging environmentally friendly transportation, and it has received several awards for its creative public transportation strategy. Additionally, the system has assisted in enhancing the city's air quality and reducing traffic congestion.

- SmartBike: A PBS called SmartBike was launched in Delhi, India, in 2019. The Delhi Metro Rail Corporation (DMRC) developed the system with the intention of offering city commuters an environmentally friendly and cost-effective form of transportation. Users of the SmartBike system can hire bicycles from a fleet using a smartphone app. By setting up an account on the app or by physically visiting a SmartBike station, users may sign up for the SmartBike system. Once enrolled, customers may use the app to find and unlock bicycles. Users pay a small cost for the duration of the trip, and they may return the bike to any SmartBike station that is close to where they are going. The SmartBike system has been well-received by users and has been effective in encouraging environmentally friendly transportation in Delhi. Since it is inexpensive and practical, the system is a well-liked choice for errands and everyday commuting. The DMRC intends to eventually extend the SmartBike system to more areas of the city.

- MIBICI: A bike-sharing programme called MIBICI was introduced in Mumbai, India, in 2019. The Mumbai Metropolitan Region Development Authority implemented the system as part of its initiatives to promote environmentally friendly transportation and lessen traffic in the city. Users of a mobile app can hire bicycles from MIBICI's fleet using the app; 500 bicycles were initially available at 25 stations across the city when the system first began operating. Since then, the number of bicycles has increased to nearly 3,000 at over 300 stations. For a small cost, users can hire bicycles for the duration of their trip, and they can return them to any MIBICI station that is close to their final destination. Residents of Mumbai have embraced MIBICI, which has aided in promoting environmentally friendly transportation in the city. The system has been commended for its simplicity of use and low cost, and it has assisted in reducing traffic congestion and improving air quality.

- MYBYK: A PBS called MYBYK was introduced in Ahmedabad, Gujarat, in 2017. The Ahmedabad Municipal Corporation created the system with the intention of fostering environmentally friendly and sustainable transportation in the city. Users of the MYBYK system can rent a fleet of bicycles using a smartphone app or by going to one of the authorised MYBYK stations spread out across the city. By setting up an account on the mobile app or by physically visiting an MYBYK station, users can sign up for the system. Once logged in, customers may use a smart card or the mobile app to unlock a bicycle from a station. The cost of the ride is low, and users may return the bicycle to any MYBYK station that is close to their destination. In Ahmedabad, MYBYK has been effective in encouraging environmentally friendly transportation, which has decreased traffic congestion and enhanced air quality. The system has earned several awards for its contribution to India's sustainable urban growth and has also been praised for its creative public transportation strategy.

- PEDL: Zoomcar has a bike-sharing programme called PEDL (Pedal Easy). The PEDL system was introduced in 2017 and is now accessible in a number of Indian cities, including Pune, Chennai, Mumbai, Kolkata, and Delhi. The PEDL system includes a fleet of bicycles that consumers may hire using a smartphone app. By entering their personal data and payment information, users may download the PEDL app and sign up for the service. After signing up, users may use the app to find nearby PEDL cycles and unlock them by scanning a QR code on the bike. Users pay a small cost for the entire journey, which may be finished by leaving the bicycle at a specific PEDL station. In cities all around India, the PEDL system has been effective in encouraging environmentally friendly and sustainable transportation, and users have given it high marks. The system appeals as a practical and economical choice for quick journeys within the city.

- Yulu: A dockless bicycle sharing programme called Yulu was introduced in India in 2018. The firm seeks to offer a first-mile/last-mile connection and sustainable and economical transportation solution. Smart locks, GPS, and IoT technologies are all included in the fleet of bicycles that Yulu rents out. Customers may use a smartphone app to find and unlock the bicycles. Yulu uses a dockless system, so customers can leave their bicycles anywhere in a specified Yulu zone as long as they are not blocking pedestrian or vehicular traffic. The business also sells Yulu Miracle electric bicycles, which include a battery pack and can go up to 60 km on a single charge. Yulu has had success encouraging environmentally friendly transportation in a number of Indian cities, including Bangalore, Pune, and Delhi. To grow its operations and promote environmentally friendly transportation, the firm has cooperated with a number of local government agencies and private organizations. In addition, Yulu has started offering a sustainable alternative to gasoline-powered automobiles with its Yulu Move electric scooter sharing programme.

- Yaana: In Bengaluru, India, Yaana, a bike-sharing network, was introduced in 2021. The platform was created to offer city commuters a practical and reasonably priced mobility option. Yaana encourages people to ride bicycles rather than automobiles or other vehicles that add to traffic congestion and pollution in order to promote sustainable mobility. Users may register for the Yaana service by downloading the app and entering their personal and payment information. Users of the app may use a QR code to find the closest Yaana bike and unlock it. After that, they may ride the bike to their destination and leave it in any authorized parking space nearby. Currently available in a few Bengaluru areas, Yaana intends to eventually provide its services in more cities. The platform has been commended for its user-friendly user interface (UI), reasonable cost, and excellent customer feedback.

10.5.3 Use Cases of the Bike Sharing System

The research paper [11] examines the environmental benefits of bike sharing by analysing big data on bike sharing usage. The researchers found that bike sharing can reduce carbon dioxide emissions, air pollution, and traffic congestion, and promote sustainable transportation, with the greatest benefits seen in densely populated urban areas. Due to its many advantages, bike-sharing programmes are becoming more and more common in smart cities throughout the world. These programmes enable the short-term rental of bicycles, making them an easy and cost-effective mode of transportation for local travel. Here are some examples of how bike-sharing programmes are used in smart cities:

- Improved mobility: Bike-sharing programmes offer a practical substitute for conventional forms of transportation like automobiles and buses. They provide a more time and money-efficient means of short-distance transport, easing traffic congestion and enhancing mobility in metropolitan locations.
- Health advantages: Cycling is a low-impact workout that has several advantages for your body, such as better cardiovascular health, stronger muscles, and less stress. A better lifestyle is encouraged through bike-sharing programmes, which motivate people to pedal more frequently.
- Benefits for the environment: Bike sharing programmes help cut carbon emissions and are ecologically benign. They aid in decreasing traffic by encouraging riding as a practical means of transportation, which lowers pollutants and improves air quality.
- Data collection: Bike-sharing programmes generate a lot of information, such as demographics, traffic numbers, and usage trends. The system may be improved with the use of these data, which can also shed light on urban mobility trends and help with traffic planning. Overall, bike-sharing programmes are a creative and environmentally friendly response to the mobility problems in smart cities. We may anticipate seeing more use cases and advantages as they gain popularity in the coming years.

10.5.4 Challenges in Bike Sharing Systems

Researchers and urban planners are both interested in the connection between the natural and built environments and the use of bike-sharing programmes. The study in [12] explores this connection in further detail and focuses on how urban planning and infrastructure might affect how well-liked and effective bike-sharing programmes are. One of the main conclusions of the study is that the use of these programmes is significantly influenced by the natural environment, including terrain, climate, and vegetation. Places with flat terrain and pleasant temperatures, for instance, often have greater rates of bike-sharing utilization than places with steep topography and severe climates. Additionally, the presence of greenery and plants in urban areas can enhance riding enjoyment and promote the usage of bike-sharing programmes. The effectiveness of programmes is greatly influenced by the built environment, especially the transportation infrastructure. The study discovered that communities with designated bike lanes and traffic-calming features like speed bumps and roundabouts typically have greater rates of use for bike-sharing programmes. Additionally, communities with efficient public transit systems and integrated bike-sharing programmes that operate with other means of transportation frequently have better outcomes.

The layout and positioning of bike-sharing stations is a significant element that affects the success of bike-sharing systems. The study discovered that stations adjacent to hubs for public transit and popular tourist destinations typically had greater utilization rates. Additionally, having access to safe bike storage options like bike racks and lockers may increase customers' comfort levels and propensity to utilize the programmes. The study also emphasizes how crucial user behaviour and attitudes are to gauging the effectiveness of bike-sharing initiatives. People's decision to utilize bike-sharing programmes or not is heavily influenced by aspects including convenience, accessibility, and safety worries. For instance, the availability of helmets and safety gear, as well as the bike-sharing system's usability and accessibility, can have a big influence on user behaviour. Overall, the study emphasizes the intricate connection between the use of BSSs and the built and natural environments. It highlights the necessity for city planners to take a variety of elements into account when developing and executing BSSs, including the physical environment, transportation infrastructure, station design, and user behaviour. Cities may do this to promote sustainable and healthy transportation alternatives for their citizens while maximizing the advantages of bike-sharing programmes.

10.6   LITERATURE REVIEW

In this section we are going to discuss some of the previous work done in the field of BSSs to get a general idea as to which parameters are needed to train our model. The research in [13] examines the impact of the built environment and weather on bike-sharing demand using a station-level analysis of commercial bike-sharing in Toronto. The results of the study indicate that the built environment (land use, land value, and population density) and weather (temperature, precipitation, and wind) have a significant impact on bike-sharing demand, with temperature and precipitation having the strongest effects, and land use and population density having a moderate effect. Paper [14] is a study that examines the various factors that influence the demand for bike-sharing services, including demographic factors, weather, accessibility, and pricing. It concludes that these factors play a significant role in determining the success of bike-sharing programmes and that a better understanding of them is crucial for the effective planning and management of BSSs. Paper [15] is a study that uses machine learning techniques to predict the demand for bicycle rentals. The authors use both random forests and multiple linear regression to model the data, and find that both methods are effective in forecasting demand. The study concludes that the combination of both methods improves the accuracy of the predictions and can be useful for bike-sharing companies to optimize their resources.

The conclusion of paper [16] is that machine learning models are effective in predicting bike availability in a BSS. The authors used a dataset of bike rentals, weather, and other factors to train and test their models, and found that random forest, gradient boosting, and neural network models performed well in predicting bike availability. They suggest that this kind of model can be useful for bike-sharing companies to optimize their resources and improve the user experience by ensuring that bikes are available at the right place and at the right time. Additionally, the authors suggest that future work could include more features in the model and also more sophisticated models such as deep learning to enhance the accuracy of the predictions.

In paper [17] the authors use a generalized extreme value count model (GEV) to predict the demand for BSSs in real time. The GEV model is a statistical model that is commonly used for modelling the behaviour of extreme events in count data, such as the number of bike rentals in a specific location and time. It is a type of probability distribution that can be used to describe the distribution of rare events, such as extreme weather conditions or high bike rental demand. The authors use the GEV model to make predictions on future demand, and they also show that the model is able to accurately predict demand, with prediction errors being lower than other commonly used methods.

Paper [18] examines the impact of weather and calendar events on BSSs, specifically the trip patterns of bike rentals at different stations. The study uses data on bike rental patterns, weather conditions, and calendar events to analyse how these factors affect bike rental demand. The author finds that weather conditions such as temperature, precipitation, and wind speed have a significant impact on bike rental demand, with higher demand on days with good weather and lower demand on days with poor weather. The study also finds that calendar events such as holidays and festivals have a positive effect on bike rental demand, with increased demand on days when these events are taking place. The author suggests that this information could be used by BSS operators to optimize fleet management and make adjustments to their pricing strategies in response to weather and calendar events.

Paper [19] discusses the use of data mining techniques for predicting demand for BSSs in a large urban area. The study analyses historical data on bike rentals and weather conditions to train a model that can make predictions on future demand. The paper uses different data mining techniques such as decision tree, random forest, and multiple linear regression to analyse the data and compare the performance of each technique. The study finds that the random forest method is the best performing model with the highest accuracy. The authors also investigate the correlation between weather conditions, time of day, day of the week, and bike demand. The study shows that weather conditions, especially temperature and precipitation, have a significant effect on bike rental demand, with higher demand on days with good weather and lower demand on days with poor weather. Also, the study indicates that time of day and day of the week have a strong impact on bike rental demand, with higher demand on weekends and during the daytime. The authors suggest that this information could be used by BSS operators to optimize fleet management, predict demand, and make adjustments to their pricing strategies in response to weather conditions, time of day, and day of the week. Table 10.1 shows a comparison of existing models and our model which is developed to predict bike count.

TABLE 10.1

Comparison between Different Models

Dataset	Methodology	Performance	Reference
Data of Washington	Multiple linear regression models and random forest with Gradient boosting machine (GBM) packets	Highest accuracy is achieved in random forest with rate of 82%	[15]
Data of San Francisco Bay Area	Univariate models: random forest and LS boost and multivariate model: partial least squares regression (PLSR)	Least mean absolute error (MAE) is achieved in random forest: 0.37 bikes/station	[16]
Washington–Arlington–Alexandria	GEV count models	The presented system was perform well with a 5% margin of error and also a good arrival and departure prediction with 75% accuracy	[17]
Seoul Public Data Park website of South Korea	Apply data mining techniques first on data then train five models.	As per the result presented with the models support vector machine (SVM), EGBT, LR, BT, etc. where gradient boosting machine performed well in comparative analysis	[19]

10.7 EXPLORATORY DATA ANALYSIS

It is essential to thoroughly analyse each feature and its link with the goal attribute – in this example, the bike count – in order to acquire a full grasp of the information at hand. By using this method, we can evaluate how much each attribute affects the target variable and, as a result, spot any noteworthy patterns or trends in the data. Furthermore, while examining the link between characteristics and the goal attribute, it is critical to take the data dispersion into account. We can determine the degree of variability in the dataset and spot any potential outliers or anomalies that can influence our research by looking at the data distribution. A thorough examination of each feature and its relationship to the goal attribute, together with a study of the data distribution, will allow us to develop a comprehensive knowledge of the dataset and make judgements based on the findings of our analysis.

10.7.1 DATASET INFORMATION

The famous UC Irvine Machine Learning Repository (https://archive.ics.uci.edu/ml/datasets/Bike+Sharing+Dataset) provided the bike sharing dataset that was employed in this work. This dataset, which documents the process of renting bikes, is significantly impacted by a variety of environmental and seasonal elements, including the weather, precipitation, the day of the week, the season, and the time of day. The dataset offers a thorough and in-depth two-year historical record of the Washington, DC, USA, Capital Bikeshare system. The public may see these data, which have been compiled both daily and every two hours, on the Capital Bikeshare website (https://capitalbikeshare.com/system-data). The dataset also contains weather and seasonal data that were taken from a different source (http://www.freemeteo.com) in addition to the bike rental data. This additional information is essential to comprehending the connection between the rental process and environmental factors, enabling more precise analysis and insights. Overall, the availability of this comprehensive dataset, which covers a range of variables influencing bike rentals, offers academics and practitioners in the fields of machine learning and data analysis a useful tool.

10.7.2 ATTRIBUTE INFORMATION

The following are the attributes of a bike rental dataset:

- instant: index number for each record
- dteday: date of the rental
- season: classification of the season – spring, summer, fall, or winter
- yr: year of the rental – 2011 or 2012
- mnth: month of the rental – 1 to 12
- hr: hour of the rental – 0 to 23
- holiday: indicates if the day is a holiday or not
- weekday: day of the week
- workingday: indicates if the day is neither a weekend nor a holiday
- weathersit: classification of the weather situation – 1-clear, 2-misty, 3-snowy, or 4-rainy
- temp: normalized temperature in Celsius, divided by 41 (maximum value)
- atemp: normalized feeling temperature in Celsius, divided by 50 (maximum value)
- hum: normalized humidity, divided by 100 (maximum value)
- windspeed: normalized wind speed, divided by 67 (maximum value)
- casual: count of casual users
- registered: count of registered users
- cnt: count of total rental bikes, including both casual and registered users.

Here cnt is our target attribute and the remaining ones are all independent attributes. Data are already normalized and contain no null values. All the attributes have numeric values.

10.7.3 INSIGHTS PROVIDED BY THE DATA

From Figure 10.1, we can say that during the weekdays, from Monday to Friday, between 8 am in the morning and 5 pm in the evening, the bike count is at its peak. The obvious inference that can be made is that many people go to their work and leave the office at that time. However, during the weekends, the bike count is high in the afternoon as people engage in leisure activities such as biking in parks or visiting friends and family.

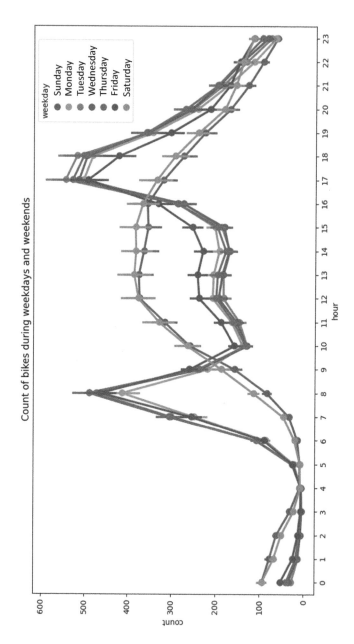

FIGURE 10.1 Graph of hour vs count for different days.

Based on the information presented in Figure 10.2, we can draw the conclusion that, during weekends, the majority of bike users are casual riders. This implies that casual riders tend to utilize bikes as a means of leisure and recreational activities during their free time at weekends. It can also be inferred that the frequency of bike usage by casual riders is higher during weekends as compared to weekdays, possibly due to a more relaxed schedule and availability of time for non-work-related activities.

Based on the information provided in Figure 10.3, it can be concluded that individuals who have registered to use the service tend to utilize it more frequently on weekdays compared to weekends. Specifically, the data suggest that there is a noticeable decrease in usage among registered users during the weekends. It is also observed that those users are riding bikes more in the afternoon period on weekends as compared to weekdays.

According to Figure 10.4, there is a noticeable decrease in bike count during snowy weather. This observation is consistent with the usual argument that people are less likely to ride bikes when weather conditions are too cold, wet, or snowy.

After carefully analysing Figure 10.5, it is evident that there is a noticeable pattern in the bike count data. Specifically, we can see that the number of bikes being used increases significantly during the months of June to September (Figure 10.6).

According to the correlation matrix, registered users and casual users have a considerable influence on the demand for bike rentals as seen by their high positive association with the number of bikes. The number of bikes also significantly changes with temperature, indicating that meteorological factors, particularly temperature, are important in determining the demand for bike rentals.

10.8 EXPERIMENTAL ANALYSIS

Our main goal is to examine and evaluate how well various regression models estimate the number of bikes using a given dataset. We've used a variety of regression models, including linear regression, ridge regression, Huber regressor, ElasticNet, decision tree regressor, random forest, gradient boosting regressor, and extra trees regressor, to achieve this. Accuracy score and mean squared error (MSE) have been computed as measures to assess each model's performance. The MSE calculates the difference between the anticipated and real bike counts, while the accuracy score evaluates how well the model can predict the bike count. By splitting the dataset into training and testing sets, we were able to determine the accuracy score and MSE for each model. The model is trained using the training set, and its effectiveness is assessed using the testing set. Based on the testing set, the accuracy score and MSE for each model are computed. The model that predicted the bike count the most accurately may be identified by comparing its accuracy score and MSE values to those of the other models. This research could support stakeholders in making well-informed resource allocation and demand planning choices, including bike-sharing firms and municipal planners.

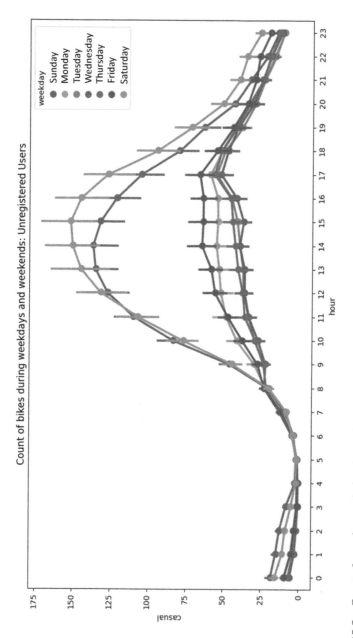

FIGURE 10.2 Count of casual users during the week.

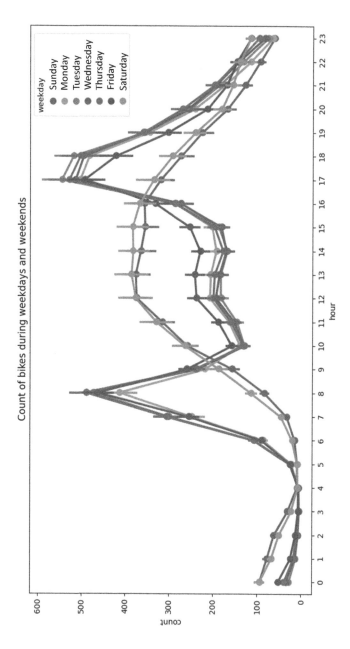

FIGURE 10.3 Count of registered users during the week.

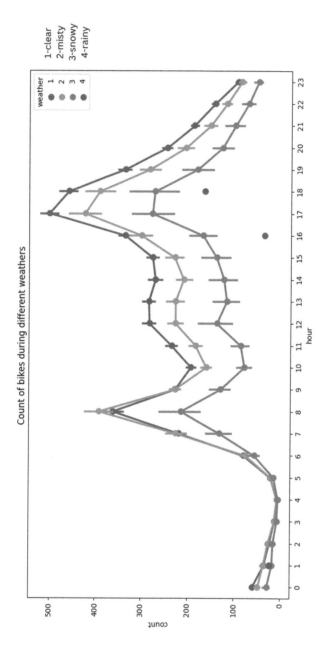

FIGURE 10.4 Count of bikes according to the weathers.

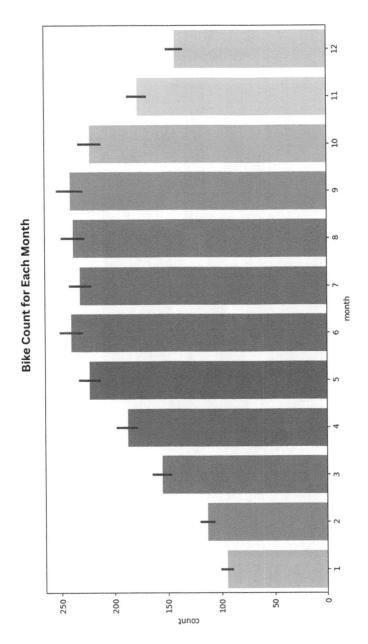

FIGURE 10.5 Bike count for each month.

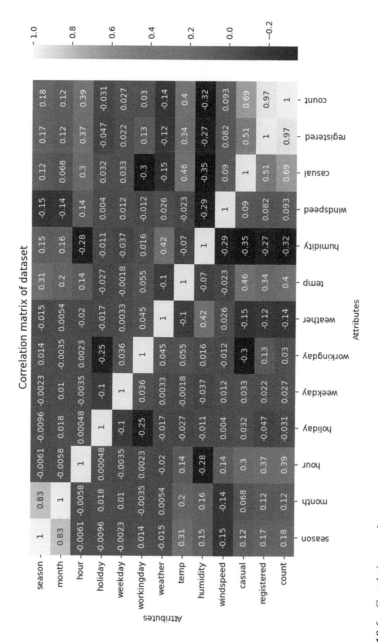

FIGURE 10.6 Correlation matrix.

10.9   RESULTS AND DISCUSSION

According to Table 10.2, the models for linear regression and ridge regression have model scores of 0.3, which indicate a poorer accuracy in predicting the number of bikes. The MSE for both models is above 23,000, which further suggests that the forecasts contain a significant inaccuracy. In comparison to the earlier models, the Huber regressor model has a significantly lower model score of 0.27 and a little higher MSE of 24,406. ElasticNet yielded a model score of 0.21, which is lower than that of the earlier models, and an MSE of 26,371, which denotes a bigger prediction error. The decision tree regressor model, on the other hand, demonstrated a substantially higher model score of 0.69, suggesting a greater accuracy in predicting the number of bikes. This model's MSE is 9,274, which is significantly less than that of the earlier versions. The random forest model provides a score that is even higher, 0.81, showing a better level of accuracy in predicting the number of bikes. This model's MSE is 5,631, which is considerably less than that of the earlier versions. Both the extra trees regressor and gradient boosting regressor models have a score of 0.78, showing great accuracy in predicting the number of bikes. The gradient boosting regressor model has an MSE of 7,252, but the extra trees regressor model has an MSE of 6,498.

Overall, the dataset analysis revealed that the decision tree regressor, random forest, gradient boosting, and extra trees regressor models outperformed the linear regression, ridge regression, Huber regressor, and ElasticNet models in terms of accuracy and precision in predicting the number of bikes. The findings imply that the bike count in this dataset may be better predicted using the decision tree regressor, random forest, gradient boosting, and extra trees regressor models. The following are some of the reasons why random forest gives the best accuracy:

- Ensemble learning: The random forest regressor uses an ensemble of decision trees, which reduces the variance and overfitting that can occur in single decision trees. The average of multiple models results in a more robust prediction.

TABLE 10.2
Model Scores and MSE

Model Name	Model Score	MSE (Mean Square Error)
Linear Regression	0.30	23493.17
Ridge	0.30	23492.64
Huber Regressor	0.27	24406.09
ElasticNet	0.21	26371.88
Decision Tree Regressor	0.69	9274.34
Random Forest Regressor	0.81	5631.53
Gradient Boosting Regressor	0.78	7252.63
Extra Trees Regressor	0.78	6498.95

- Bagging technique: The bagging technique used in the random forest regressor randomly selects a subset of the training data and builds a decision tree on it. This leads to diversity among trees and improved performance.
- Handling missing values: The random forest regressor is robust to missing values in the data, as it can handle them automatically. This can lead to a better accuracy when compared to other models.
- Handling outliers: The random forest regressor can handle outliers effectively, as it builds multiple decision trees that are averaged together, reducing the effect of outliers.
- Feature importance: The random forest regressor also provides feature importance, which can be useful for identifying which features are contributing most to the prediction. This can help in reducing noise and overfitting in the model.

10.10 SUMMARY

The random forest regressor provides a robust and accurate prediction by combining the strengths of decision trees, reducing variance and overfitting, handling missing values and outliers effectively, and providing feature importance.

REFERENCES

[1] Park, Eunil, Angel P. Del Pobil, and Sang Jib Kwon. "The role of Internet of Things (IoT) in smart cities: Technology roadmap-oriented approaches." *Sustainability* 10.5 (2018): 1388.

[2] Verma, J. P., Mankad, S. H., & Garg, S. (2019). A graph based analysis of user mobility for a smart city project. In *Next Generation Computing Technologies on Computational Intelligence: 4th International Conference, NGCT 2018*, Dehradun, India, November 21–22, 2018, Revised Selected Papers 4 (pp. 140–151). Springer Singapore.

[3] Jensen, Søren B. "Free City Bike Schemes." City of Copenhagen, Conference Proceedings. 2000.

[4] DeMaio, Paul. "Bike-sharing: History, impacts, models of provision, and future." *Journal of Public Transportation* 12.4 (2009): 41–56.

[5] Furness, Zack. *One less car: Bicycling and the politics of automobility.* Temple University Press, 2010.

[6] Larsen, Janet. *Bike-sharing programs hit the streets in over 500 cities worldwide.* Washington, DC: Earth Policy Institute, 2013.

[7] Marshall, Aarian. *Americans are falling in love with bike share.* 2018.

[8] Shaheen, Susan, and Stacey Guzman. "Worldwide bikesharing." *Access Magazine* 1.39 (2011): 22–27.

[9] Shirky, Clay. *Here comes everybody: The power of organizing without organizations.* Penguin, 2008.

[10] Shir, B., Prakash Verma, J., & Bhattacharya, P. Mobility prediction for uneven distribution of bikes in bike sharing systems. *Concurrency and Computation: Practice and Experience* 35.2 (2023): e7465.

[11] Zhang, Yongping, & Zhifu Mi. Environmental benefits of bike sharing: A big data-based analysis. *Applied Energy* 220 (2018): 296–301.

[12] Mateo-Babiano, Iderlina, et al. "How does our natural and built environment affect the use of bicycle sharing?." *Transportation Research Part A: Policy and Practice* 94 (2016): 295–307.

[13] El-Assi, Wafic, Mohamed Salah Mahmoud, and Khandker Nurul Habib. "Effects of built environment and weather on bike sharing demand: A station level analysis of commercial bike sharing in Toronto." *Transportation* 44 (2017): 589–613.

[14] Eren, Ezgi, and Volkan Emre Uz. "A review on bike-sharing: The factors affecting bike-sharing demand." *Sustainable Cities and Society* 54 (2020): 101882.

[15] Feng, You Li, and Shan Shan Wang. "A forecast for bicycle rental demand based on random forests and multiple linear regression." *2017 IEEE/ACIS 16th International Conference on Computer and Information Science (ICIS)*. IEEE, 2017.

[16] Ashqar, Huthaifa I., et al. "Modeling bike availability in a bike-sharing system using machine learning." *2017 5th IEEE International Conference on Models and Technologies for Intelligent Transportation Systems (MT-ITS)*. IEEE, 2017.

[17] Sohrabi, Soheil, et al. "Real-time prediction of public bike sharing system demand using generalized extreme value count model." *Transportation Research Part A: Policy and Practice* 133 (2020): 325–336.

[18] Kim, Kyoungok. "Investigation on the effects of weather and calendar events on bike-sharing according to the trip patterns of bike rentals of stations." *Journal of Transport Geography* 66 (2018): 309–320.

[19] Sathishkumar, V. E., Jangwoo Park, and Yongyun Cho. "Using data mining techniques for bike sharing demand prediction in metropolitan city." *Computer Communications* 153 (2020): 353–366.

11 Building Future-Proof, Sustainable, and Innovative Startups in Africa Powered by Smart Cities

Kithinji Muriungi
Purdue University, Indiana, USA

Kiyeng P. Chumo
Moi University, Kenya

11.1 INTRODUCTION

Green computing, also extensively referred to as "sustainable computing," focuses on reducing the environmental impact of computing and promoting sustainability in the technology industry. It focuses on the design, development, and use of computer systems and technologies that are environmentally sustainable through the reduction of energy consumption, waste, and carbon emissions throughout the lifecycle of computing systems, from manufacturing to disposal [1]. Environmental challenges can be addressed to some extent by using computing technologies to promote sustainable practices through monitoring energy consumption and supporting renewable energy resource utilization.

Smart cities are made up of people, by people, and for people [2]. Sustainability, liveability, efficiency, and safety are all measures that need to make sense to people living in cities. Furthermore, people who drive in a sustainable way are the city's ultimate source and beneficiaries. As smart cities become the key driving force in the knowledge-based economy, there is a need to assess the viability of the companies they power in terms of their innovativeness and, more importantly, their sustainability. Implementing smart cities in Africa, such as Konza Technopolis in Kenya, The New Capital City in Egypt, Kigali Innovation City in Rwanda, and Eko Atlantic in Nigeria, have significantly influenced the growth and development of technology and business processes supporting the Fourth Industrial Revolution (4IR). With the development of smart cities, infrastructure costs are expected to decrease while increasing innovation and entrepreneurial culture in rapidly developing cities. According to Kummitha [3], smart cities are expected to fuel public–private partnerships (PPP) through centralized digital governance enabling entrepreneurship to scale through emerging technologies and data, as seen in Figure 11.1.

DOI: 10.1201/9781003388814-11

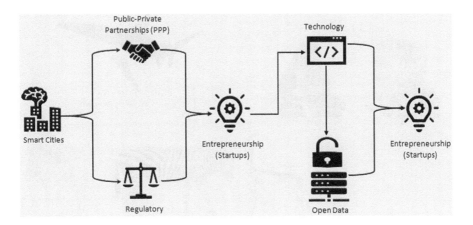

FIGURE 11.1 How smart cities are powering startups.

The expectation through this research is to demystify why smart cities in Africa have the potential to power the startups which are sustainable and future-proof with optimal innovations taking place. The justification for this looks more deeply into how smart cities are built and why their model is vital in ensuring the sustainability of these innovative startups. In the context of this chapter, future-proof, sustainable, and innovative startups are those that can evolve and surpass the test of time by adapting to customer needs through progressive development and growth. These innovative startups influence technology, business processes, and people in their design, development, and growth life cycles, making them highly adaptable to ever-changing ecosystems. This refers to using a user-centered methodology and design thinking approach to develop innovative ideas.

Startup establishments are categorized based on the startup's value and intention in the startup ecosystem. Two major startup categories considered in this context are the unicorn and zebra. A unicorn is a startup with an evaluation of over USD 1 billion and focuses mainly on growth for profit. A zebra is a startup smaller than unicorns and valued at less than USD 1 billion. Even though zebras can make profits, their primary intentions are sustainability and the promotion of social development. Examples of zebras are those startups working on solutions for quality education, poverty eradication, zero hunger, and better health among other social-economic solutions. There are other categories under what is famously known as "Entrepreneurship Zoo," where pioneer researchers used animal names to categorize startups: sharks, cash cows, white elephants, gazelles, gorillas, and others (Figure 11.2).

Three key pillars drive the development of smart cities: smartness, safety, and sustainability, emphasizing consideration of the pillars differing from country to country depending on the priority area. Some of the fundamental challenges faced by the African continent, which are expected to be addressed with intelligent city development, include sustainability and safety issues. The smart cities concept has taken root in Europe and North America [5, 6]. The leading African countries that have made significant progress in achieving smart city concepts include Kenya, Rwanda, Egypt, and South Africa. However, there needs to be more research on smart city development in developing countries.

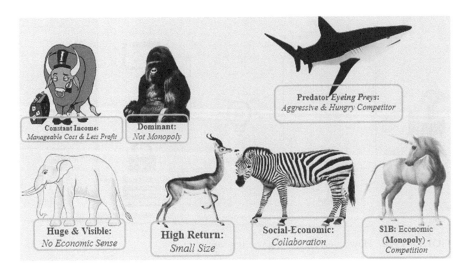

FIGURE 11.2   Entrepreneurship zoo: startup categorization [4].

According to Yanrong et al. [7], a smart city needs to resolve the following challenges to ensure its sustainability:

- *Governance*: A smart city needs to be inclusive while embracing diversity
  through open networks, infrastructure, and data for optimal innovation. It
  needs to facilitate visualization, citizen engagement, and integrated management structures.
- *Financing*: Energy saving in a smart city demonstrates how efficient the city
  is. Efficiency is among the metrics used to measure the success of a smart
  city. PPPs, tax increment financing, crowdfunding, performance contracts,
  green bonds, and private investment are all methods into which a smart city
  can ensure sustainable financing is achieved.
- *Business Model*: Pilot projects and mature innovative businesses and startups need to adopt innovative and flexible business models such as pay-as-
  you-go (PayGO) models, which are highly effective for Internet of Things
  (IoT) solutions and cloud-based solutions to enhance sustainability business
  operations and growth. More intelligent procurement and revenue creation
  from data are ways in which smart businesses can be innovative while striving to become sustainable.
- *Services*: Some of the potential smart services for smart cities include public security and safety for real-time policing, smart learning solutions, smart
  grid technology (for data center tier level considerations), smart health solutions, and smart traffic systems. An intelligent traffic system helps reroute
  traffic, avoid congestion, and maximize work time for people, businesses,
  and processes in a smart city.
- *Technology*: Cloud computing, the IoT/Internet of Everything (IoX), broadband connectivity, smart personal wearables, consumer devices, and big data

analytics are all technological challenges that, when appropriately utilized, can solve local challenges by providing relevant solutions. Technology challenges can either be operational, infrastructural, or even human resource or talent-based, which can be resolved by capacity building, among other mechanisms.

- *Government Policies*: Local personal data processing policies are among the challenges intelligent cities face due to the ever-changing privacy and data policies due to advanced security issues globally.

Yanrong et al. [7] state that it is, therefore, significantly clear that smart cities need to power startups through the provision of connectivity for both machines and humans, facilitate open-data platforms for innovation, and a "future internet lab" for start-ups and entrepreneurs to facilitate the innovation ecosystem. Connectivity facilitates the development of solutions that are not only safe for machines but for humans as well. Additionally, connectivity enhances the livability aspect of the smart city by facilitating an ecosystem suitable for humans and machines. Similarly, efficiency in machines and systems is directly linked to the operational efficiency of humans using those systems and machines. A future internet lab would provide a tinkering environment for the innovators and entrepreneurs in the smart city to provide appropriate, relevant, and locally viable solutions.

Monitoring, evaluating, and assessing startups in a smart city is critical in ensuring progress, growth, and development are experienced. The features of the intelligent city assessment framework relevant to the development of startups include smart city services, smart city strategy (infrastructure then ecosystem), value assessment, business models, legal and regulatory policies, stakeholders (government, industry, academia), governance, information and communication technology (ICT) infrastructure, and funding. Appropriate assessment of these framework elements ensures the sustainability of the present and future startups in a smart city ecosystem (Figure 11.3).

For proper alignment between the contributing startups in building solutions powered by smart cities, there is a need to consider the following prospective choices to support the smart city frameworks, policies, and strategies [9]:

- *Purpose Choice*: Business city (e.g., New York), eco-city (e.g., Dubai), citizen city, smart cocktail city.
- *Ecosystem Choice*: Technopolis (Konza), datapolis, standard city, beacon city, bot city (political and sociological transformation), or community city.
- *Society Choice*: Gentry city; inclusive city.
- *Governance Choice*: Top-down; bottom-up; co-modernization city.
- *Ecology Choice*: Survival city; organic city; liveable city (healthy city, organic city, agri city).
- *Privacy Choice*: Data deal (civic, control, business, privacy, concession, co-profiling).
- *Programming Choice*: New deal (problem-solving and one-to-one customization); PPP trends.

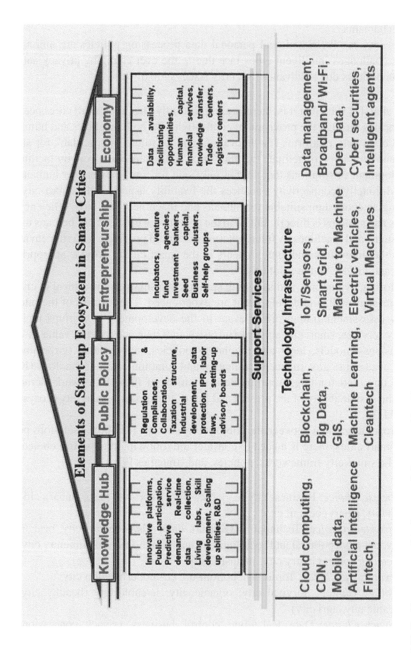

FIGURE 11.3　Entrepreneurship framework in smart cities [8].

This chapter explores the developments in innovations and established startups within the smart city environment and how the ever-emerging technologies have supported and will support the same innovative startups. The objective is to analyze the existing stream of research, including the work done by Eckinger and Sanders [10], carried out in this area to understand the context and provide insight into the present and future startups and scale-ups. The focus was on Africa; however, there is a need to review the global research to compare and contrast and provide relevant insights related to the research topic.

Technology is critical in any innovation ecosystem, especially in smart cities. There are other factors besides technology alone, even in a technopolis. Implementing appropriate trending, emerging, and future technologies in building startups powered by smart cities is significant in ensuring that the developed startups are sustainable, scalable, and replicable in and out of smart cities. Different technologies are responsible in various significant ways in a startup: planning, implementing, developing, scaling, or even sustaining. The availability of global trending, emerging, and future technologies varies based on an intelligent city's nature and development status. These technologies are essential in the startup and scaleup phases of growth and development. Technology availability in smart cities varies depending on the geographical location, partnerships, infrastructure strategy, development framework, and startup technology focus, among other factors. For example, a trending or emerging technology in one smart city based in a particular region could be considered a future technology in another smart city ecosystem. These considerations affect adoption and local contextual startups' needs impacting the smart city strategic drivers.

Trending technological tools are great in utilization and implementation due to their maturity or instant benefits realized in a smart city ecosystem. Examples of trending technological tools are not limited to the blockchain, cloud computing, IoT, AI/ML, 5G, VR/AR, big data and data science, e-mobility, nanotechnology, robotics, and drones among others [11]. Unlike other tools such as emerging and future technologies, trending technological tools are mostly available and adopted in most regions globally. Emerging technological tools either have a potential that has yet to be fully realized or are currently under development. Similar to the considered technological tools under the trending technologies, some in that category still qualify as emerging technologies, especially in Africa. Standards are often a key aspect during considerations and categorizations of these technologies. In Africa, technological standards, applicability, security, and the nature of the local challenges qualify some global trending technologies to be referenced as emerging technologies, especially in AI, 5G, cloud computing, and the IoT.

Future technological tools are those whose business potential considerations have yet to be realized, and the application of the technology is not precise. Other than the trending and emerging technologies, future technologies still need to be commercially viable and often are entirely under the research and development phase. Examples of these technologies include quantum computing, web 3.0, 6G, and 7G [12]. No commercially viable 6G or 7G solutions are present even though significant research and development work has been done, especially in 6G [13].

In a smart city, the application of emerging, trending, and future technologies in the development of innovative, futureproof, and sustainable solutions and startups varies based on the need, application context, technical specifications, business potential, target market, governance, policies, socio-economic situation, impact scaling metrics, and others. Some advanced and sophisticated technologies demand high-level technical human resources and talents to explore intelligent cities, solutions, or startups' full potential. A technology-driven startup has different tools, needs, and requirements than a social-based startup in a smart city. According to Thomas, Passaro, and Quinto [14], technology is a powerful tool in a smart city ecosystem in fostering the digital transformation journey facilitated by startups.

11.2 LITERATURE REVIEW

11.2.1 DEVELOPMENT OF THE STARTUP SECTOR IN AFRICA

The average entrepreneurial rate in Africa is among the highest and fastest growing, resulting in a massive startup ecosystem valued at over USD 7 billion [15]. The lack of employment opportunities for the youth primarily drives this. The rising investment by some governments and private sector interest in telecommunication technologies infrastructure coupled with liberalization has motivated young African entrepreneurs, leading to increased startups in the technology sector. This has, in turn, attracted both local and international investors. South Africa, Nigeria, Egypt, and Kenya are the most favorable destinations for startups as of 2022, representing over 89% of the continent's investment.

The global outlook shows that African startups have grown tremendously in the last decade up to 2022. Some of these startups are not only making a social, economic, and technological impact locally but also globally. Few of these startups founded in the last decade have already reached unicorn status, putting Africa on the global map and motivating a dozen more to revolutionize further and disrupt some sectors and markets. Figures 11.4 and 11.5 compare startups founded globally between 2013 and 2020, which shows tremendous growth in the startup ecosystem over the period.

Several African startups have qualified to reach the unicorn level due to adopting appropriate combinations of innovative business models in their technology solutions. Most of these startups have their origin in Africa and solve global challenges. Figure 11.6 shows the ranking of the startups in 2021.

Innovative and sustainable business models have been identified as among the main drivers and enablers of a few unicorns in Africa. A successful startup needs a clear growth vision, mission, and a suitable business model. The combination of both enables startups to transform from "startup" to "scale-up" [16]. This approach works with startups of any type (gazelle, zebra, unicorn), especially those beyond "the valley of death" (see Figure 11.7).

Multi-stakeholder engagement to support startups is essential in ensuring they become sustainable and scalable. Figure 11.8 shows an overview of the Kenya Innovation Ecosystem representing the entrepreneurial life cycle.

FIGURE 11.4 Startups founded by city by the year 2013. (Map by Zara Matheson, Martin Prosperity Institute. Source: https://www.seedtable.com.)

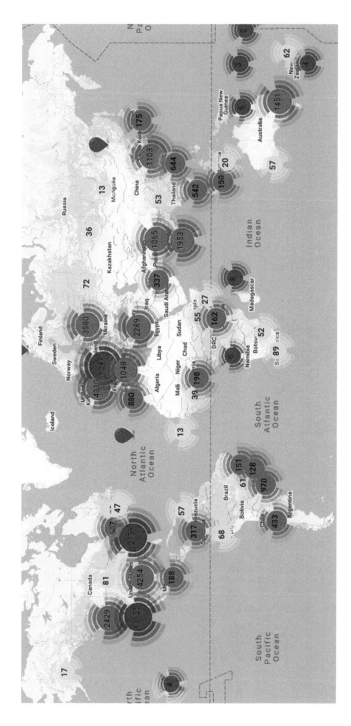

FIGURE 11.5　Startups founded by city by the year 2020.

(Source: https://www.startupblink.com.)

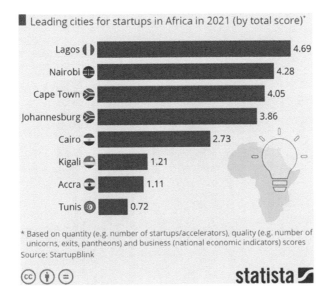

FIGURE 11.6 Leading cities for startups in Africa in 2021 (by total score). (Source: StartupBlink.)

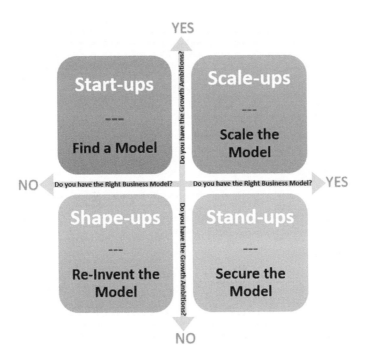

FIGURE 11.7 Startups, shapeups, standups, and scaleups.

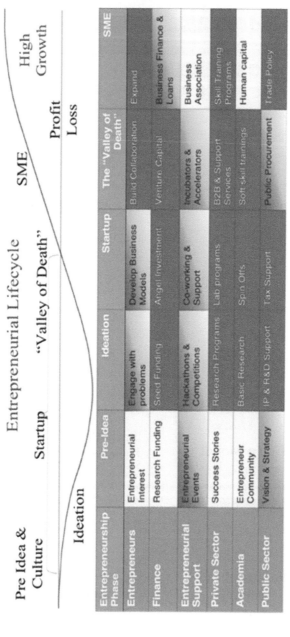

FIGURE 11.8 Entrepreneurial life cycle.

(**Source:** [17].)

The colors used represent the following:

- **Light Gray cells** are well-supported;
- **White cells** are partially but insufficiently supported;
- **Gray cells** are primarily unsupported, but efforts have been initiated;
- **Black cells** are mainly unsupported.

The ITU has developed the innovation ecosystem shown in Figure 11.9 to minimize the black color in the entrepreneurial life cycle.

11.2.2 ACHIEVING FUTUREPROOF AND SUSTAINABLE STARTUPS

Digital transformation (digital experience) is the next big thing in business evolution, changing and disrupting how people, processes, and technology interact. This shift will ultimately reshape business, people, and processes, with a more significant impact experienced progressively as we build toward the future. Digital experience presents the understanding and integration of technology ecosystems, data growth, and its impact on the knowledge-based economy, providing more significant opportunities to startups powered by intelligent cities.

The optimal digital experience will empower people for optimal productivity and efficiency, and for optimized operations for maximum efficiency and outcome through learner processes, innovative solutions, and business models (Figure 11.10).

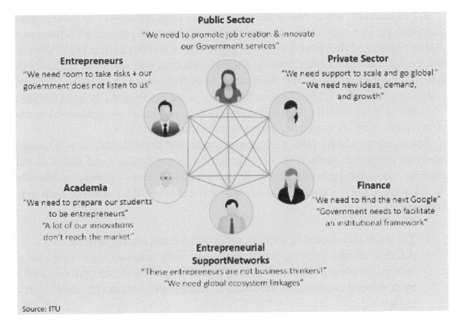

Source: ITU

FIGURE 11.9 Stakeholders in an innovation and entrepreneurship ecosystem.

(Source: [17].)

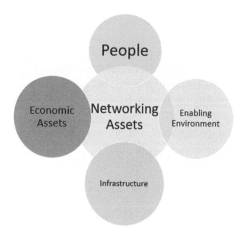

FIGURE 11.10 Elements constituting the digital experience.

(Source: The World Bank [18].)

Artificial intelligence (AI), often referred to as a general-purpose technology with a vast potential to advance inclusive knowledge-economy growth, presents more significant opportunities to startups in Africa. Despite the potential challenges that AI might face, increasing innovation and productivity in society is one of its ultimate goals. AI can be deployed in many sectors, enabling startups to build local solutions more quickly and that are adaptable to the local context.

IoT capabilities, which enable the processing of large amounts of data demanding maximum performance, open up a significant need for smart city infrastructures such as data centers and the cloud. Using big data analytics and blockchain in IoT systems will enable startups to offer trusted, reliable, efficient, and real-time solutions for everything for government and businesses, from filing taxes to customer services. Integrating the IoT, AI, and cloud technologies at the intelligent city level creates a "digital city" that smartly senses and responds to particular environmental needs, safety, public-city services, and enterprise and industrial activities.

According to ASSAf [19], a successful smart city implementation in a country needs an integrated national framework that meets the fundamental needs of that country, enhancing inclusive economic (knowledge-economy) growth sustainably. To achieve a country's inclusive and sustainable economy, the following need to be considered:

- The smart city should be locally contextualized and clearly defined.
- Address all the existing local challenges with relevant local solutions without emulating standard frameworks and models developed majorly by developed countries in the world to address their smart city needs and models.
- Local intelligent cities should have a global market link but offer localized solutions to local problems and challenges instead of competing with the developed countries in the world that are also implementing intelligent cities.

- Clearly define vital stakeholders' roles and responsibilities for the intelligent city national framework and policies.
- Be flexible, adaptable, locally aligned, and amendable to 4IR local needs.
- Integrate all fundamental building blocks of smart cities, including holistic STEAM education [20], citizen rights, city enablers (connectivity, power, and water), rural versus urban, shared objectives, goals, vision, and mission.

For startups to successfully build innovative solutions powering the knowledge economy and supported by intelligent city strategies, there is a significant need for collaboration, partnership, sustainable strategies, and adequate capacity building through advanced and sophisticated research and development methodologies. Figure 11.11 shows a model adopted by Living Labs to facilitate "citizen-driven innovations" [21].

11.2.3 The Smart Cities Landscape in Africa

Under smart cities, the design of infrastructure and spaces is aimed at ease of use, safety, efficiency, and liveability by their citizens. By 2050, up to 68% of the world's population will live in cities [22]. Therefore, it is paramount to design our cities to accommodate the postulated influx and ensure they are economically and socially sustainable. This can be achieved by adopting technologies to help manage the influx by adopting the intelligent cities concept. Smart cities depict areas with everything interconnected, from financial transactions to social payments, and are more accessible in intuitive and intelligent ways [23]. There is, therefore, innovative, automated, high-quality life and sustainability.

Most leading intelligent city initiatives are in Europe, with Asia following closely [24]. Barcelona is a smart city intending to be self-sufficient, resulting in savings of EUR 75 million and creating over 45,000 jobs [25]. The city is focused on reducing pollution, sharing transport resources such as bikes, and reducing noise. Singapore has employed technologies such as the IoT and Wi-Fi to develop its smart city.

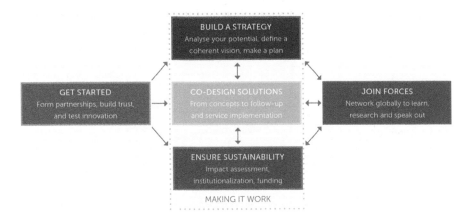

FIGURE 11.11   Citizen-driven innovations model.

(Source: Living Labs.)

Intelligent grid technologies, urban traffic flow, intelligent lighting, and smart Wi-Fi access exist. Sensors have been fitted, and officials are informed when people smoke in unauthorized and inaccessible zones or throw rubbish from high-rise buildings [26]. London has a roadmap planned to make it the most intelligent city globally. The initiative is to stimulate innovation from the technology sectors through the Civic Innovation Challenge for better communities and citizen engagement through unrolling civic platforms while addressing inequality regarding access to the internet and technologies. Part of the plan includes free Wi-Fi access in public areas and 5G connectivity to all selected areas. The success of all these intelligent cities has been achieved by engaging citizens to create more intelligent city ideas by making the data publicly available and encouraging input [22]. In this case, the foundations for achieving smart cities include open and secure city data, data capability for citizens, user-centered designs, connectivity, smarter streets, and intelligent city collaboration.

The International Monetary Fund estimates that spending 1% on technology and engineering infrastructure results in a 1.4% GDP growth within four years [27]. Technological, successful growth and development in Africa will depend on developing economies and creating integrated social-economic, environmental, political, and legal factors to enable an "enabling environment." All these factors collectively enhance a feasible digital transformation economy. Figure 11.12 shows an overview of scenarios while implementing sustainable development goals (SDGs), either within the same, faster, harder, or smarter scenarios.

The smarter scenarios maximize collective intelligence, diverse teams, optimal knowledge acquisitions with proper learning alignments, and early investment. To manage fundamental risks while working on smarter scenarios, smart cities are expected to explore innovation ecosystems, participatory design, regulatory sandboxes, and public–private–people partnerships and collaborations. Smart cities are also expected to provide mechanisms to manage economic value without overriding societal values. The ability to advance technology in intelligent cities correlates with ethical, transparent, responsible, and trusted data utilization and powerful but potentially harmful technology like AI. According to D'Orville [27], Singapore created an Advisory Council on the Ethical Use of AI and Data, while

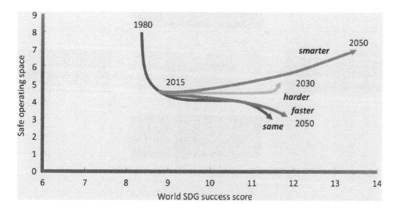

FIGURE 11.12   SDG digital transformation feasibility scenarios.

Japan's "Society 5.0" has already created a roadmap that envisions a "human-centric" society concerning data and AI.

According to Eskelinen et al. [21], Africa had the highest urban population growth ten years ago despite being a developing country. The trend is still the same for the next ten years and, even after three decades, as the United Nations had predicted (Figure 11.13).

Despite most African cities' acute problems, this has offered a natural collaboration platform to provide societal solutions. Urbanization provides diverse capabilities and digital resources that are difficult to find in rural or ordinary cities. The concentration of critical mass creates opportunities through diverse innovations, technology adoption, services, and business processes and models [20]. Cities connect external and foreign markets, influencers, and even investors. Ideally, cities' perceptions are dominant, especially in innovative and entrepreneurial activities. Smart cities enforce this perception by intensively spurring innovation cost-effectively by providing fundamental infrastructure products and services required and offering a low-risk environment in testing ideas for sustainability.

The inspirations and motivations for governments to transform existing cities into smart cities are fueled by the threats of the current demographic. These demographics include rural–urban migration, population growth, public health, safe living conditions, sustainable development, and mitigating pollution. Additionally, the unlimited potential solutions that can be developed through technological innovation, digital transformation, and citizen inclusion in modern and intelligent city management fuels the government [28]. Urban managers and planners have to face the ever-growing urban population and the citizens of the current existing cities. The World

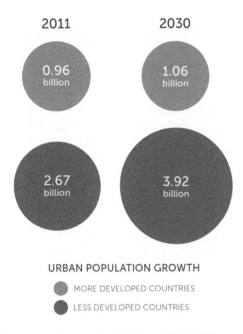

FIGURE 11.13 Urban population growth in 2011 and 2030 [22].

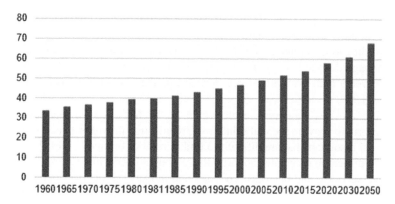

FIGURE 11.14   World's urban population growth (% of total world population) between 1960 and 2050 [15].

(Source: World Bank.)

Bank predicts that the world's urban population will experience an increase to 68% by 2050 compared to an average of 59% in 2023, as shown in Figure 11.14.

With the global population growth trend, the urban population for about three decades will almost double. The World Bank envisions that about 85% of the world's population will live in urban cities by 2100. This is unlike the less than one billion people in urban cities in 1950.

Table 11.1 displays the comparative urban population size, the total population growth percentage rate, and the situation of large urban agglomerations.

The 4IR fundamentally changes how people live, work, and relate to both machines and fellow humans. 4IR presents a new phase in human advancement, powered by technology integrating the digital, physical, and biological spheres in a manner that forges both positive promise and menace. Digital transformation has enormous potential to benefit societies, economies, and environments. Through smart city implementation, digital transformation can be realized. Different architectures and frameworks have been developed to address the best approach and methodology while implementing a smart city [29]. Nokia Future X Architecture incorporates the following layers in creating a SmartCity-as-a-Platform: multi-cloud, high-performance networking; digital value platforms; dynamic security; and business applications – present in all the initial four layers. These layers provide mission-critical, flexible, and reliable connectivity (high-performance networking); diversity of applications (multi-cloud); common enablers and analytics (digital value platforms); and entire value chain continuous innovation providing NextGen products and services (business applications) (Figure 11.15).

Emerging technologies promote the systematic, holistic integration of a smart, innovative, safe, efficient, and ecological city into an urban and country-wide model and framework. Emerging technologies are easing the stimulation of the transformation and innovation of city public services through adequate digitalization of all the essential services. Therefore, information technology and engineering become the core innovation enablers in advancing the impact of industry on different businesses. With these excellent engineering and technological advances disrupting business

TABLE 11.1
World Development Indicators: Urbanization between 1990 and 2016 [15]

	Urban Population					Population in Urban Agglomerations of more than 1 million (% of total Population)		Population in the Largest City (% of urban Population)	
	Millions		Percentage of World Population		Growth (%)				
	1990	2016	1990	2016	2016	1990	2016	1990	2016
World	2272	4037	43	54	2.00	18	24	17	16
East Asia & Pacific	619	1318	34	57	2.30	18	20	17	12
Europe & Central Asia	568	653	68	72	0.70			15	17
Latin America & Caribbean	312	504	71	80	1.30	34	38	25	23
Middle East & North Africa	140	281	55	65	240	23	26	28	26
North America	209	294	75	82	1.00	42	46	10	8
South Asia	284	587	25	33	2 50	11	15	10	11
Sub-Saharan Africa	139	400	27	39	4.0	12	15	29	27
Low income	77	212	23	32	390	10	12	35	31
Lower middle income	559	1164	30	40	2.60	12	16	17	16
Upper middle income	890	1690	43	65	2.10	17	29	14	12
High income	746	970	74	81	0.80			20	19

Source: World Bank [15].

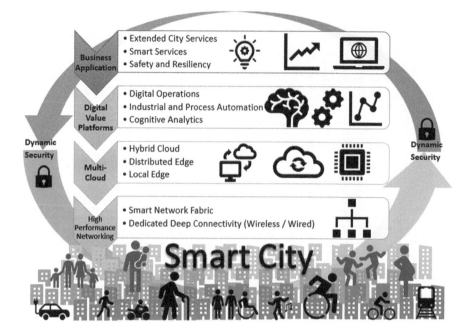

FIGURE 11.15   Future X architecture for smart city-as-a-platform [29].

processes and, ultimately, people and profits, digital policies and governance need closer monitoring by the relevant personnel.

As a fundamental building block of 21st-century infrastructure, the internet supports economic progress and businesses, drives innovation, and helps governments address the digital transformation agenda. However, responsible, secure, and reliable governance of the internet infrastructure is a difficult challenge, especially with the emerging technologies that threaten the already available legacy infrastructure and systems. The oversight role of a critical internet framework led by the non-profit Internet Corporation for Assigned Names and Numbers and other consortiums have become more complex recently due to the high number of domain name registrations (332.4 million) since 2017 [27]. Disruptive innovations are pushing the boundaries of the technology landscape, rapidly changing the socio-economic perspectives. These innovations generated by disruptive technologies initiate immediate change in work, business, and society.

A hierarchical representation of how the internet powered by fiber optics revolutionizes different industries by enabling technologies while facilitating the ultimate digital experience or digital transformation is depicted in Figure 11.16.

11.2.4 Startups in Smart Cities

Connectivity improvement and internet access in intelligent cities are critical for startups and the city's digital economy growth and development. Internet access in

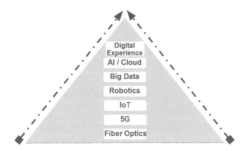

FIGURE 11.16 Hierarchy of technological digital transformation.

the intelligent city is a significant enabler and facilitator of innovative, inclusive, sustainable, and scalable startups. This ultimately requires the investment to support digital connectivity infrastructure locally and globally, such as 5G networks, fiber optic networks, and other ultra-high-speed internet-related connectivity technologies and infrastructures. The G-20 stipulation on affordable and universal internet access for all by 2025 [27], with a particular focus on poverty eradication and distance learning, will benefit African startups generally and facilitate faster adoption and growth of smart cities. Universal access to an affordable internet provides a direct and indirect benefit to local African startups to build and test faster and for their digital solutions to be relevant locally. This has the potential for startups to build solutions for the government and for the general public to fundamentally reshape public services which align and respond directly to citizen needs.

It should also be noted that the traditional business models need help to fit into the current digital business, either powered by the modern city or a typical smart city. Smart cities provide a complex and sophisticated framework that many startups driven by the 4IR need help to cope with. According to Dorantes-Gonzalez [30], the typical business models, such as the Lean Canvas and the Business Model Canvas, do not factor in open innovation, problem definition, creative problem solving, business metrics, and disruptive strategies. Besides open innovation approaches, disruption strategy, and startup metrics, innovative business models must address unique value propositions and introduce tools and resources for problem-solving. A startup being powered by intelligent cities and specifically meant to resolve specific challenges in smart cities should be able to justify, if not all, most of the items to be able to demonstrate its suitability in the intelligent city but, most significantly, its ability to survive the ever-changing business arena due to innovative and sustainable models. In startups, adopting intelligent cities might be more challenging than thinking, but learning ability and adaptability are needed.

Intelligent technologies have great potential to transform cities. However, the limitation is that they come with potentially hidden high implementation and operation costs. Defining sustainable, efficient, liveable, and safely achievable intelligent city policies, frameworks, and strategies necessitates an understandable interpretation of the strengths, weaknesses, opportunities, and threats (SWOT) facing intelligent cities.

A given feature of a smart city can depict multiple roles in the SWOT analysis, depending on the perspectives. The data are an excellent example of such an intelligent city feature, representing multiple roles. Data can be seen as a weakness, an opportunity, and a threat depending on the scenario and perspective. Data provide a real opportunity for startups powered by smart cities to exploit in the future by understanding the context and operating within the legal framework provided by the smart city governance body in a particular country.

Similarly, data can also establish a weakness for startups in smart cities that need to be more capable of using it. Data can also be a threat when considering privacy and security interests that originated from the rich data generated through smart cities and shared with startups. Figure 11.17, presented by the Organisation for Economic Co-operation and Development [31], shows intelligent cities' significant strengths, weaknesses, opportunities, and threats.

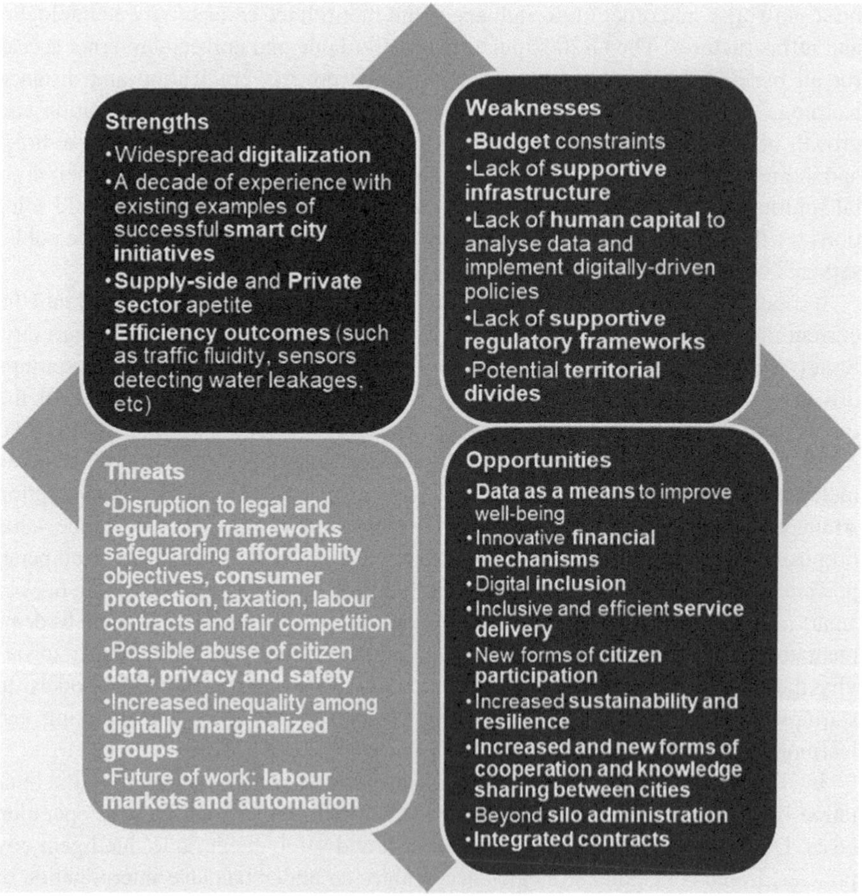

Strengths
- Widespread **digitalization**
- A decade of experience with existing examples of successful **smart city initiatives**
- **Supply-side** and **Private sector** apetite
- **Efficiency outcomes** (such as traffic fluidity, sensors detecting water leakages, etc)

Weaknesses
- **Budget** constraints
- Lack of **supportive infrastructure**
- Lack of **human capital** to analyse data and implement digitally-driven policies
- Lack of **supportive regulatory frameworks**
- Potential **territorial divides**

Threats
- Disruption to legal and **regulatory frameworks** safeguarding **affordability** objectives, **consumer protection**, taxation, labour contracts and fair competition
- Possible abuse of citizen **data, privacy and safety**
- Increased inequality among digitally marginalized groups
- Future of work: **labour markets and automation**

Opportunities
- **Data as a means** to improve well-being
- Innovative **financial mechanisms**
- Digital **inclusion**
- Inclusive and efficient **service delivery**
- New forms of **citizen participation**
- Increased **sustainability and resilience**
- Increased and new forms of cooperation and knowledge sharing between cities
- Beyond **silo administration**
- **Integrated contracts**

FIGURE 11.17   SWOT analysis of smart cities [31].

11.2.5 Green Computing for Startups in Smart Cities

Startups can help the green smart city ecosystem and promote the green environment in various ways. In so doing, they enable the generation of convenient, adaptable, and affordable solutions for city inhabitants. Areas where startups have sprung up have provided solutions which include:

- *Sustainable transportation* where startups have provided solutions that help reduce traffic congestion and air pollution by promoting projects in sustainable transportation in smart cities through the provision of alternative modes such as bike-sharing and carpooling services, electric vehicles, and smart parking systems [32].
- *Energy-efficient buildings* are achieved through startups which focus on building solutions that use energy-efficient lighting, heating, ventilation, and air conditioning (HVAC) systems. These solutions make buildings more sustainable because of minimized energy consumption and greenhouse gas emissions [33].
- *Renewable energy* startup initiatives can help promote the use of renewable energy in smart cities by providing solutions such as solar panels, wind turbines, and energy storage systems. These solutions can help to reduce reliance on fossil fuels and promote a more sustainable energy system [34].
- *Waste management* startups promote a circular economy by reducing landfill waste in smart cities by providing solutions for innovative technologies such as waste-to-energy, recycling, and composting [35].
- *Green data centers* are startups aimed at reducing energy consumption and carbon emissions by promoting green data centers in smart cities by providing energy-efficient servers, cooling systems, and renewable energy sources [36].

Africa has not been left behind in the green revolution, and startups focusing on green computing in smart cities have emerged with success. The following are some notable startups that have emerged:

- *Solaris Offgrid* is a startup that provides solar-powered off-grid solutions for rural communities in Africa. Their solutions include solar-powered mini-grids, home solar systems, and energy-efficient appliances that are designed to reduce carbon emissions and promote sustainability [37].
- *Ujuzi Code* is yet another startup that provides software development and IT services for businesses in Africa. They focus on using green computing technologies such as cloud computing, virtualization, and energy-efficient servers to reduce energy consumption and promote sustainability [38].
- *Afrilabs* is a pan-African network of technology innovation hubs that promotes innovation and entrepreneurship in Africa. They focus on supporting startups working on solutions for smart cities, including green computing technologies that can help reduce carbon emissions.
- *Ecobuild* is a startup that provides energy-efficient building solutions for businesses and homes in Africa. Their solutions include energy-efficient

lighting, insulation, and HVAC systems that can help to reduce energy consumption and promote sustainability.

- *Rensource* is a startup that provides clean energy solutions for businesses in Africa. Their solutions include solar-powered mini-grids, energy-efficient appliances, and battery storage systems that can help to reduce carbon emissions and promote sustainability.

These startups are making a significant impact on promoting sustainability and reducing carbon emissions in Africa's smart cities. By leveraging innovative solutions and technologies, they are helping to create more sustainable and liveable cities.

11.3 DISCUSSION

11.3.1 CORE PILLARS OF THE SMART CITY ECOSYSTEM THAT POWER STARTUPS

The 21st-century smart city ecosystems are primarily built by considering various factors; however, they still maintain two core pillars: community and infrastructure, which demonstrate that the developed smart city ecosystem emphasizes the philosophy that such ecosystems are socio-technical and built by the people, for the people, and of the people [2]. Community fosters a sense of belonging to people in the city. In contrast, physical and virtual infrastructure fosters the spirit of the abundance of resources, which is critical, especially in capital-dependent solutions and startups. The smart ecosystem can be scaled in diverse aspects, sectors, and concepts with adequate support from the community and the availability of the required infrastructure. An appropriate infrastructure enhances the scalability of startups and solutions without negatively affecting the smart city ecosystem's outcome or impact. A combination of appropriate communities and infrastructure makes the smart city ecosystem adaptive, sustainable, and innovative to the changing needs of customers and solutions. The adaptability of the ecosystem ensures its sustainability is available through the development of a future-proof approach being in place. Chen and Dahlman [39] state that another core factor, particularly when considering the infrastructure pillar, is the efficiency and safety mechanism, which highly impacts people, business processes, and technologies.

The community pillar in a smart city ecosystem fosters a sense of belonging or a humanistic sense in the ecosystem [40]. In a smart city, the community binds people, businesses, and stakeholders together, enhancing the processes and technologies utilized in the ecosystem. Communities are built based on the needs of the startups and other entities in the ecosystem. Some communities are built considering the social-economic, technological, environmental, political, or even legal needs. According to Melé [40], a community (people-oriented) pillar, in the humanistic sense, can be achieved holistically through wholeness, growth and development, substantial knowledge, dignity, sustainable stewardship, common social good, and transcendence. A sense of belonging facilitates the humanization of startups, which enhances humanism in innovation ecosystems, eventually driving startups to add economic sense to the smart city ecosystem.

The intelligent city infrastructural pillar in the ecosystem refers to resources such as physical, technological, and intangible (knowledge and skill set). In this case a technopolis (e.g. Konza Technopolis) is a smart city representing an entity where startups are hosted and act as an infrastructural reference in an intelligent ecosystem [41]. A smart city's infrastructural pillar represents a technology, administration, and innovation hub while serving the physical infrastructural demands of the startups. Despite the social outcomes offered by the community pillar [42], the infrastructure pillar in an intelligent city ecosystem offers technological and economic values.

Developing a startup, an incorporated business, or an innovation hub differs from developing and orchestrating a smart city ecosystem. A startup, a business, or an innovation hub has a particularly focused organizational vision, mission, goal, and value, among other strategic elements, which inform the general direction individuals and organizations are expected to take. A startup can be referred to as a fundamental element in an intelligent city ecosystem since it can fit into either an incorporated business or an innovation hub [43]. Most startup solutions have a narrow and limited scope that often informs their value chain delivery to end-users in any relationship framework. These narrow scopes are sometimes defined by their value to the end-user through a solution in the form of a product or a service. These focused products and services get scaled as the startup becomes an established company that grows and adopts different focus areas to keep the startup sustainable and scalable [44]. Most startups exit either through merging or acquisition by established companies, significantly where the startup solves a specific need for the established company's value chain. Besides a startup being adopted through a merger or acquisition by an established company, most startups fit well in innovation hubs, where they benefit through accelerators and incubators [45]. Innovation hubs differ from innovation ecosystems in various ways. Narrow focus areas just define some innovation hubs. Innovation hubs, in most cases, have a limited scope from which they operate. Generally, the operational principles of an innovation hub vary from that of an innovation ecosystem. Hypothetically speaking, an innovation hub can fit into an innovation ecosystem. However, an innovation ecosystem cannot fit into an innovation hub. Taking a broader definition of hub and ecosystem, a hub could mean a central connection point, whereas an ecosystem comprises more than what the system has to offer. Taking inspiration from the natural ecosystem, an innovation ecosystem is more sophisticated and complex than a hub [46].

Startups can still grow and develop outside an ecosystem; however, a startup that belongs in an ecosystem has a more competitive advantage [47]. According to Feng et al. [43], exceptional startups can also grow from mere startups to innovation ecosystems. Startups are the fundamental elements that give life to the ideas, problems, challenges, and solutions driven by individual entrepreneurs and innovators in an intelligent city innovation ecosystem. Innovators play a vital role in an ecosystem since they are the masterminds and the architects of challenges that offer opportunities of greater interest to the ecosystem's stakeholders. Through the support of other stakeholders, these innovators create products and services that ultimately add social-economic, environmental, and even political values. Startups are easy to form and work with due to their agility, flexibility, and adaptability, which in most cases are lacking in established companies. The simplicity of startup management is a competitive advantage compared to the high chain of protocol commands [45]. This management simplicity facilitates multidisciplinary problem-solving capabilities, which have overlapping roles, especially at the beginning of a startup, making the decision-making shorter and faster. This faster response is a crucial asset in managing crisis, faster product development, and quick failure recovery, which confirms the viability of the tested and validated idea or solution. A robust innovation ecosystem nurtures its startups to be prepared to fail quickly for iterations for testing, validation, and viability purposes. This mindset is also critical in responding to time-sensitive challenges and opportunities that do not require much planning. The most vital startups, in this case, understand how power sprints work in their favor.

The sustainability of scale-ups in an innovation ecosystem varies [48]. Some of the recommendations which are likely to sustain the scale-ups, according to Kwak and Lee [49], include the emphasis on conducting a solid groundwork on governance, policy, and services to create a flourishing ecosystem. This ecosystem serves as a service platform for incubating and delivering long-term solutions to society through successful, adaptive, and innovative ideas and technologies for innovative solutions and startups. Thomas, Passaro, and Quinto [14] argue that dynamic strategic approaches are significant in ensuring that a startup can scale and still be sustained at the scale-up phase. In an innovation ecosystem, pioneer startups which have scaled into becoming established companies should choose to remain and actively participate in the ecosystem in growing others. This results in a never-ending cycle where the starters grow other startups. Other than relying on external networks to develop fundamental technologies, the network of innovators supported by all ecosystem partners should assist each other through open innovations in developing new technologies and solutions that primarily serve the entire ecosystem.

Regarding financing, notable technical-based investors should choose to remain and keep supporting other kinds of technical startups by offering guidance to fellow trusted investors. Additionally, other than startups and scale-ups existing through mergers and acquisitions, the innovation ecosystem should make it easy for startups and scaleups to pursue initial public offerings [14]. Cultural diversity in terms of talents, skills, and experiences through the inclusion of different nationalities, practices, and ideas will also facilitate sustainability due to high innovation and out-of-the-box problem-solving skills and building state-of-the-art startups, which are easier to scale and sustain. Additionally, innovation ecosystems need to adopt remarkable "subtle legal peculiarity" by engaging policymakers in structuring digitally fit policies, especially in non-compete agreements and non-disclosure agreements (NCAs/NDAs) for positive competition purposes, which leads to collaboration.

Startups can significantly impact on improving urban lifestyles by leveraging innovative solutions and technologies [24]. The startups are found in all sectors and domains and can contribute to improving urban lifestyles. By leveraging technology and innovation, startups can help to create more sustainable and liveable cities for all. Some of the leading areas where startups have successfully been deployed include the following.

- *Mobility startups* are key and can improve urban lifestyle by providing alternative and sustainable modes of transportation, such as bike sharing, carpooling, and ride-hailing services. These services can help to reduce traffic congestion and air pollution, making cities more liveable [50].

- *Health technology (Healthtech)* startups focus on enhancing urban lifestyles by providing access to healthcare services through telemedicine, wearable devices, and mobile apps. These services can help reduce healthcare costs and improve urban residents' overall health and well-being [51].
- *Property technology (PropTech) startups* provide smart building solutions that promote sustainability and enhance the quality of life for residents of smart cities. These solutions can include energy-efficient lighting, smart thermostats, and smart home security systems [52].
- *Financial technology (Fintech) startups* can improve urban lifestyles by providing financial services that are more accessible and affordable to urban residents. These services can include mobile banking, microloans, and digital payments, which can help to reduce financial exclusion [53].
- *Education technology (Edtech) startups* focus on improving urban lifestyles by providing access to education and training through online platforms, mobile apps, and digital content. These services can help to bridge the digital divide and provide opportunities for lifelong learning and skill development [54].

11.3.2 STARTUPS AS THE DRIVER OF THE KNOWLEDGE ECONOMY AND INNOVATION FRAMEWORK: PEOPLE, PROCESSES, AND TECHNOLOGY

In the 21st century, which is popularly referred to as the information age, consideration of the knowledge economy by government is essential [55]. Developing an intelligent ecosystem requires a holistic and systemic partnership that brings together public and private partnerships to address ecosystem matters and handle economic interests. Unlike a physical resource-dependent economy, a knowledge-based economy depends on an intangible valuable resource whose value can be mistaken. Legal protection in a knowledge economy considers different measures predominantly strengthened by different governing bodies. Knowledge economy assessment, evaluation, and monitoring metrics vary compared to the other forms of the economy. According to Andreu and Rosanas [56], with more significant opportunities in it, the knowledge economy possesses greater management challenges, which are more demanding in humanistic matters than before. Managing knowledge with economic value is expected to be challenging, especially with the current technological advancement where privacy and security are significant points of concern.

Open innovations enhance sustainable and scalable innovation ecosystems [57]. Despite the greater good that this model is likely to create, security plays a significant role in ensuring safety in physical space and cyberspace. Open innovations thrive due to the optimal contribution and high value of the institutional knowledge quickly disseminated from individuals within organizations forming part of the ecosystem [58]. Community facilitates a great sense of belonging for individuals who foster unique experiences in the innovation ecosystem. Developing a knowledge-based economy is a great way to foster unique innovation ecosystems in countries [55]. Therefore, communities, as a pillar in the intelligent city innovation ecosystem, play a significant role in its development and sustainability.

An intelligent city ecosystem's growth and development take time [59]. They require strategically sustained PPPs that may require decades to succeed under constant and regular quality monitoring. These collaborations and partnerships require the participation of all stakeholders and players in government, academia, and industry. The growth and development of a flourishing smart city ecosystem require transformative government ecosystem players to establish and grow or discover and mobilize powerful, innovative, startup champions from the developed network of potential partners in a region where the smart city ecosystem is being established. To provide strategic direction for innovation ecosystems, governments must also pinpoint sectors with developmental priorities to provide macroeconomic, social, and political benefits. A balance should also be found in their economic development practices and human capital establishment strategies, offering incentives for innovative startups in urgently required sectors [60]. For example, funds and investments can be channeled to activities and startups at various innovation phases of the value chain, depending on the high-potential startups.

Similarly, startups, businesses, or companies can be established to operate and provide solutions in the selected strategic sectors. In a smart city ecosystem, it is significant to acknowledge that startups will thrive only if they organically further the economic involvement of both an innovative startup champion and the concerned government. These champions are responsible for catalyzing the development of smart city innovation ecosystems by assisting in the creation of innovation hubs and innovation ecosystems with technical and infrastructural capabilities and human resource talent pools. These champions are also essential in stimulating and supporting research and development activities. They also assist in bridging the gap between research and commercialization success. This gap is a critical challenge hindering the long-term viability of startups powered by smart city innovation ecosystems, negatively affecting the national economic sectors that the solutions are expected to support. With these startups and innovation champions, there is a significant rise in the possibilities of building a successful startup in a smart city innovation ecosystem; without them, the possibilities of failure are very high and almost inevitable.

11.3.3 Strategy Implementation and Quality Monitoring of Startups in a Smart City

Implementing a successful smart city innovation ecosystem strategy and maintaining sustainability [61] requires establishing effective and efficient quality assessment, evaluation, and monitoring policies and systems in the respective innovation ecosystem [62], since the smart city innovation ecosystem represents the integration and combination of scientific, engineering, and technological ventures and their business application. Hence, smart ecosystems serve as research and development commercialization catalysts by transforming technological advances into marketable products and services. Accordingly, the smart ecosystem value chain that thrives in giving rise to serial innovators includes research, development, commercialization, and the production and dissemination of new innovative solutions. Among the areas of focus for implementing a successful innovation ecosystem strategy is the ability to build capabilities, support and develop research and development activities, and enable commercialization. This creates an end-to-end value chain that benefits all the partners and stakeholders in the ecosystem [63].

Building capacity is the responsibility of all key stakeholders in the ecosystem. Each partner is responsible for a particular strategic area that requires proper capacity [64]. These strategic areas of focus include but are not limited to advanced technical research, governance and legal policies, finance and business, and socio-economic areas that affect both private and public domains. Strategic capacity, especially in advanced technology research and development, refers to the capability of creating and safeguarding intellectual property (IP) and having accessibility to the capital resources and human expertise required to develop it commercially [65].

Transformative governance strategy implementation in an innovation ecosystem is essential for establishing stable macroeconomic, political, and legal environments [64]. This aligns people, processes, and technologies implemented and ensures a balanced, stable, and sustainable ecosystem throughout the evolution. Governance and policy ecosystem strategies improve the availability, standard, and productivity of generalized inputs, infrastructure, and organizations legally engaged and involved in the ecosystem for everyone's safety. This sets up the general rules and incentives governing competition, such as investment incentives, antitrust laws, IP protection rules, and others, which also facilitates cluster growth and development through active participation and collaboration in the ecosystem. Generally, the transformative government's strategic role is to improve the innovation environment and ecosystem rather than to intervene directly in any competitive process.

End-to-end management of innovations, solutions, and startups is critical to the success of a smart city ecosystem [66]. Effective pipelines and a funnel establishment are specific mechanisms to grow and develop the most impactful and deserving startups, which lead to successful, established businesses and companies. Effective pipelines should elaborate innovation sourcing, problem curating, prioritization mechanisms, solution exploration or testing the innovation hypothesis, incubation, and accelerators. An optimal integration mechanism for benefiting ecosystems through the programs enables startups to scaleups and ultimately to an incorporated company.

Like any intricate ecosystem, the prosperity of established startups within a smart city framework relies on a multitude of diverse factors [67]. Some of these factors are fueled by the availability of problems and challenges with equal strengths and opportunities. The balance or imbalance of problems and opportunities leads to either the demand and supply being equally stable or unstable [68]. For example, to have a balanced innovation ecosystem, many available problems need to have equally many viable opportunities to create solutions to avoid the stakeholder not being overwhelmed or underwhelmed. Within the smart city innovation ecosystem of demand and supply, there are entities that offer, entities that receive, those engaged in both offering and receiving, and those actively seeking. The value chain in the intelligent city innovation ecosystem engages resources, financial capital, physical assets, knowledge (tangible and intangible), branding and connectivity, and innovations [60]. The ever-demanding smart city innovation ecosystem with limited supply sets the ecosystem for survival of the fittest, rewarding the winners and always considering positive competitive advantages. In an ecosystem, everything is interconnected and interdependent. Therefore, the success or failure of a significant entity or component in the value chain might have a catastrophic effect on the entire ecosystem. According to Chapin III, Kofinas, and Folke [69], it is paramount for all players in the ecosystem to practice stewardship for appropriate resilience-based innovation resource management in the ever-changing ecosystem.

11.3.4 DIVERSITY, EQUITY, INCLUSION, AND ACCESSIBILITY OF STARTUPS IN A SMART CITY ECOSYSTEM

Diversity can best be explained through the nature of a natural ecosystem. Ecosystems are hard to replicate anywhere else due to the many unwritten system rules and cultures. Diversity is always a strength in a smart city ecosystem, especially when specific models and frameworks are in place to enhance appropriate inclusion, diversity, equity, and accessibility (IDEA). These models factor in metrics like compliance by developing policies and defining regulatory requirements. Creating awareness while building a diverse pipeline and tracking and assessing its effectiveness improves compliance. Unique pipeline models' effectiveness is realized when they are accepted, and there is a culture of belonging. This removes models' execution barriers. Removal of execution barriers makes the practices consistently applicable and easily monitored. This creates a diverse and inclusive ecosystem brand through commitment to the greater good and collaboration. Such practices ultimately create sustainable outcomes and impact ecosystem actions.

In a smart city ecosystem, IDEA efforts can broadly aim at demographics (such as gender, age, race, religion, and people with disabilities), spatial communities and regions, and industrial innovation research and development promotion in traditional industries [70]. According to Klingler-Vidra [70], smart city innovation ecosystems can create synergies through inclusive innovation activities, especially by promoting women in engineering and technology-related engagements. Policies are another area where synergies can be created through broad demographic engagements in policy development. The gaps that diversity, equity, and inclusion (DEI) needs to address in the innovation include the initiatives of the LGBTQ+ and people with disabilities, consistent social media campaign engagements to create awareness, robust data, measurement, and coherent collaboration across diverse entities who are ordinarily not invited into such ecosystems and where a limited range of policy types is used.

According to Fasnacht [71], as he explains in his book *Open Innovation Ecosystems*, open innovation fosters collaboration and attracts strong partnerships, which amass knowledge and speed up the rate of innovation and growth in an ecosystem. With the open nature of the ecosystem, there is a high sense of belonging for many partners and individuals, which facilitates the agenda of DEI. Similarly, Rego and Gergen [72] state that in fostering an inclusive innovation ecosystem, breaking and limiting the "zero-sum game" through the development of leadership bridges is critical in creating a culture that nurtures community growth through appropriate socio-economic mechanisms. A zero-sum game is "anti-DEI" since it fosters situations where a person's gain or benefit is equivalent to another person's loss, resulting in a net change in wealth or economic change that is zero. In the same regard, Mullin et al. [73], concerning organizational responsibility to leadership competency, expand the DEI concept by adding an accessibility aspect into it, which changes from DEI to inclusion, diversity, equity, and accessibility (IDEA), which rethinks performance management, governance, and mentorship through the IDEA lens. This concept emphasizes that there is no DEI if it is not accessible by all the stakeholders in the ecosystem.

11.4 CONCLUSION

To achieve sustainable solutions developed by startups in smart cities, it is vital to allow the participation of citizens. Equally important is supporting startups focused on providing solutions that aid in better growth, development, and management of the intelligent city. Smart cities need to enhance the perspective that "all the money is there" through multi-stakeholder financing, especially investors in technology companies, to facilitate adequate capital to develop ideas into startups and scale them globally. Of significance is to ensure that most of the incubated and accelerated unicorn-formed startups in Africa within smart cities choose to stay in Africa and grow other startups. Strong partnerships, collaborations, and supporting networks will ultimately facilitate skills and talent development in smart cities, empowering startups to create new solutions using existing infrastructure and resources. Diversity and inclusion facilitated by the intelligent cities bringing together industry and academia will facilitate the development of highly qualified scientists, engineers, technologists, and designers, through locally and organically developed capacity building due to the accommodation of diverse nationalities, cultures, practices, and ideas. Additionally, smart cities will need to adopt "subtle legal peculiarity" similar to what has been seen in Silicon Valley by engaging policymakers in structuring digitally fit policies, especially in NCAs/NDAs. Lastly, local African startups should be willing to start small and either be zebras or unicorns but to ensure that their solutions are sustainable, safe, efficient, and liveable in a smart city.

REFERENCES

[1] Raza, K., Patle, V. K., & Arya, S. (2013). A review on green computing for eco-friendly and sustainable IT. *Journal of Computational Intelligence and Electronic Systems*, *1*(1), 3–16. https://doi.org/10.1166/jcies.2012.1023

[2] Golubchikov, O. (2022). *People-Smart Sustainable Cities*. Ssrn.com. Retrieved from https://ssrn.com/abstract=3757563

[3] Kummitha, R. (2019). Smart Cities and entrepreneurship: An agenda for future research. *Technological Forecasting and Social Change*. https://doi.org/10.1016/j.techfore.2019.119763

[4] Melenciuc, S. (2018). *Unicorns, zebras, white elephants, or gorillas? Which one describes your business? BR invites you to learn the business zoo*. Business Review. https://business-review.eu/business/unicorns-zebras-white-elephants-or-gorillas-br-invites-you-to-embark-on-a-tour-of-the-business-zoo-191336

[5] Paskaleva, L. (2009). Enabling the smart city: the progress of city e-governance in Europe. *International Journal of Innovation and Regional Development*, *1*(4), 405.

[6] Mora, L., Deakin, M., & Reid, A., (2020). Smart-city development paths: Insights from the first two decades of research. In the *International conference on smart and sustainable planning for cities and regions*, Berlin, pp. 403–427.

[7] Yanrong, K., Lei, Z., Cai, C., Yuming, G., Hao, L., & Ying, C. et al. (2014). *EU-China Smart and Green City Cooperation "Comparative Study of Smart Cities in Europe and China"- White Paper.* Huet, Jean-Michel: https://www.bearingpoint.com/files/smart-cities-the-key-to-africas-third-revolution.pdf

[8] Mitra, S., Kumar, H., Gupta, M. P., & Bhattacharya, J. (2023). Entrepreneurship in smart cities: Elements of Start-up Ecosystem. *Journal of Science and Technology Policy Management, 14*(3), 592–611.

[9] Cathelat, B., (2019). *Smart Cities: Shaping the Society of 2030.* UNESCO Publishing.

[10] Eckinger, C., & Sanders, M. (2019). User Innovation and Business Incubation. USE Working Paper Series, 19(16).

[11] Kramer, W. J., Jenkins, B., & Katz, R. S. (2007). *The role of the information and communications technology sector in expanding economic opportunity.* Cambridge, MA: Kennedy School of Government, Harvard University, pp. 22, 1–45.

[12] Rawal, B. S., Ahmadand, S., Mentges, A., & Fadli, S. (2022). Opportunities and challenges in metaverse the rise of the digital universe. In *International Conference on Metaverse* (pp. 3–17). Springer, Cham.

[13] Khutey, R., Rana, G., Dewangan, V., Tiwari, A., & Dewamngan, A. (2015). Future of wireless technology 6G & 7G. *International Journal of Electrical and Electronics Research, 3*(2), 583–585.

[14] Thomas, A., Passaro, R., & Quinto, I. (2019). Developing entrepreneurship in digital economy: The ecosystem strategy for startups growth. *Strategy and Behaviors in the Digital Economy*, 1–20.

[15] World Economic Forum. *These four countries are leading Africa's start-up scene.* World Economic Forum. (2022, August 24). Retrieved from https://www.weforum.org/agenda/2022/08/africa-start-up-nigeria-egypt-kenya-south-africa/

[16] Ritter, T., & Pedersen, C. (2022). *An Entrepreneur's Guide to Surviving the "Death Valley Curve."* Harvard Business Review. Retrieved 12 May 2022, from https://hbr.org/2022/04/an-entrepreneurs-guide-to-surviving-the-death-valley-curve

[17] ITU (2017). Bridging the Digital Innovation divide: *A toolkit for Strengthening ICT Centric Ecosystem.*

[18] Mulas, V., Minges, M., & Applebaum, H. (2015). *Boosting Tech Innovation Ecosystems in Cities: A Framework for Growth and Sustainability of Urban Tech Innovation Ecosystems.* World Bank, Washington, DC. http://hdl.handle.net/10986/23029. License: CC BY-NC-ND 3.0 IGO.

[19] Academy of Science of South Africa (ASSAf) (2020). The Smart City Initiatives in South Africa and Paving a Way to Support Cities to Address Frontier Issues Using New and Emerging Technologies. http://dx.doi.org/10.17159/assaf.2019/0059

[20] Wainaina, I., Muriungi, K., Chumo, K., & Mulongo, D. (2022). *STEM Vs. STEAM Approach Dilemma in the Innovation Ecosystem for Smart Cities Sustainability.* IEEE Smart Cities.

[21] Eskelinen, J., Robles, A. G., Lindy, I., Marsh, J., & Muente-Kunigami, A. (Eds.). (2015). *Citizen-driven innovation.* World Bank Publications.

[22] United Nations (UN), Department of Economic and Social Affairs, Population Division (2019). *World Urbanization Prospects: The 2018 Revision (ST/ESA/SER.A/420).* New York: United Nations.

[23] United Nations Economic Commission for Europe (UNECE) (2020d). *AgingAging in Sustainable and Smart Cities.* UNECE Policy Brief on Aging No. 24. Available from http://www.unece.org/fileadmin/DAM/pau/age/Policy_briefs/ECE_WG-1_35.pdf

[24] Caragliu, A., Chiara, D., and Nijkamp, P., (2011). "Smart Cities in Europe." *Journal of Urban Technology, 18*(2), 37–41.

[25] Mora, L., & Bolici, R. (2016). The development process of Smart City strategies: The case of Barcelona. In J. Rajaniemi (Ed.), *Re-city: Future City - Combining Disciplines* (pp. 155–181). Tampere: Juvenes Print.

[26] Achieng, M., Ogundain, O, Makola, D, and Iyamu, T. (2021). The African Perspective of Smart City: Conceptualization of Context and Relevance.

[27] D'Orville, H. (2020). Urbanization, innovation, and governance: The quest for sustainable development. *Cadmus*, *4*(2), 308–319.

[28] Fagadar, C., Trip, D., Darie, G., & Badulescu, D. (2021). Smart Cities and the European Vision. The Annals of the University of Oradea. *Economic Sciences*, *30*, 49–60. http://dx.doi.org/10.47535/1991AUOES30(1)004

[29] Weldon, M. K. (2016). *The Future X Network: a Bell Labs perspective*. CRC Press.

[30] Dorantes-Gonzalez, D. (2017). A Business Model Frame For Innovative Startups.

[31] Organization for Economic Co-operation and Development (OECD) (2020). Smart Cities and Inclusive Growth. Available online: https://www.oecd.org/cfe/cities.OECD_Policy_Paper_Smart_Cities_and_Inclusive_Growth.pdf

[32] Skala, A. (2022). Sustainable transport and mobility—Oriented innovative startups and business models. *Sustainability*, *14*, 5519. https://doi.org/10.3390/su14095519

[33] Kim, D., Yoon, Y., Lee, J., Mago, P.J., Lee, K., & Cho, H. (2022). Design and implementation of smart buildings: A review of current research trend. *Energies*, *15*, 4278. https://doi.org/10.3390/ en15124278

[34] Vukovic, N.A., & Nekhorosheva, D.E. (2022). Renewable energy in Smart Cities: Challenges and opportunities by the case study of Russia. *Smart Cities*, *5*, 1208–1228. https://doi.org/10.3390/smartcities5040061

[35] Tozlu, A., Özahi, E., & Abuşoğlu, A. (2016). Waste to energy technologies for municipal solid waste management in Gaziantep. *Renewable and Sustainable Energy Reviews*, *54*, 809–815.

[36] Uddin, M., Shah, A., & Memon, J. (2014). Energy efficiency and environmental considerations for green data centres. *International Journal of Green Economics*, *8*(2), 144–157.

[37] Wim, J.K., & Emmanuel, G.M. (2010). Increasing sustainability of rural community electricity schemes—case study of small hydropower in Tanzania. *International Journal of Low-Carbon Technologies*, *5*(3), 144–147. https://doi.org/10.1093/ijlct/ctq019

[38] Bharany, S., Sharma, S., Khalaf, O.I., Abdulsahib, G.M., Al Humaimeedy, A.S., Aldhyani, T.H.H., Maashi, M., & Alkahtani, H. (2022). A systematic survey on energy-efficient techniques in sustainable cloud computing. *Sustainability*, *14*, 6256. https://doi.org/10.3390/su14106256

[39] Chen, D. H., & Dahlman, C. J. (2005). The knowledge economy, the KAM methodology and World Bank operations. World Bank Institute Working Paper, (37256).

[40] Melé, D. (2016). Understanding humanistic management. *Humanistic Management Journal*, *1*(1), 33–55.

[41] Linde, L., Sjödin, D., Parida, V., & Wincent, J. (2021). Dynamic capabilities for ecosystem orchestration: A capability-based framework for smart city innovation initiatives. *Technological Forecasting and Social Change*, *166*, 120614.

[42] Bandera, C., & Thomas, E. (2018). The role of innovation ecosystems and social capital in startup survival. *IEEE Transactions on Engineering Management*, *66*(4), 542–551.

[43] Feng, N., Fu, C., Wei, F., Peng, Z., Zhang, Q., & Zhang, K. H. (2019). The key role of dynamic capabilities in the evolutionary process for a startup to develop into an innovation ecosystem leader: An indepth case study. *Journal of Engineering and Technology Management*, *54*, 81–96.

[44] Bereczki, I. (2019). An open innovation ecosystem from a startup's perspective. *International Journal of Innovation Management, 23*(8), 1940001. https://doi.org/10.1142/S1363919619400012

[45] Marcon, A., & Ribeiro, J. L. D. (2021). How do startups manage external resources in innovation ecosystems? A resource perspective of startups' lifecycle. *Technological Forecasting and Social Change, 171*, 120965.

[46] Browning, T. R. (2014). Managing complex project process models with a process architecture framework. *International Journal of Project Management, 32*(2), 229–241.

[47] Isenberg, D., & Onyemah, V. (2016). Fostering scale up ecosystems for regional economic growth. In *Global Entrepreneurship Congress* (pp. 71–97), Tagore LLC.

[48] Rowe, A., Dong, L., Landon, J., & Rezkalla, E. (2019). Scaling Start-ups: Challenges in Canada's Innovation Ecosystem. In *ISPIM Conference Proceedings* (pp. 1–17). The International Society for Professional Innovation Management (ISPIM).

[49] Kwak, Y. H., & Lee, J. (2021). Toward sustainable smart city: Lessons From 20 years of Korean programs. *IEEE Transactions on Engineering Management, 70*(2), 740–754. doi: 10.1109/TEM.2021.3060956

[50] Guyader, H., Friman, M., & Olsson, L.E. Shared Mobility: Evolving Practices for Sustainability. *Sustainability* 2021, *13*, 12148. https://doi.org/10.3390/su132112148

[51] Chitungo, I., Mhango, M., Mbunge, E., Dzobo, M., Musuka, G., & Dzinamarira, T. (2021). Utility of telemedicine in sub-Saharan Africa during the COVID-19 pandemic. A rapid review. *Human Behavior and Emerging Technologies, 3*(5), 843–853. https://doi.org/10.1002/hbe2.297

[52] Gunge, V. S., & Yalagi, P. S. (2016). Article: Smart Home Automation: A Literature Review. *IJCA Proceedings on National Seminar on Recent Trends in Data Mining RTDM, 2016*(1), 6–10.

[53] Chuka I., Kenechukwu O., Eze F., Samuel M.T., Anthony E., Godwin I., & Josaphat U. J. (2022). Financial inclusion and its impact on economic growth: Empirical evidence from sub-Saharan Africa, *Cogent Economics & Finance, 10*, 1. https://doi.org/10.1080/23322039.2022.2060551

[54] Kinshuk, D.H., & Sampson, D., & Chen, N.-S. (2013). Trends in educational technology through the lens of the highly cited articles published in the Journal of Educational Technology and Society. *Educational Technology & Society, 16*, 3–20.

[55] Byat, A. B., & Sultan, O. (2014). The United Arab Emirates: fostering a unique innovation ecosystem for a knowledge-based economy. *The Global Innovation Index, 101–111.*

[56] Andreu, R., & Rosanas, J. M. (2010). Manifesto for a better management. A rational and humanistic view. *Research paper.*

[57] Costa, J., & Matias, J. C. (2020). Open innovation 4.0 as an enhancer of sustainable innovation ecosystems. *Sustainability, 12*(19), 8112.

[58] Pulford, L. (2011). The global ecosystem for social innovation. *Social Space,* 112–113.

[59] Rabelo, R. J., & Bernus, P. (2015). A holistic model of building innovation ecosystems. *If-Papers Online, 48*(3), 2250–2257. https://doi.org/10.1016/j.ifacol.2015.06.423

[60] Gereffi, G. (2019). Global value chains and international development policy: Bringing firms, networks and policy-engaged scholarship back in. *Journal of International Business Policy, 2*(3), 195–210. https://doi.org/10.1057/s42214-019-00028-7

[61] Mora, L., Deakin, M., Aina, Y. A., & Appio, F. P. (2019). Smart City Development: ICT Innovation for Urban Sustainability. In *Encyclopedia of the UN Sustainable Development Goals: Sustainable Cities and Communities.* Cham: Springer.

[62] Adner, R. (2006). Match your innovation strategy to your innovation ecosystem. *Harvard Business Review, 84*(4), 98. https://hbr.org/2006/04/match-your-innovation-strategy-to-your-innovation-ecosystem

[63] O'Leary, D. E. (2019). Technology life cycle and data quality: Action and triangulation. *Decision Support Systems*, p. 126, 113139.

[64] Tiwana, A. (2013). *Platform ecosystems: Aligning architecture, governance, and strategy.* Newnes.

[65] Velibeyoglu, K., & Yigitcanlar, T. (2010). An evaluation methodology for the tangible and intangible assets of city-regions: the 6K1C framework. *International Journal of Services Technology and Management*, *14*(4), 343–359.

[66] Lang, N., von Szczepanski, K., & Wurzer, C. (2019). The emerging art of ecosystem management. Boston Consulting Group. Hendersen Institute, 1.

[67] Audretsch, D. B., & Belitski, M. (2017). Entrepreneurial ecosystems in cities: Establishing the framework conditions. *Journal of Technology Transfer*, *42*(5), 1030–1051.

[68] Syrbe, R. U., & Grunewald, K. (2017). Ecosystem service supply and demand–the challenge to balance spatial mismatches. *International Journal of Biodiversity Science, Ecosystem Services & Management*, *13*(2), 148–161.

[69] Chapin III, F. S., Kofinas, G. P., & Folke, C. (Eds.). (2009). *Principles of ecosystem stewardship: resilience-based natural resource management in a changing world.* Springer Science & Business Media.

[70] Klingler-Vidra, R. (2019). Global review of diversity and inclusion in business innovation.

[71] Fasnacht, D. (2018). Open innovation ecosystems. In *Open Innovation Ecosystems* (pp. 131–172). Springer, Cham.

[72] Rego, L., & Gergen, C. (2017). Fostering Inclusive Innovation Ecosystems. In *Breaking the Zero-Sum Game*. Emerald Publishing Limited.

[73] Mullin, A. E., Coe, I. R., Gooden, E. A., Tunde-Byass, M., & Wiley, R. E. (2021, November). Inclusion, diversity, equity, and accessibility: From organizational responsibility to leadership competency. *Healthcare Management Forum*, *34*(6), 311–315.

12 A Recommended Pesticide Approach to a Smart City for Sustainable Agriculture and Food Production

Himanshi, Ritwik Duggal, and Abhishek Kumar
National Institute of Technology Himachal Pradesh,
Hamirpur, India

Nagendra Pratap Singh
National Institute of Technology, Jalandhar, India

Rohit Thakur
National Institute of Technology Himachal Pradesh,
Hamirpur, India

12.1 INTRODUCTION

Pest attacks on crops can be defined as a living organism which is thriving where it is not wanted and causes substantial harm to plants that are grown for food [1]. This is often the reason for a failed season for farmers due to ruined crops, and studies show that these pest attacks are incremental over the years [2]. To deal with this, in recent years there have been methods used to change the biology of the plants to make them more resistive [3, 4] or by changing the environment of the plant fields [5]; however, these methods require intensive changes and hence are not possible for every farmer to come up with. The alternatives rely on plant disease recognition using image processing [6] or by using machine/deep learning methods [7], discussed in more detail below. Despite being effective these are mostly limited to a small number of crops and therefore farmers rely on a broad spectrum of pesticides.

Every year a huge number of crops are ruined due to either infestation by pests or by degradation of soil due to the use of excessive pesticides [8]. Pesticide application also requires a lot of financing, which many small farmers cannot afford, hence many resort to the use of cheap pesticides which can be very harmful to the soil [9] and

DOI: 10.1201/9781003388814-12

degrade it even further. Other farmers unable to afford the pesticides often fall into a debt trap. These issues are particularly prevalent in agriculture. The idea of a green city is highly fashionable right now. However, the food chain is becoming contaminated because of farmers' unintentional and excessive usage of pesticides [10]. The overuse of pesticides causes the soil to lose its natural fertility, and frequently most of these pesticides are not even necessary because most farmers are unaware of the pests that infest their crops. In addition, pesticides have an impact on the nature of the soil because different soil types respond differently to the nature of the pesticide. The use of recommended pesticides, instead, could be very helpful in dealing with this problem, as it would help prevent soil degradation and be economic.

There have been many approaches made in this field, such as the whiteflies which were detected using the relative difference in pixel intensities algorithm in [11], which distinguishes between the whitefly and aphid. As proposed in [12], detecting whiteflies is done by measuring the size of the whiteflies and counting them using a background subtraction of images containing whiteflies [13]. The main issue with these traditional methods is that they are inaccurate when dealing with multi-class pests.

A setup aimed at detecting whiteflies with a gray level co-occurrence matrix (GLCM) feature extraction yields very low accuracy with multi-class classification; additionally, the use of GLCM is decreasing nowadays due to its low efficiency [14]. Parallel processing is more cost-effective, and deep learning yields more accurate and better results than GLCM. To address the issue of multi-class classification, in the proposed study, a machine learning model is trained, with pesticide recommendations based on pest recognition rather than traditional single-class classification methods and fertilizer recommendations based on soil and climate conditions.

Our method ensures high accuracy even when detecting multiple classes of pests infesting tomato leaves along with real-time deployment, using a dataset of images [15] to train the model using k-nearest neighbors (KNN). This method can be developed further and potentially help to detect multiple pests in multiple crops.

12.2 COMPARATIVE ANALYSIS OF VARIOUS "SUSTAINABLE AGRICULTURAL AND FOOD PRODUCTION PROJECTS"

There are various projects referred to as "sustainable agricultural and food production projects". This section presents a comparative analysis of various methods along with the proposed method and their pros and cons (Table 12.1).

12.3 IMPACT OF SMART AGRICULTURAL TECHNIQUES FOR GREEN CITY DEVELOPMENT

The terms "green computing" and "smart cities" are sometimes used interchangeably to refer to projects that aim to lower energy consumption, increase the use of renewable energy sources, and enhance resource efficiency. However, sustainable

TABLE 12.1

Comparative Analyses of Various "Sustainable Agriculture and Food Production Projects"

Reference	Crop/Pest	Method	Accuracy	Pros	Cons
[7]	Rice crops	ResNet101V2	86.799	High accuracy, intensive image processing	Only for rice crops, takes longer training time and excessive resources
[6]	Cocoa fruit	Image processing	70	Resource efficient, takes less time to train, fast working	Limited to cocoa fruit, low accuracy, low elasticity
[13]	Whiteflies	Image processing and Support Vector Machine	97	High accuracy, resource efficient	Limited to whiteflies, low elasticity, doesn't work on the crop just the whiteflies
[52]	Cotton	Deep learning	96	High accuracy	Takes longer training time depending on the resources, limited to cotton crops, doesn't take multiple diseases into account
[53]	Tomato (a) Gray mold; (b) Canker; (c) Leaf mold; (d) Plague; (e) Leaf miner; (f) Whitefly; (g) Low temperature; (h) Nutritional excess or deficiency; (i) Powdery mildew	Faster R-CNN with VGG-16	83	Reliable accuracy, more enhanced pixelwise analysis, takes multiple diseases into account	Heavy resources are needed, takes long training time
Proposed solution	Tomato (a) Tetranychus urticae; (b) Bemisia argentifolii; (c) Zeugodacus cucurbitae; (d) Trips palmi; (e) Myzus persicae; (f) Spodoptera litura; (g) Spodoptera exigua; (h) Helicoverpa armigera	Algorithms: (a) SVM; (b) Ensemble; (c) KNN; (d) Convolutional Neural Network	61.4 84.1 85.9 87.5	Requires less computational power, identifies multiple classes, i.e. 8	Model needs to be more accurate, dataset needs to be increased

urban development also involves the encouragement of the creation of green areas, healthy living situations, and the availability of fresh food that is healthy. Methods of sustainable agriculture have the potential to play a key part in the promotion of these features of environmentally responsible urban development.

One of the ways in which smart farming practices may help in the development of environmentally friendly cities is via the use of the technologies of precision farming. Farmers can make the most efficient use of resources such as water, fertilization, and pesticides by using sensor technology and performing data analysis in real time. This not only enhances crop yields and decreases waste, but it also lessens the detrimental effects that farming methods have on the surrounding ecosystem. For instance, precision irrigation systems may drastically cut down on the amount of water used, while precision spraying can limit the amount of potentially dangerous pesticides that are used.

Fostering urban agriculture and community gardening is another way that sustainable agriculture may contribute to the development of environmentally conscious cities. These practices not only allow us to produce food locally and to increase the availability of fresh products, but they also contribute to the development of green spaces in urban areas. The urban heat island effect may be mitigated, air quality can be improved, and biodiversity can be promoted via the practice of urban agriculture and gardening.

In addition to these advantages, the implementation of innovative agricultural practices has the potential to aid in the creation of a food system that is both more sustainable and more robust. Farmers can swiftly react to changing climatic circumstances and change their practices by making use of cutting-edge technology such as machine learning and real-time data analysis. This has the potential to help alleviate the negative effects that climate change will have on food production, decrease emissions of greenhouse gases, and enhance the health of the soil.

We can create urban landscapes that are more liveable and sustainable by including smart agriculture in the greater framework of developing smart cities. This will allow us to place a higher priority on the health and well-being of both people and the earth. Innovative agricultural practices have the potential to contribute to the construction of a more sustainable and liveable urban future, as well as to the creation of a food system that is more egalitarian and resilient; they can also assist in promoting green areas and healthy living conditions.

12.4 BACKGROUND AND RELATED WORKS

The research issues at play here are:

- *Excessive use of pesticides and the resulting damage to the environment* [16]: Excessive use of pesticides in agriculture may have a severe effect on the environment, including the contamination of soil and water supplies, the killing of beneficial insects, and the possible creation of pests that are resistant to pesticides. Our research contributes to a more accurate and focused approach to the application of pesticides, which in turn helps reduce the total quantity of pesticides that are needed and lessens the impact on the environment.

- *Ineffective and time-consuming techniques of pest management* [17]: Many farmers are forced to deal with ineffective and time-consuming methods of pest control that are also expensive. Our research aims to enhance pest control in tomato plants by establishing a system that is both more efficient and automated in terms of pest identification and management. This will be accomplished via the use of machine learning algorithms and real-time data to identify pests and offer suitable treatment approaches.
- *Difficulty gaining access to agricultural experience* [18]: Because of the difficulty in gaining access to agricultural expertise and resources in urban settings, it may be difficult for people to successfully manage their gardens and crops. Our research aims to solve this problem by building user-friendly software that can give assistance and suggestions for sustainable agricultural practices and pest control. This will make it simpler for people to produce their own food and contribute to a food system that is more sustainable.
- *The rising demand among consumers for locally produced and organic products* [19]: There is a growing demand among consumers for locally grown and organic produce; nevertheless, this desire may be challenging to satisfy in metropolitan settings. Our research focuses on this topic by establishing a system for urban agriculture that is more environmentally friendly and productive. This method makes use of innovative technology and techniques for precision farming in order to cultivate crops in confined spaces and satisfy customer demand for fresh, locally farmed products.
- *The influence of climate change on food production* [20]: The influence of climate change on food production is considerable, with shifting weather patterns and severe weather events influencing crop yields and quality. Our research could provide a solution to this problem by coming up with a smart city strategy for environmentally friendly farming that takes into consideration the effects of climate change and makes use of technology to adapt to shifting environmental circumstances.
- *Food security and availability of fresh produce* [21]: A lack of access to fresh produce in many metropolitan settings is one factor that might lead to food insecurity and poor health consequences. Our research could provide a solution to this problem by adopting a "smart city" strategy for urban agriculture. This strategy would concentrate on expanding access to fresh food in disadvantaged regions by making use of technologies such as hydroponics and vertical farming in order to cultivate crops in a reduced amount of space.

These are only a few instances of the research challenges that are connected to the subject that we are researching. Our study has the potential to make a substantial contribution to the area of environmentally responsible agriculture and food production if we focus on resolving problems and coming up with specific answers.

12.4.1 ANALYSIS OF PESTICIDES

A pesticide is any substance which we use to kill or control certain forms of plant and animal life (known as pests). Pesticides include herbicides, insecticides, fungicides, disinfectants, larvicides, and compounds used to control mice and rats.

12.5 IMPACT OF PESTICIDES ON THE ENVIRONMENT

Pesticides have a severe environmental impact since they can pollute soil, water, turf, humans, and other living things [22]. Pesticides adversely influence the proliferation of beneficial soil micro-organisms and their associated biotransformation in the soil [23]. Long-term and excessive pesticide usage can cause adverse negative impacts on soil ecology, resulting in the erosion of beneficial and plant probiotic soil microflora [24]. The recommended pesticides in the proposed work are:

1. *NeemAzal (Azadirachtin)*: This is a botanical insecticide generated from the seed kernel of the neem tree, produced by a watery extraction process. The chemical group of NeemAzal is Tetranortriterpenoid. The active ingredient of NeemAzal is 40 g/liter of azadirachtins [25]. Non-target species and helpful insects such as Encarsia wasps are largely unaffected by NeemAzal. Its formulations are broken down by microorganisms and water very quickly so it is not a water hazard material and has low toxicity and is non-mutagenic. It is free of aflatoxin and solvent residue. It is an anti-feeding pesticide with a delayed action. It helps in feeding inhibition and reduction in fecundity and breeding ability [25].
2. *Avermectin*: This is a naturally occurring or semisynthetic macrocyclic lactone endectocide [26]. Avermectin degrades rapidly in light and soil and does not translocate in the environment, reducing the environmental impact on non-target organisms [27]. Avermectin exerts its therapeutic effects by binding to glutamate-gated chloride channels, causing flaccid paralysis and parasite death (Figure 12.1).
3. *Flubendiamide*: This comes as a white suspension liquid or as granules. It has a slight odor. It is extremely harmful to certain aquatic invertebrates but only moderately toxic to birds, bees, earthworms, and ladybirds. It is

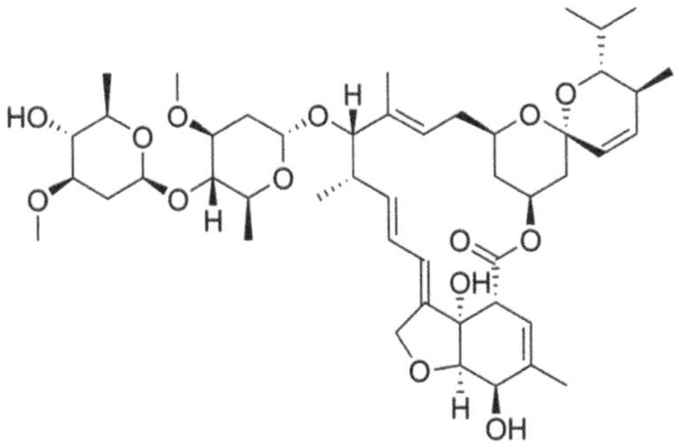

FIGURE 12.1 Chemical structure of Avermectin [26].

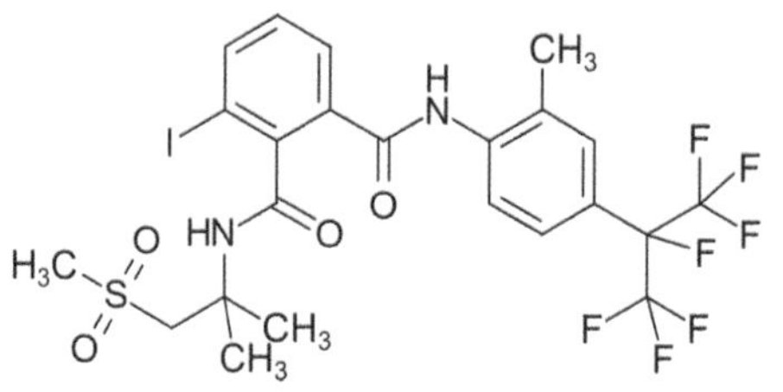

FIGURE 12.2   Chemical structure of Flubendiamide [28].

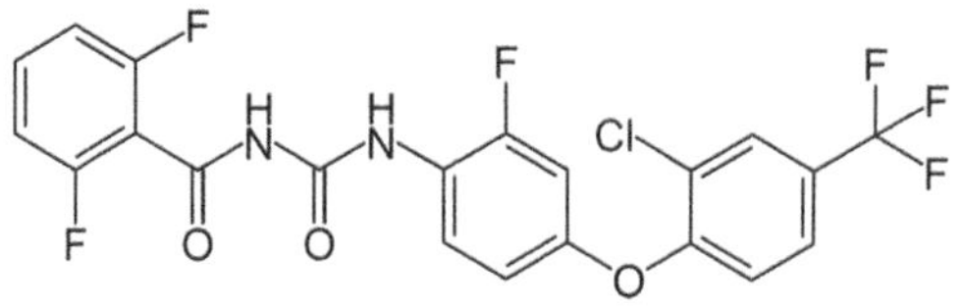

FIGURE 12.3   Chemical structure of Flufenoxuron [31].

quickly degraded in soil by indirect photolysis. It acts mainly by being eaten by larvae. It is slow acting and kills larvae in 4–5 days [28] (Figure 12.2).

4. *Flufenoxuron*: This is a white crystalline powder used to suppress phytophagous mites in their immature stages [29]. If we release flufenoxuron in water, it will be absorbed into suspended solids and sediment. It inhibits chitin formation and thus kills pests by regulating acyl urea insect growth. It is effective in both a residual spray and baits formulation treatments [30]. The chemical structure of flufenoxuron is given in Figure 12.3.

5. *Bt (Bacillus thuringiensis)*: This is a biological pesticide found in the guts of caterpillars. Insecticidal action occurs by crystal proteins known as delta endotoxins produced by Bt strains. Mammals are not poisoned by Bt toxins. The majority of the Bt toxin is rapidly destroyed by microbes, but some is absorbed by organic tissue. According to several studies, poisons do not stay in soil. When crystal proteins are ingested by insects, their alkaline digestive tracts make them soluble and become incorporated into the insect's stomach cell membrane which paralyzes them. The insect stops eating and dies as a result of starvation [32].

6. *Malathion*: This is also known as organophosphate and is a synthetic phosphorous compound [33]. Exposure to malathion can cause respiratory distress, gastrointestinal distress, neurological problems, skin reactions, and eye distress. Malathion is usually degraded rapidly by soil bacteria and, when in water, it is degraded by hydrolysis [34]. It affects insects' nervous systems. When a healthy nerve sends a signal, a chemical messenger travels from one nerve to the next, and it comes to a halt when an enzyme is released into the space between the nerves. Malathion prevents nerve signals from stopping, causing nerves to signal to each other without stopping, causing the insect to be unable to move or breathe properly and eventually die [35]. The chemical structure is given in Figure 12.4.

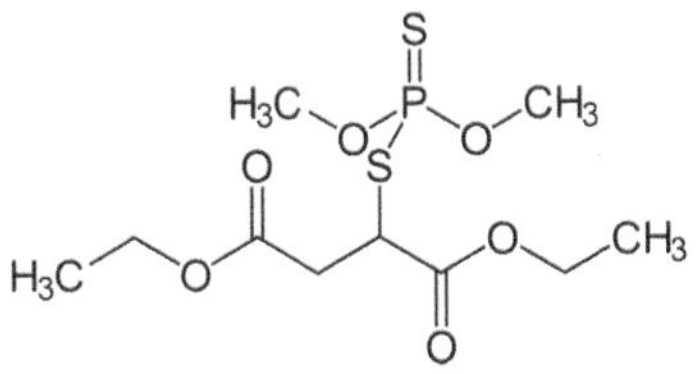

FIGURE 12.4 Chemical structure of Malathion [33].

7. *Pyrethroids*: These are man-made pesticides and like the natural pesticide pyrethrum. They include tetramethrin, resmethrin, bio allethrin, etofenprox, silafluofen, and many other insecticides [36, 37]. Pyrethroids are less toxic to mammals and birds but highly toxic to fish if applied directly to water. These insecticides bind tightly to soil and organic matter so they do not penetrate soil. Pyrethroids' solubility in water is very poor [36]. In insects, pyrethroids block channels in axonal layers from closing. The toxin keeps the channels in their open state due to which nerves cannot repolarize, making the axonal membrane completely depolarized. This results in muscular paralysis and the death of insects [37, 38].

12.6 PROPOSED METHODOLOGY

The statistical analysis is computationally fast and requires less memory to implement. An algorithmic approach, as shown in Figure 12.5, is proposed in this chapter. The image dataset was classified using a common statistical method. All the test images were pre-processed by converting the RGB image to grayscale image. Following the scaling of the grayscale image, the texture features used a GLCM. After this, a comparative study of four models was considered: a K-nearest neighbor algorithm, a support vector machine algorithm, an ensemble bagged tree algorithm, and a CNN to classify images and annotate them with the predicted pest and pesticide for crop protection.

12.6.1 DATASETS

The dataset was taken from [39] (https://data.mendeley.com/datasets/s62zm6djd2/1) and was originally extracted from the National Bureau of Agricultural Insect Resources which had 609 original images, though after image enhancement it was extended to 4263 images. The images are in JPEG format. The dataset is not uniformly distributed so we performed preprocessing. It consists of eight common tomato pest images: (1) Tetranychus urticae TU, (2) Bemisia argentifolii BA, (3) Zeugodacus cucurbitae ZC, (4) Thrips palmi TP, (5) Myzus persicae MP, (6) Spodoptera litura SL, (7) Spodoptera exigua SE, and (8) Helicoverpa armigera HA.

12.6.2 RGB TO GRAYSCALE IMAGES

In general, an image is commonly represented by three components: red, green, and blue (RGB). An RGB image uses 24 bits/pixel, that is, 8 bits for each component [40].

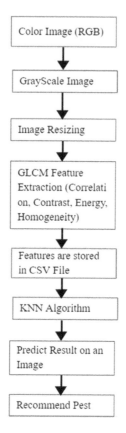

FIGURE 12.5 Work flow of proposed GLCM based recommended system.

A color image needs more computational power and memory. It is common for an image to lose shadows, contrast, sharpness, and the structure of the original after general conversion [41]. Thus using the method of the addition of chrominance and luminance and RGB approximation, reduction an algorithm is developed to convert the color image to grayscale:

$$Grayscale = 0.299R + 0.587G + 0.114B$$

12.6.3 IMAGE RESIZING

Image resizing is also known as image interpolation, and it is the most basic requirement for image preprocessing. It is used for creating all the images in a dataset at a similar size without reducing fine details. There are many common algorithms to perform this, such as nearest neighbor interpolation, bilinear, and bi-cubic [42]. In the bi-cubic interpolation method we generally weight an average of 4x4 neighboring pixels. Pixels which are adjacent provide a higher weight and merge output quality and processing time. For this reason, it is commonly used in various software like Photoshop and in default image resizing techniques in OpenCV and MATLAB [43].

12.6.4 FEATURE EXTRACTION

In feature extraction, our primary goal is to reduce data points to have a subset of the most impacting data points in the images [44]. Texture has been measured using a variety of techniques, including a co-occurrence matrix, fractals, Gabor filters, histograms of gradients, and a local binary pattern [45].

Regarding the planned work, a GLCM is used for texture extraction. Texture classification researchers frequently use a GLCM matrix. This is a statistical approach of the second order that figures out the spatial arrangement of the gray level in an image [46]. The co-occurrence matrix is used to compute numerical features that can represent the texture more compactly, such as angular momentum, contrast, correlation, entropy and homogeneity.

- *Contrast*: Contrast determines the local variations that exist in an image. If the intensity of a picture varies widely, the pixel will be localized away from the principal diagonal of the co-occurrence matrix and contrast will be high.
- *Correlation*: Correlation computes the linear structure in an image. Its high value indicates a number of linear structures in the image.

$$\text{Correlation} = \sum_{a,b=0}^{N-1} P_{a,b} \frac{(i - \mu_a)(j - \mu_b)}{\sqrt{(\sigma_a^2)(\sigma_b^2)}}$$

- *Entropy*: Entropy determines the randomness of the intensity distribution in an image. It is greater when there is a large number of pixels with similar intensities, and vice versa.

$$\text{Entropy} = \sum_{a,b=0}^{N-1} P_{a,b} \left(-\ln P_{a,b} \right)$$

- *Homogeneity*: A homogeneous image has a co-occurrence matrix with a merger of low and high-intensity pixels. On the other hand, a heterogeneous image means there is an even distribution of pixel intensities.

$$\text{Homogeneity} = \sum_{a,b=0}^{N-1} \frac{P_{a,b}}{1 + (a - b)^2}$$

12.7 THE CNN MODEL

TensorFlow and Keras are used to build a deep learning model with a dataset of 4211 images in Python on a Jupyter notebook. The model is partitioned into two datasets: training and testing. An image data generator is used to increase the parameters, with the image shape set to (150, 150, 3). The sequential model with activation function "Relu" is applied to the three convolution layers. As it is a multiclass classification problem, loss is calculated as categorical cross-entropy. The performance is evaluated using accuracy matrices. The model recognized 4179 images from 8 classes.

Layer (type)	Output Shape	Parameters #
conv2d (Conv2D)	(None, 147, 147, 32)	1568
max_pooling2d (MaxPooling2D)	(None, 73, 73, 32)	0
conv2d_1 (Conv2D)	(None, 70, 70, 64)	32832
max_pooling2d_1 (MaxPooling2	(None, 35, 35, 64)	0
conv2d_2 (Conv2D)	(None, 32, 32, 128)	131200
max_pooling2d_2 (MaxPooling2	(None, 16, 16, 128)	0
flatten (Flatten)	(None, 32768)	0
dense (Dense)	(None, 256)	8388864
activation (Activation)	(None, 256)	0
dropout (Dropout)	(None, 256)	0
dense_1 (Dense)	(None, 1)	257
activation_1 (Activation)	(None, 1)	0

Total parameters: 8,554,721
Trainable parameters: 8,554,721
Non-trainable parameters: 0

FIGURE 12.6 Summary of the deep learning model.

The total number of parameters discovered was 8,554,721 and they are all trainable. A summary of the model is depicted in Figure 12.6.

12.8 WEB APP

In industry there are many libraries and frameworks, which are extremely valuable for industrializing the machine learning model for its real-life application. These include Dash, Streamlet, and Flask. To increase the outreach and compare our analysis with real data, we have put our bet on Dash Platform.

Dash apps make use of the opportunity to use a graphical user interface (GUI) written in Python, R, and Julia. Data scientists use Dashboards for complex analytics for easy use to a non-tech person in the corporate world.

The deep learning model is saved using the model.save() method of the Keras framework. The model is saved in H5 format. It is a popular method for serializing and deserializing the deep learning model [47].

Consequently, a web app as shown in Figure 12.7 has been developed using Dash; the model is saved and loaded in H5 format. The GUI of the application is designed with the help of HTML, CSS, and Bootstrap components of the Dash library. A file

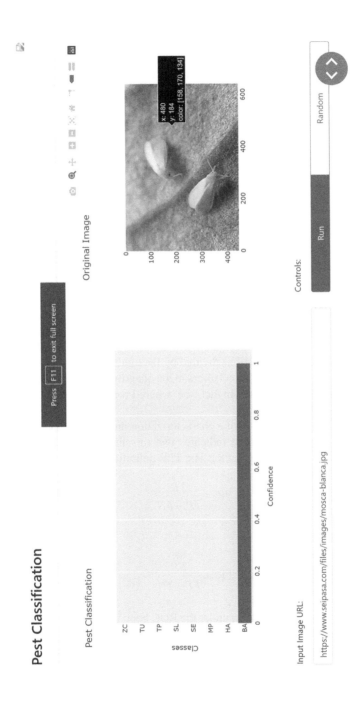

FIGURE 12.7 A web app for pest classification.

containing a link to random images extracted from the Internet is provided to the application. The designed application is deployed on the local server at port number 8090.

12.9   RESULTS

The data are collected and then fed into the machine learning model. Classification techniques include SVM, ensembles, KNN, and convolutional neural network (CNN). The same dataset has been used to train and test every model. The dataset is randomly split into two portions, one for testing and the other for training. The receiver operating characteristic curve (ROC) and the area under its curve (AUC) are used to compare the SVM, Ensemble, and KNN algorithms.

To start, the phrases below are explained to make clear the exhibition estimations used for the correlation of calculations:

- *True Positive (TP)*: The algorithm accurately predicts the positive class, when the outcome is truly positive.
- *True Negative (TN)*: When the model correctly predicts the negative class, the result is considered to be truly negative.
- *False Positive (FP)*: A case of mistaken identity is a result that occurs when the model forecasts the positive class incorrectly.
- *False Negative (FN)*: A false negative outcome occurs when the model suggests the negative class incorrectly.
- *Accuracy*: Accuracy = (TP + TN) / Total, where the Accuracy ratio is the number of accurate predictions divided by the total number of data tests.
- *Precision*: The term "precision" refers to the quantity of observable, positive pieces of evidence that were in fact true, where Precision = TP / (TP + FP).

An illustration of the presentation of a characterization model at each grouping edge is called a ROC curve. This curve indicates two thresholds: the true positive rate (TPR) and the false positive rate (FPR). The TPR definition is

$$TPR = \frac{TP}{TP + FN}$$

When it comes to FPR,

$$FPR = \frac{FP}{FP + TN}$$

The AUC calculates the total region under the ROC bend in two dimensions between (0,0) and (1,1). The overall execution percentage across all conceivable characterization constraints is represented by AUC. One way to interpret AUC is the possibility that the system places an arbitrarily good prognosis more than an arbitrarily negative one. The value of AUC ranges between 0 and 1. When a system suggests 100% incorrect results then the AUC is 0.0; when 100% correct expectations is suggested then the AUC is 1.0 [48].

Support Vector Machine: An SVM is a supervised AI algorithm that resolves two-bunch arrangement challenges by using order computations. After giving them an SVM model layout for the named preparation information for each class, they are ready to sort the additional content. It is a quick and accurate characterization calculation that performs well with few data to deal with. These information focuses are sent into an SVM, which then produces the hyperplane – which, in two measurements, is just a line – that isolates the labels the best. The choice limit is this line. The hyperplane, which in this case is a line, has the greatest distance from each tag's nearest component.

The ROC curve of the SVM is shown in Figure 12.8. Additionally, the AUC is included; its value is 0.95. The true positive rate is found to be nearly constant between 0.01 and 0.64.

A potent machine learning technique known as ensemble learning has proven to have distinct advantages in several applications. The ensemble is an algorithm designed using several distinct models that operate concurrently and whose results are integrated with a judgment fusion technique to yield a single solution to a particular problem.

Ensemble Bagged Trees: Bootstrap aggregation is abbreviated to "bagging." Averaging together multiple estimates is one approach to lowering an estimate's variance.

$$F(x) = \frac{1}{M} \sum_{m=1}^{M} f_m(x)$$

M is the number of different trees.

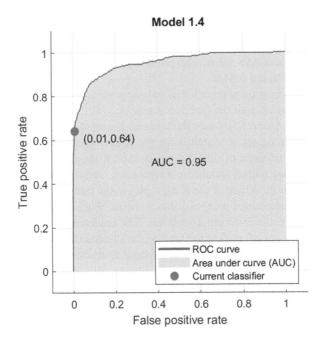

FIGURE 12.8   ROC curve of the SVM model.

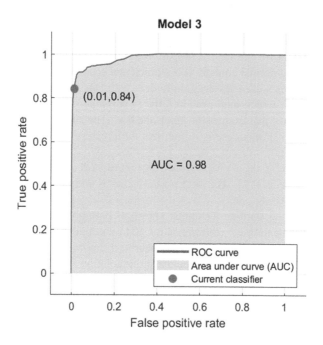

FIGURE 12.9 ROC curve of the ensemble bagged tree model.

To obtain data subsets for the base learners' training, bagging employs bootstrap sampling. Bagging aggregates the output of training sets by polling for categorization and normalizing for regression [49].

Figure 12.9 shows the ROC curve of the Ensemble bagged tree. It also contains the AUC, which has a 0.98 value. The genuine positive rate is seen to be practically consistent from 0.01 and 0.84.

KNN is a classification approach that only approximates the function locally and delays all computation until the function is assessed [50].

K-Nearest Neighbors: KNN uses the training dataset to make predictions directly. The complete training set is searched for the K examples (neighbors) that are the most like the new instance (x); the output for such K clusters is then summed. This could be the average output variable in a regression, and it could be the modal class value in classification. Which of the K examples in the training sample are closest to the new input is determined using a distance metric. Euclidean distance is the most typical metric for real-valued input variables. The Euclidean distance (ED) is calculated by taking the square root of the total of the squared difference between a point location (a) and an older point (a_i) over all input characteristics (j) [51].

$$ED = \sqrt{\left(a_2 - a_1\right)^2 + \left(b_2 - b_1\right)^2}$$

Figure 12.10 represents the ROC curve of the KNN model and its area, which is calculated to be 0.93. Since the AUC is nearly equal to 1, this indicates good accuracy. The true positive rate is found to be nearly constant between 0.01 and 0.87.

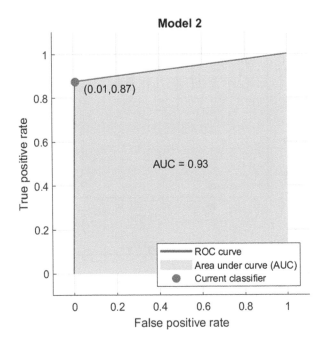

FIGURE 12.10   ROC curve of K-nearest neighbor model.

Convolutional Neural Networks: A CNN model utilizes input data to create class probabilities using a variety of hidden units, including convolutional, pooling, recurrent neural networks, and completely associated layers. A CNN is divided into two layers in the proposed work, using a large number of straight channels in each convolutional layer to obtain neighborhood highlights at various points on the previous layers' highlight maps. The successive convolutional reactions are carried out and down sampled after each convolution layer. One portion from the samples is used to train the model; the remaining data are used to test the process.

Furthermore, 100 epochs were used in the data preparation, which demonstrates the accuracy curve in Figure 12.11(a). The figure shows that the curve starts with accuracies of 0.875. The graph stabilizes after ten epochs and oscillates between 0.870 and 0.875.

Another often used graphic to analyze a deep learning model is a loss curve. This provides a sneak preview of the interaction during preparation and the environment in which the organization learns. Each information item's loss work is calculated throughout the course of a period, and the quantifiable loss measure is guaranteed at the specified age. The loss, however, is only provided for a portion of the whole dataset when the curve is plotted over cycles. Figure 12.11(b) depicts the proposed CNN model's loss vs. epoch curve. The graph shows that the loss began at 1.1921e-07 and remains constant. It then stayed below 1 as the epoch count is increased. This shows that as the count of epochs grows, the loss of information diminishes or the accuracy rises, indicating that the system is adapting to its errors and making the necessary corrections in the subsequent epoch.

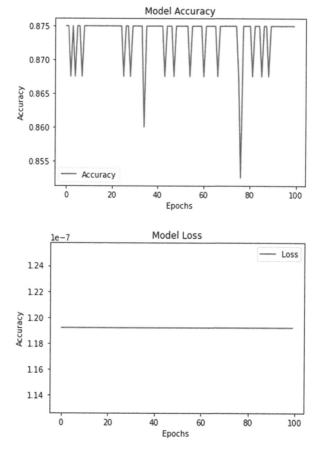

FIGURE 12.11 (a) Accuracy and epoch of the CNN model (b) graph showing the CNN model's loss and epoch.

Table 12.2 summarizes the outcomes of all three algorithms. Based on the table, CNN is the best method for predicting pesticides. The distinction between CNN and KNN is also quite close. Ensemble bagged trees and SVM should be avoided because they provide the least accuracy.

A confusion matrix is a particular type of table that allows for the explicit discussion of the problem of measurable arrangement as well as the interpretation of the exhibition of a computation in the field of ML. The confusion matrix's lines deal with

TABLE 12.2

Comparative Results of Machine Learning Models

	SVM	Ensemble	KNN	CNN
Accuracy (%)	61.4	84.1	85.9	87.5

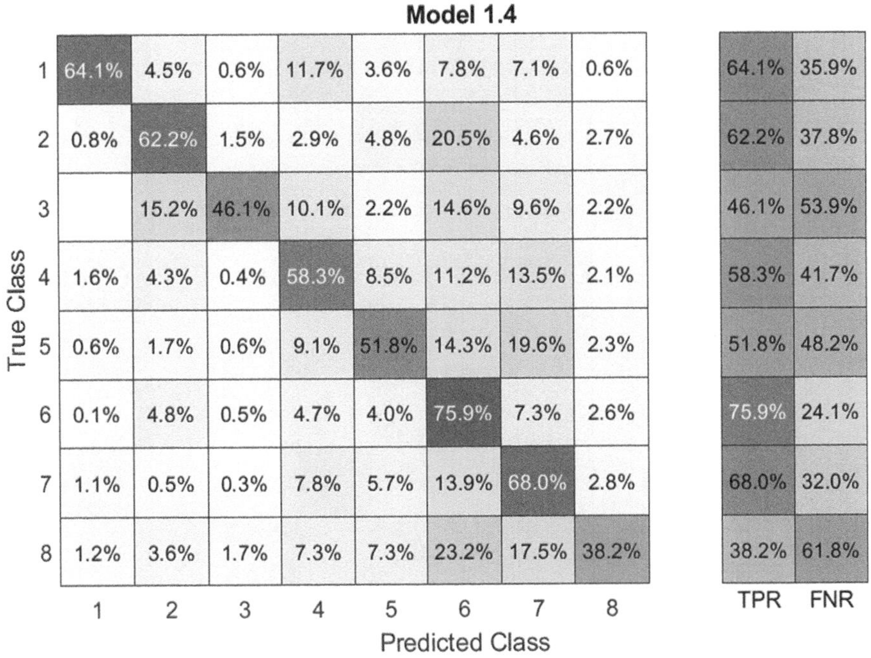

Model 1.4

True Class	1	2	3	4	5	6	7	8		TPR	FNR
1	64.1%	4.5%	0.6%	11.7%	3.6%	7.8%	7.1%	0.6%		64.1%	35.9%
2	0.8%	62.2%	1.5%	2.9%	4.8%	20.5%	4.6%	2.7%		62.2%	37.8%
3		15.2%	46.1%	10.1%	2.2%	14.6%	9.6%	2.2%		46.1%	53.9%
4	1.6%	4.3%	0.4%	58.3%	8.5%	11.2%	13.5%	2.1%		58.3%	41.7%
5	0.6%	1.7%	0.6%	9.1%	51.8%	14.3%	19.6%	2.3%		51.8%	48.2%
6	0.1%	4.8%	0.5%	4.7%	4.0%	75.9%	7.3%	2.6%		75.9%	24.1%
7	1.1%	0.5%	0.3%	7.8%	5.7%	13.9%	68.0%	2.8%		68.0%	32.0%
8	1.2%	3.6%	1.7%	7.3%	7.3%	23.2%	17.5%	38.2%		38.2%	61.8%
	1	2	3	4	5	6	7	8			

Predicted Class

FIGURE 12.12 Confusion matrix for support vector machine model.

cases in hypothetical classes, while its segments deal with examples in actual classes. The name comes from the ease with which it can be determined whether the framework is conflating two classes. With two measurements (actual and predicted) and identical configurations of "categories" in the two measures, this is a special kind of possibility table. The number of FP (1,0), FN (0,1), TP (0,0), and TN (0,0) are displayed in a table.

The detailed confusion matrix for the SVM model is shown in Figure 12.12. Since total erroneous results outnumber total real results, we should refrain from using the SVM model for pesticide recommendation.

The comprehensive confusion matrix with parameters for the Ensemble Bagged Tree model is shown in Figure 12.13. We should not use this model to suggest pesticides because the total number of erroneous results exceeds the total number of accurate outcomes. Figure 12.14 shows the KNN's confusion matrix and demonstrates how many more true results overall there are than incorrect outcomes. This demonstrates that the KNN model should be chosen over others when recommending pesticides.

When the information was tested, the SVM model had an accuracy of 61.4%, the Ensemble Bagged Tree model had 84.1%, the KNN model had 85.9%, and the CNN model had 87.5%. CNN, Ensemble, and KNN performed admirably, producing nearly accurate results, with CNN slightly outperforming Ensemble and KNN. It is also possible to conclude that SVM is unsuitable for pesticide recommendation. Table 12.3 highlights the findings from our suggested approach, which indicates insecticides based on reputable, well-known pests.

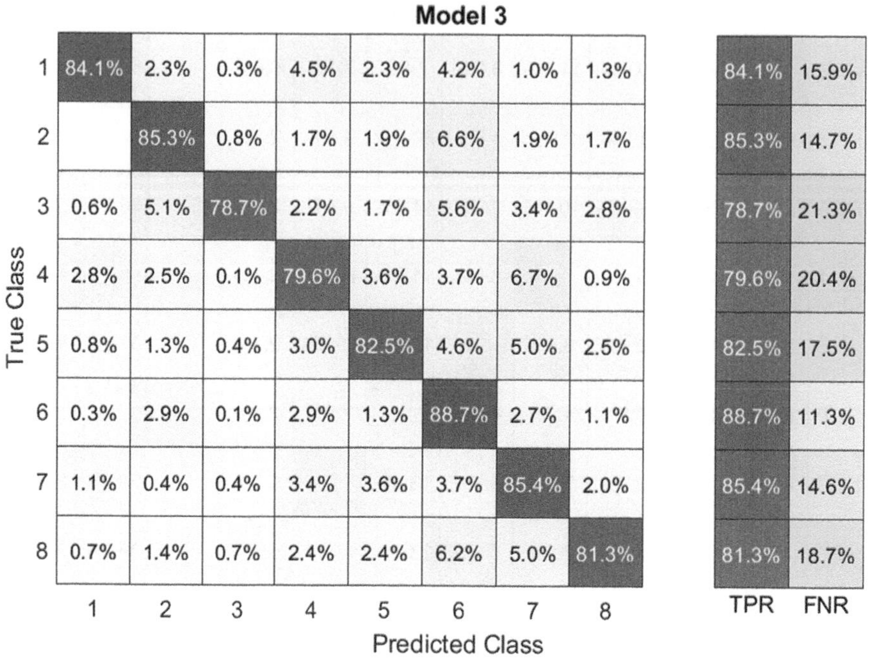

Model 3

	1	2	3	4	5	6	7	8		TPR	FNR
1	84.1%	2.3%	0.3%	4.5%	2.3%	4.2%	1.0%	1.3%		84.1%	15.9%
2		85.3%	0.8%	1.7%	1.9%	6.6%	1.9%	1.7%		85.3%	14.7%
3	0.6%	5.1%	78.7%	2.2%	1.7%	5.6%	3.4%	2.8%		78.7%	21.3%
4	2.8%	2.5%	0.1%	79.6%	3.6%	3.7%	6.7%	0.9%		79.6%	20.4%
5	0.8%	1.3%	0.4%	3.0%	82.5%	4.6%	5.0%	2.5%		82.5%	17.5%
6	0.3%	2.9%	0.1%	2.9%	1.3%	88.7%	2.7%	1.1%		88.7%	11.3%
7	1.1%	0.4%	0.4%	3.4%	3.6%	3.7%	85.4%	2.0%		85.4%	14.6%
8	0.7%	1.4%	0.7%	2.4%	2.4%	6.2%	5.0%	81.3%		81.3%	18.7%

True Class / Predicted Class

FIGURE 12.13   Ensemble bagged tree model confusion matrix.

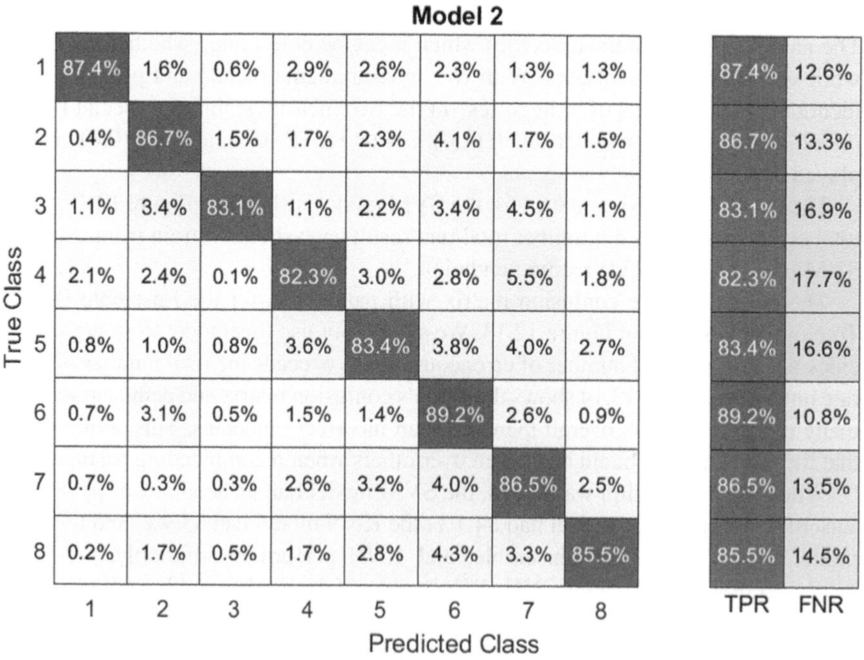

Model 2

	1	2	3	4	5	6	7	8		TPR	FNR
1	87.4%	1.6%	0.6%	2.9%	2.6%	2.3%	1.3%	1.3%		87.4%	12.6%
2	0.4%	86.7%	1.5%	1.7%	2.3%	4.1%	1.7%	1.5%		86.7%	13.3%
3	1.1%	3.4%	83.1%	1.1%	2.2%	3.4%	4.5%	1.1%		83.1%	16.9%
4	2.1%	2.4%	0.1%	82.3%	3.0%	2.8%	5.5%	1.8%		82.3%	17.7%
5	0.8%	1.0%	0.8%	3.6%	83.4%	3.8%	4.0%	2.7%		83.4%	16.6%
6	0.7%	3.1%	0.5%	1.5%	1.4%	89.2%	2.6%	0.9%		89.2%	10.8%
7	0.7%	0.3%	0.3%	2.6%	3.2%	4.0%	86.5%	2.5%		86.5%	13.5%
8	0.2%	1.7%	0.5%	1.7%	2.8%	4.3%	3.3%	85.5%		85.5%	14.5%

True Class / Predicted Class

FIGURE 12.14   Confusion matrix for KNN model.

TABLE 12.3
Pests and Their Respected Pesticides

Pest	Pesticides
Bemisia Argentifolii	Avermectin and Pyriproxyfen
Helicoverpa Armigera	Organophosphates and Synthetic Pyrethroids
Myzus Persicae	NeelAzal
Spodoptera Exigua	Bacillus Thuringien-sis
Tetranychus Urticae	Pyrethroids and Pyrethrins (3A)
Zeugodacus Cucurbitae	Malathion with Hydrolysed protein
Spodoptera Litura	Flubendiamide, Spinosad & Chlorfenapyr
Thrips Palmi	Flufenoxuron, Imidacloprid & Chlorfluazuron

12.10 CONCLUSION

Controlling pests and using pesticides is crucial for sustainable agriculture. The suggested approach would be helpful in the transition to green cities. Four machine learning models were utilized in the suggested methodology: Ensemble Bagged Tree, SVM, KNN, and CNN. In the suggested work for pest-specific pesticide recommendation, the CNN-based model, which had an accuracy and precision of 87.5%, had the greatest accuracy and precision. Due to the use of GLCM rather than CNN, most earlier models were unable to categorize multi-class categorization among different pest classes. The proposed model and developed system will outperform the expectation for real-life datasets. Enormous effort has been made to tackle issues due to various constraints such as variability in the number of images, image quality, size of pest in the image, and the camouflaged nature of various pests. After trying many processes and algorithms to withstand the diversity of the nature of pests, that is, pest of eight classes is fixed in the proposed work.

In future, work will also be done to promote cultural techniques, rather than using any pesticides at all, to protect crops from pests. This will lead to new agricultural developments that do not require the use of any pesticides, making them more affordable for small and medium-sized farmers and people who farm in their terraces and gardens. This will also result in increased soil fertility because there won't be any pesticide deterioration. Further, multiple crops have not been considered in the present work and therefore may be so considered in the future.

REFERENCES

[1] Juuko I, Amwine JP, Magembe H, Kikabi IK, Naggenda J. Pests and disease advisory and notification system for food crop farmers in Uganda. 2022.

[2] Maxmen A. Crop pests: under attack. *Nature*. 2013; 501(7468): S15–7.

[3] Zhu F, Zhou YK, Ji ZL, Chen XR. The plant ribosome-inactivating proteins play important roles in defense against pathogens and insect pest attacks. *Frontiers Plant Science*. 2018; 9. Available from: https://www.frontiersin.org/articles/10.3389/fpls.2018.00146

[4] Naskar S, Roy C, Ghosh S, Mukhopadhyay A, Hazarika LK, Chaudhuri RK, et al. Elicitation of biomolecules as host defense arsenals during insect attacks on tea plants (*Camellia sinensis* (L.) Kuntze). *Applied Microbiology and Biotechnology*. 2021; 1–13.

[5] Marini L, Ayres MP, Jactel H. Impact of stand and landscape management on forest pest damage. *Annual Review of Entomology*. 2022; 67: 181–99.

[6] Areni IS, Tamin R, et al. Image processing system for early detection of cocoa fruit pest attack. *Journal of Physics: Conference Series*. vol. 1244. IOP Publishing; 2019. p. 012003.

[7] Burhan SA, Minhas S, Tariq A, Nabeel HM. Comparative study of deep learning algorithms for disease and pest detection in rice crops. 2020: 1–5.

[8] Sharma A, Kumar V, Shahzad B, Tanveer M, Sidhu GPS, Handa N, et al. Worldwide pesticide usage and its impacts on ecosystem. *SN Applied Sciences*. 2019; 1: 1–16.

[9] Toby BL, Duleeka U, David KG, Michael E. Suicide by pesticide poisoning in India: a review of pesticide regulations and their impact on suicide trends. Bonvoisin et al. *BMC Public Health*. 2020; 20: 251.

[10] Tang FH, Lenzen M, McBratney A, Maggi F. Risk of pesticide pollution at the global scale. *Nature Geoscience*. 2021; 14(4): 206–10.

[11] Huddar SR, Gowri S, Keerthana K, Vasanthi S, Rupanagudi SR. Novel algorithm for segmentation and automatic identification of pests on plants using image processing. *2012 Third International Conference on Computing, Communication and Networking Technologies (ICCCNT12)*. 2012: 1–5.

[12] Rupesh G, Mundada Rupesh G, Mundada. Detection and Classification of Pests in Greenhouse Using Image Processing. *IOSR Journal of Electronics and Communication Engineering*. 2013; 5(6): 57–63.

[13] Srinivas K, Usha B, Sandya S, Rupanagudi SR. Modelling of edge detection and segmentation algorithm for pest control in plants. 2011: 293–5.

[14] Gondal D, Khan Y Early pest detection from crop using image processing and computational intelligence. *FAST-NU Research Journal* ISSN: 2313-7045. 2015; 2: 1.

[15] Preethi RG, Priya R, Sanjula SM, Lalitha DS, Vijaya B. Agro based crop and fertilizer recommendation system using machine learning. *European Journal of Molecular & Clinical Medicine*. 2020; 7(4): 2043–51.

[16] Mahmood I, Imadi SR, Shazadi K, Gul A, Hakeem KR. Effects of pesticides on environment. *Plant, Soil and Microbes*. 2016; 1: 253–69.

[17] Li Y, Wang H, Dang LM, Sadeghi-Niaraki A, Moon H. Crop pest recognition in natural scenes using convolutional neural networks. *Computers and Electronics in Agriculture*. 2020; 169: 105174.

[18] Travaline K, Hunold C. Urban agriculture and ecological citizenship in Philadelphia. *Local Environment*. 2010; 15(6): 581–90.

[19] Jensen JD, Christensen T, Denver S, Ditlevsen K, Lassen J, Teuber R. Heterogeneity in consumers' perceptions and demand for local (organic) food products. *Food Quality and Preference*. 2019; 73: 255–65.

[20] Arora NK. Impact of climate change on agriculture production and its sustainable solutions. *Environmental Sustainability*. 2019; 2(2): 95–6.

[21] Diekmann LO, Gray LC, Baker GA. Growing 'good food': Urban gardens, culturally acceptable produce and food security. *Renewable Agriculture and Food Systems*. 2020; 35(2): 169–81.

[22] Aktar WM, Sengupta D, Chowdhury A. Impact of pesticides use in agriculture: their benefits and hazards. *Interdisciplinary Toxicology*. 2009.

[23] Hussain S, Siddique T, Saleem M, Arshad M, Khalid A. Chapter 5 Impact of Pesticides on Soil Microbial Diversity, Enzymes, and Biochemical Reactions. *Advances in Agronomy*. 2009.

[24] Kalia A, Gosal S. Effect of pesticide application on soil microorganisms. *Archives of Agronomy and Soil Science*. 2011; 57(6): 569–96.

[25] Kauer R, Friedel M, Doring J, Meissner G, Stoll M. Organic/Biodynamic: Integrated, organic and biodynamic viticulture: A comparative 10-year study. *Wine & Viticulture Journal*. 2018; 33(1): 38, 40–2.

[26] Pitterna T, Cassayre J, Hüter OF, Jung PM, Maienfisch P, Kessabi FM, et al. new ventures in the chemistry of avermectins. *Bioorganic & Medicinal Chemistry*. 2009: 17 (12): 4085–95.

[27] Bruce HA, William VHJ, Peter WG. Environmental effects of the usage of avermectins in livestock. *Veterinary Parasitology*. 1993; 48 (1–4): 109–25.

[28] Authority EFS, Crivellente F, Hart A, Hernandez-Jerez AF, Bennekou SH, Pedersen R, et al. Establishment of cumulative assessment groups of pesticides for their effects on the thyroid. *EFSA Journal*. 2019; 17(9).

[29] Kamel A, Al-Dosary S, Ibrahim S, Ahmed MA. Degradation of the acaricides abamectin, flufenoxuron and amitraz on Saudi Arabian dates. *Food Chemistry*. 2007: 100(4): 1590–3.

[30] Gong HS, Liao LW, Hu WB, Yin GQ. Progress of research on the inspection of pesticide residues in food by GC-MS. *Food & Machinery*. 2013: 05.

[31] Albaseer SS. *Factors controlling the fate of pyrethroids residues during post-harvest processing of raw agricultural crops: An overview*. Elsevier; 2019.

[32] Rawat K, Yadav AK. Advances in biological techniques for remediation of heavy metals leached from a Fly Ash Contaminated Ecosphere. *Pollutants and Water Management: Resources, Strategies and Scarcity*. 2021: 151–71.

[33] Malathion NZ. *Africa Journal of Physical Sciences* ISSN: 2313-3317. 2020; 4.

[34] Tchounwou PB, Patlolla AK, Yedjou CG, Moore PD. Environmental exposure and health effects associated with Malathion toxicity. *Toxicity and Hazard of Agrochemicals*. 2015; 51: 2145–9.

[35] Lewis K, Malathion. 2016.

[36] Dereumeaux C, Saoudi A, Goria S, Wagner V, De Crouy-Chanel P, Pecheux M, et al. Urinary levels of pyrethroid pesticides and determinants in pregnant French women from the Elfe cohort. *Environment International*. 2018; 119: 89–99.

[37] Soderlund DM, Bloomquist JR. Neurotoxic actions of pyrethroid insecticides. *Annual Review of Entomology*. 1989; 34(1): 77–96.

[38] Mohsin M, Shafi J, Naz SI, Aslam S, Hassan S. Variability in susceptibility status of malaria vectors and other Anopheles species against different insecticides in district Faisalabad, Central Punjab, Pakistan. 2021.

[39] Huang M-L, Chuang TC. A database of eight common tomato pest images; 2020.

[40] Padmavathi K, Thangadurai K. Implementation of RGB and grayscale images in plant leaves disease detection–comparative study. *Indian Journal of Science and Technology*. 2016; 9(6): 1–6.

[41] Saravanan C. Color image to grayscale image conversion. In: *2010 Second International Conference on Computer Engineering and Applications*. vol. 2. IEEE; 2010. pp. 196–9.

[42] Parsania PS, Virparia PV. A comparative analysis of image interpolation algorithms. *International Journal of Advanced Research in Computer and Communication Engineering*. 2016; 5(1): 29–34.

[43] Asamwar RS, Bhurchandi KM, Gandhi AS. Interpolation of images using discrete wavelet transform to simulate image resizing as in human vision. *International Journal of Automation and Computing*. 2010; 7(1): 9–16.

[44] Ping Tian D, et al. A review on image feature extraction and representation techniques. *International Journal of Multimedia and Ubiquitous Engineering*. 2013; 8(4): 385–96.

[45] Jain S. Brain cancer classification using GLCM based feature extraction in artificial neural network. *International Journal of Computer Science & Engineering Technology*. 2013; 4(7): 966–70.

[46] Kekre H, Thepade SD, Sarode TK, Suryawanshi V. Image retrieval using texture features extracted from GLCM, LBG and KPE. *International Journal of Computer Theory and Engineering*. 2010; 2(5): 695.

[47] Agrad H, Mourhir A. Deep Learning Meets Fashion-A Look Into Virtual Try-On Solutions; 2020.

[48] Huang J, Charles LX. Using AUC and accuracy in evaluating learning algorithms. *IEEE Transactions on knowledge and Data Engineering*. 2005; 17(3): 299–310.

[49] Smolyakov V. Ensemble learning to improve machine learning results. *Stats and Bots*. 2017.

[50] Yoshikawa T, Losing V, Demircan E. Machine learning for human movement understanding. *Advanced Robotics*. 2020; 34(13): 828–44.

[51] Brownlee J. K-nearest neighbors for machine learning. *Machine Learning Mastery*. 2016; 15.

[52] Li Y, Yang J. Few-shot cotton pest recognition and terminal realization. *Computers and Electronics in Agriculture*. 2020; 169: 105240.

[53] Fuentes A, Yoon S, Kim SC, Park DS. A robust deep-learning-based detector for real-time tomato plant diseases and pests' recognition. *Sensors*. 2017; 17(9): 2022.

13 Digital Twin Technologies for Green Smart Cities

Akankshya Subhadarshini
College of Engineering and Technology, Bhubaneswar, India

Alok R. Prusty
DGT, NSTI, Bhubaneswar, Odisha, India

13.1 INTRODUCTION

A digital twin is a technique to virtually signify a physical process or phenomenon by means of real-time data. This modernisation technique prepares industries and organisations to oversee their products and processes so they are able to fill the gap that exists between project design and the implementation phase. A digital twin helps in preparing a miniature model of a smart city that has been digitally duplicated. With the help of building information modelling systems that make it possible to track every aspect of a building, it was initially applied to the construction of buildings and other infrastructure. This approach helps us to foresee any potential flaws in the content and was quickly expanded to include additional services like the sewage system, energy, roads, and education. The virtual metropolis in the digital twin is identical to the real city, much like the SimCity video game [1]. A digital twin is computer software that simulates a tangible thing, such as a building, or a service using data from the current world. It can be applied to improve all stages of design, construction, operation, and administration in addition to making performance predictions for products and processes [2].

They both conceptualise the merger of virtual space and physical space, making the digital twin in the broadest sense a type of cyber physical system (CPS). Even so, the digital twin system's comparison focuses on amazing resolution scenarios, which are one of its distinguishing aspects, and it has more data and models. This furthermore improves customer understanding by offering a digital feel to the product. The future of information systems has a new reality in this digital twin. A manufacturing execution system will eventually be the new pattern that incorporates supervisory control and data acquisition, (sProduct Life Management) may emerge. Systems become digital twin application subsystems when they are implemented with a digital twin as their kernel [1, 2]. Data integration technology is strongly influenced by the Internet of Things technology to increase the stakes and enormous possibilities in industry, medicine, green infrastructure, and most domestic services.

DOI: 10.1201/9781003388814-13

The massive involvement of critical data in digital twins leads to better efficacy in complex systems without a physical presence, which makes it a hot cake in the marketplace. This chapter outlines the domains in which digital twin technology is being applied, notably from the standpoint of a smart city with healthier and more ecological architectural standards.

13.2 LITERATURE REVIEW

In [1], the author shows that it is feasible to deduce from this study that the lean and green supply chain will continue to develop and adapt to future Industry 4.0 requirements. The foundation of Industry 4.0 is already focused on lean and green qualities. Lean and green methodologies will, in general, advance Industry 4.0. There are several traits that will change: product and process design, manufacturing scheduling and management, logistics and supplier collaboration, information exchange, and energy and customer values. The ability to adapt operations to the needs of each customer is one of the fundamental advances. In [2], the author describes how the advancement of the smart city is based on climate change and urban technological advancements. Smart city policies are frequently built on technological orthodoxies that are conceptually and practically shallow as decision makers search for a technology cure. The goal of this study is to address the conceptual immaturity associated with the complete digitisation of the city by pursuing a dual argument concerning the implications for democracy and the environment. In [3], The author has conducted research to figure out the level of environmental sustainability and liability, as well as the degree to which the principal stakeholders in a city or state are required to collaborate effectively. Recapitulating in terms of technology, we require better architectural plans, data gathering techniques, sensor configurations that are both effective and efficient, and finally a way to interface technology with reality. Economic considerations, such as business practices, licensing, and market impact will always take precedence in this process. Several frequent worries, like cybersecurity, confidentiality, and reliability, point to further research in this field. Utilising a reducing conceptual design, the author in [4] offers a thorough investigation into collaborative edge computing for (wireless-optical broadband access network)-supporting knowledge economy. As a new model for building smart cities, the digital twin consortium would redefine local government structures and laws and infuse continual energy for urban evolvement, according to [5]. In numerous prominent cities worldwide, digital twin cities are being built. It isn't necessary to go into detail on each topic addressed by sustainable communities because the author in [6] is focused on those that were relevant to urban design. The idea is also supported by current research on eco-urban design ideas and techniques.

13.3 MANY DIFFERENT KINDS OF DIGITAL TWINS

Depending on how enlarged the product is, many types of digital twins are available. The main difference between such twins is in the implementation. It is typical for many types of digital twins to engage with one another in a method or system. Focusing on inherent variations and applications, several digital twins can be analysed. The fundamental building block of digital twin technology and the most straightforward illustration of a feature lies in the twin parts. Even though these usually pertain to significantly less significant portions, part twins are very similar to interacting components.

13.3.1 WEALTHY TWINS

A property is generated when two or more parts execute well together. With investment twins, you can examine how these interactions produce a plethora of performance data that can be analysed and transformed into useful knowledge.

13.3.2 SYSTEM TWINS

Structure or module twins constitute the next step in illustration because they demonstrate how various components of a system come to operate together. A network may be used to develop ideas and give insight into how things interact.

13.3.3 TWIN PROCESSES

The macroeconomics of the magnification of the process twin illustrates how components interact to build a full production facility. Do all those organisations' abilities function together as effectively as possible, or do delays in one system have an impact on other systems? The utilisation of process twins enables the identification of the precise strategies that ultimately influence overall profitability.

13.3.4 PROBLEMS CAUSED BY DIGITAL TWINS

For a virtual representation to be specified properly, there cannot be an automatic data flow from the simulation environment to the virtual representation. A 3D model is a computerised representation of an actual or hypothetical physical object. Plans for buildings, product designs, and development are all available as digital models. One important point of differentiation is the absence of automatic data transfer between the real object and the virtual counterpart.

An imprint of a thing that only goes in one way between the physical and the virtual is called a digital shadow. A change in the physical object's state causes a shift in the digital object, and vice versa. Figure 13.1 illustrates the flow of manual data, which is a combination of the physical object and digital object, also known as a digital model, flows into a digital shadow, a digital twin, and finally our final flow. These modern digital ideas are used by several manufacturing businesses. Digital models are useful for industrial design and concept development, while digital shadows are helpful for monitoring development and a digital twin is an effective device for evaluating real-time production.

This phrase "digital twin" was first created as information moved back and forth between a real-world physical entity and a digital object. The digital equivalent immediately changes whenever the real object does, and vice versa.

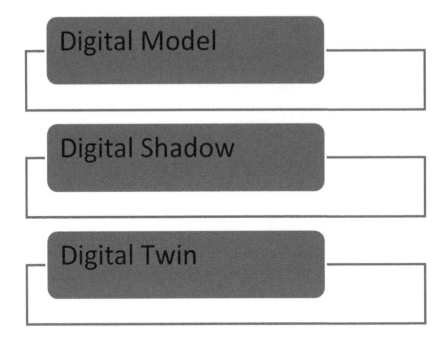

FIGURE 13.1   Digital Twin, Digital Model, and Digital Shadow.

13.4   NEW INNOVATION OF DIGITAL TWIN TECHNOLOGY

The three components that make up the digital twin platform for product lifecycle management are the finite reality, a private network, and the relevant data tier. Greater system modelling, dynamic modelling of daily life, and process improvement across the board in real time are all possible with digital twin technology [7, 8].

The current appearance of the industrial sector is going to change due to the digital twin. The way things are created, manufactured, and managed is significantly impacted by digital twins.

Figure 13.2 shows how digital twin technology is used in numerous industries, including industry, supply chains, medicine, and commerce. While reducing throughput times, it enhances and improves manufacturing efficiency [7]. The utilisation of digital twins and data from the Internet of Things (IoT) could considerably benefit the healthcare sector, including cost savings, clinical services, service and maintenance, and the provision of individualised healthcare [8].

Designing and managing a smart city using digital twins and IoT data contributes to increased economic growth, effective resource administration, a smaller ecological footprint, and an overall improvement in citizens' quality of life. Urban planners and politicians can use the digital twin model to plan smart cities by learning from diverse sensing devices and cognitive technologies. They also use the data from the digital twins to make future decisions that are well-informed [9].

The aviation sector is included under industry. In this sector, the digital twin is primarily used to maintain aircraft, forecast hazards, organise aircraft, and start

FIGURE 13.2 Application of Digital Twin Technology.

self-repair mechanisms, where the ability to process data and diagnose problems is significantly increased [8].

13.4.1 Digital Twin for Consumer Energy Service

Energy consumption analysis and forecasting are related; because of the high degree of accuracy that can be achieved between the virtual and physical worlds, it is possible to project energy use on a daily, weekly, or monthly basis. Additionally, renewable power use can be predicted, therefore allowing for the development of improved resources activities and decisions [10]. Behaviourally based energy management services use digital twin real-time monitoring services where users are assessed based on their energy consumption behaviour, enabling for the real-time correction of poor energy consumption and lifestyle choices. Utilities may achieve this by employing digital twin renewable power, which will allow them to oversee and fully integrate even while upgrading the overall service quality they provide. To do this, it is necessary to evaluate consumption patterns to compute actual live activities associated with energy behaviour. Due to their unique and independent patterns of energy use, digital twins The increasing popularity of commercialization. Modernisation and the improvement of AI tools would provide excellent methods for changing constant energy behaviour instantly. This is mostly decided by looking at a user's consumption patterns and determining whether these fall within the guidelines of healthy usage [10].

Here is an enterprise energy-saving suggestion. Since industrial operations directly account for 33% of global energy consumption, increasing energy efficiency is necessary for a low-carbon civilization to develop. In addition to having a substantial positive effect on the environment, increasing energy efficiency at the industrial scale increases firm competitiveness while reducing (Green House Gas) GHG emissions. Profitability can be greatly increased by lowering a company's energy usage profile. Contrary to this, however, organisations frequently fail to implement the various energy saving measures that are available – a situation known as the energy efficiency gap.

Researchers have stated that it is crucial to conserve energy because there are steps that may be taken to do so across production facilities. They have also talked about technological strategies for energy conservation, which has led to suggestions for specific activities that would enhance energy use in the metal, bread, and food industries.

To solve pressing societal issues, social computing encourages collaboration between humans and machines using cutting-edge socio-technical methods. A social activity that conserves energy is one of these. The recognition of an informant's web 2.0 as a significant, potent technology enabling social computing has increased the efficacy of this technique. A consequence of community networking enterprises, social robots offer both interpersonal and technical methodologies.

13.4.2 SMART GREEN CONSTRUCTION INNOVATION

Throughout a building's life cycle, green building advocates for energy efficiency, low carbon emissions, and a human-centred, peaceful coexistence with the environment [11].

Since they significantly contribute to energy conservation and emissions reduction in cities, green building design and construction are crucial for implementing sustainable development. Figure 13.3 shows the different types of class in green building design. Green, bright green, and Reducing carbon dioxide emissions has an intelligent relationship; however, this differs from green buildings and different types of logos. The number of initiatives utilising green has considerably expanded during the last ten years. Additionally, many more construction evaluation marks are now available. The implementation of technological criteria, unequal regional growth, and the market's support for green construction all remain problematic, which must be kept in mind. The key is to support the large-scale development of cutting-edge green structures and to encourage the construction of environmentally friendly metropolitan areas, continuous green building development areas, and so on. Innovative

GREEN	BRIGHT GREEN	INTELLIGENT
Air & Energy	Energy management	Converged Networks
Reduce GHG emissions	Assets management	Data collection, measurement,
	space utilisation	verification
Improve IAQ	Integrated design process	Diagnostics, sensors
Improve energy efficiency	Renewable design process	Infrastructure
Water	Healthy and comfortable	Structured cabling solution
Reduce wastewater discharge	environment	
Lower contaminant release	Green loans	Wireless systems

FIGURE 13.3 Green Building Design.

green cities present a chance for the intelligent construction sector given the current level of severe competition. An intelligent green city provides the potential for the future growth of the smart construction industry. Numerous academics have used AI to study green building. Without these organisations and methods, a system for ensuring the smooth operation of smart cities is necessary. The concept of sustainable construction improves the objective of green architecture by reducing risks. Building management is improved in terms of sustainability and financial efficiency by integrating intelligent building technologies, analytics, and cutting-edge digital services. Due to the rising awareness of businesses, merchants, and residents to environmental quality, the subjects of the green economy and sustainable growth are now very relevant. New devices are being utilised more frequently and could be extremely helpful in ubiquitous computing as a consequence of the rapid breakthroughs in networking made feasible by the Internet of Things.

A localised transportation system that uses cutting-edge technology and ICT to cut CO_2 emissions and energy consumption and to improve facilities to the people, with a focus on the most underprivileged communities including the disabled, the elderly, and families with strollers, is at once digital, smart, ecological, and green. Conversely, the development of urban parks is a green construction only, the digitisation of the Government Register Office is a digital project only, and the improvement of power generation capacity in the smart grid is (nearly) entirely smart.

Digital twins are employed more often as communities become more interconnected because of the growth of smart cities. Moreover, we will keep gathering data. All this will make it possible to conduct studies on the creation of improved AI algorithms utilising IoT sensors that are already a part of our basic city operations.

Today's construction projects frequently employ several green technologies in construction, which include:

1. Photovoltaic panels: Photovoltaic arrays may heat water for a building and provide power.
2. Turbines: The house's electricity can be produced by a wind turbine.
3. Combined heat and power systems: Geothermal systems utilise the heat from the Earth to produce power or heat water for a structure.
4. Vertical greenery: Rooftop gardens have a vegetation-covered surface that can aid in conserving water and warmth while also affording a habitat for wildlife.
5. Water collection: Systems for collecting and storing rainwater are used for agriculture and certain other uses.
6. Wastewater recycling systems: These systems reuse wastewater from drains, showers, and other sources for planting or use in toilets.
7. Lighting systems: Compared to conventional incandescent ones, lights are more energy-efficient and can assist in lowering a building's energy usage.

13.4.3 Data Centres in Digital Twin Cities

Data are the main strategic resource in digital twin cities. Every piece of data is tracked throughout its full life cycle by a data centre, particularly the mega data centre for a metropolis. One challenge brought about by the growing use of mobile communications is the creation of versatile, adaptable, yet dynamic infrastructures that can address additional challenges with smallest amount of funds and time. The integrity of data may potentially be jeopardised if extended delays are experienced when transferring data to cloud servers. Therefore, it is essential to investigate the equilibrium between efficiency and safety [2].

13.4.4   URBAN BRAIN CITIES

There have also been several new directions for research on the urban brain. As big data technologies have developed, it has become standard practice to merge data from various sources and viewpoints to produce a comprehensive, dynamic picture of urban participants. Due to the expansion of local governance in major cities, the diversity of both the major highways has also been an issue [12].

According to [6], there are different levels of users, access, business service, intermediate service, data, and support for the Highway Smart Operation and Maintenance System. Its bottom layer encompasses the system application, architecture, and communication links.

13.4.5   FLOOD SCENARIO SOLUTIONS AND FLOOD TRACKING

By establishing a set of parameters for real-time urban flood modelling and business research and assessment services, digital twin technology contributes to the growth of the smart urban environment. Continuous review and prediction before floods, fluid analysis and tracking during disasters, and review and rebuilding after catastrophic events are all part of the entire flood lifecycle. The standardised and accommodative big data surveillance of floods, the flood knowledge map, and the flood service application represent the three critical features of the urban planning flood inundation mapping and support system built on the digital twin. Detecting the big data on flooding involves an approach that employs real-time satellite, air, and earth explosion at urban and drainage dimensions [12]. This strategy is used in the background in cloud computing.

13.5   INDUSTRY 4.0 AND DIGITAL TECHNOLOGIES

The option of efficient logistic chain management in moderate- and small-scale manufacturing companies is greatly affected by Industry 4.0 innovations Ministry of Micro, Small & Medium Enterprises.

Companies all around the world are constantly looking for fresh and creative approaches to improve their brand image and gain the greatest competitive advantage. The administration of the material, information, and monetary flows between various businesses, from the supplier to the manufacturer and client, is known as supply chain management.

Investigating cutting-edge concepts like lean and green in the context of the supply chain may alter practices to produce a supply network that is more effective and sustainable. Lean and green concepts are becoming more significant. The lean paradigm places a strong emphasis on streamlining procedures across the entire supply chain, searching for areas where simplification is possible, and eliminating tasks that do not generate value. By reducing waste and addressing customer needs, lean increases quality and production while reducing costs and time.

13.5.1 Agile and Ecological Principles Are Used in the Supply Chain

Both the agile and ecological paradigms are currently being applied to supply chain operations. The goal of the lean concept is to find and eliminate any non-value-added operations from any process flow to improve it. These management systems were initially created to lower production process waste and raise the calibre of the final output [13].

Lean demands that: space be reduced, personnel be reduced, capacity be used more efficiently, system flexibility be increased, and standard parts be used. Since it aims to fulfil organisational profit and market share goals while also reducing environmental impacts and hazards, green is a crucial management paradigm. Environmental management is the foundation of the green paradigm; supply chain management uses environmentally friendly methods that can be used to reduce the environmental effect. This can be achieved by lowering the amount of green waste produced in terms of energy, water, emissions into the air, solid trash, and hazardous waste, preventing pollution, or utilising these resources more effectively. All stakeholders concerned in the logistical system, encompassing companies, facilities, vendors, logistics, assets, and consumers, are expected to be in complete contact through the phrase "Industrial 4.0." Each of these adapts its setup instantaneously based on the requirements and circumstances of the other logistical system stakeholders to include green technology. A few instances are the elimination of contamination, raw material consumption, and greenhouse gas.

Industry 4.0 is a method of conducting business that is founded on three key tenets three basic concepts:

From its accessibility and high degree of flexibility to its horizontal and vertical integration throughout the whole value creation internet backbone, equipment and processes are made possible. the ultimate development and digitization of already-existing goods and services along the course of their whole life cycles. Equipment and processes enable both horizontal and vertical integration across the whole value creation internet backbone to its accessibility and significant level of autonomy. The final evolution and digitiszation of existing products and services over the span over its entire life phases.

  i. Creative digital business models.

Infrastructure and the Internet of Things are used by the CPSs that make up Industry 4.0's information technology to connect and collaborate with people in real time. Connectivity, realised by decline, scalability, real-time abilities, adaptability, and customer orientation, are all among the cornerstones of this alliance [1].

13.6 IMPLEMENTING TECHNOLOGIES FOR DIGITAL TWINS

A digital twin app integrates technologies that enable the creation of a graphical signal, collection, and storage of actual statistics, as well as the generation of insightful information based on that data.

13.6.1 ANALYTICS OF DATA

As shown throughout the text and in scholarly publications, the phrase "advanced analytics" is an encompassing word which combines known rules. Thus, it is essential to understand and evaluate other publications. A broad field like data science, which concentrates on acquiring and presenting data for analysis to obtain a stronger insight, is where the phrase "predictive science" derives. Raw data are essential for performing data analysis.

Several steps must be completed to convert this information into data that can be utilised in algorithms and data analysis. The second phase addresses the need to acquire the necessary data, specifying the actual locations and methods for doing so. Following the collection phase, the data will undergo a processing stage where they will be sorted in accordance with predetermined criteria. Data cleaning is the last and, possibly, most crucial step.

13.6.2 ARTIFICIAL INTELLIGENCE AND MACHINE LEARNING

Artificial intelligence is the first area of focus for data analysis. The notion of creating "AI applications" earned AI its broad meaning in the 1950s, while machine learning (ML), a field within AI, is the study of techniques that allow a system to learn and act on behalf of an individual without it being overtly taught. Programs that autonomously collect data using sophisticated algorithms are made using ML.

13.6.3 SUPERVISED AND UNSUPERVISED ALGORITHMS

The most common type of ML involves large volumes of tagged data that are analysed and learned by algorithms. Image classification is one of the tasks that the algorithm must accurately be able to perform by learning from and analysing labelled data. These models are taught on datasets, then, when given test data, their accuracy in foretelling what an image will reveal is assessed. The percentage of accuracy of their forecasts is displayed. The administrator evaluates these solutions, tries to correct and re-teach all deficiencies, in order to improve the model's learning and the application's accuracy.

Another kind of ML is unsupervised learning, which does not need expensively obtained data, where the resulting data are in fact predefined for each training Predefined. Unsupervised learning algorithms evolve by categorising and emphasising patterns within data on their own, instead of relying on user reviews. Segmentation is a methodology for classifying the data.

Algorithms are trained to group together unlabelled datasets, possibly revealing hidden patterns that weren't immediately apparent [14].

13.6.4 DEEP LEARNING

Data analytics also includes deep convolutional neural networks, a branch of cognitive computing. Deep learning techniques use multi-layer neural networks with autonomous input feature extraction to make sense of structured and unstructured data instead of human retrieval. Deep learning models, which can be more accurate but take time to train due to the significantly bigger neural network models, are created by such networks using ML. A different kind of training is referred to as semi-supervised learning, which assesses how the methods may enhance their efficiency by just using data samples and much more untagged information. Although there are many more, these are the data science algorithms that are most frequently employed.

13.6.5 Visualisation of Data

Visualisation of data is the final aspect within predictive analytics. It's also characterised as a graphic illustration or visualisation of data or outcomes. Various data visualisation approaches are employed depending on the type of data. Heterogeneous actual data, which is more common, can be displayed using graphs and charts with a variety of distinct factors, such as bar or pie charts. Geospatial data can be seen using distribution maps, cluster maps, as well as common maps. These data are also gathered from the ground using GPS data [14].

13.7   DIGITAL TWIN FRAMEWORK FOR GREEN SMART CITIES

There are several traits that will change the product and process design, manufacturing scheduling and management, logistics and supplier collaboration, and information exchange. To determine widespread discrepancies through virtualisation of the physical world, a more thorough analysis is needed. According to this study, highly urbanised areas result in a process resembling that of invasion biology. Both commercial and residential projects can benefit from using green building design concepts. To lessen the negative effects on the environment and gradually conserve energy, water, and money, many architects choose to develop green commercial buildings. Certain technologies, like data science, the cloud, and ML, are used, based on the type of project. Digital twins can aid existing manufacturing processes in a line of cutting-edge replications built on real-world data sensed through IoT devices. Historic and real-time data, through data analytics, may perform the whole method and set it up on the go.

13.8   CONCLUSION

There are two additional issues that are receiving increased interest in digital twins: healthcare and smart cities. This chapter has contributed to a thorough analysis that considers not only manufacturing but also healthcare and smart cities. It has described how researchers are developing digital twins in each component, highlighting challenges, and significantly enabling technologies that will aid in research considerations. The use of real-time data by city residents enables them to learn about, automate, and better comprehend urban activities. This study has provided a methodological framework and a top city idea for digital twin modelling. The approach's logical structure supports decision-making during cities' digital transition by providing a virtual environment for what-if scenario analysis.

The power of ML algorithms and other AI techniques and technologies are antici-pated to lead to the discovery of new information about the city, which may then be utilised to guide appropriate decisions. Decisions influence behaviours and activities in the physical environment. Big data analysis and AI technologies must be applied to construct a flood expertise map. Inference and knowledge discovery based on dynamic monitoring of flood disasters and flood big data will be made easier because of the creation of a flood big data knowledge map. Real-time modelling of flood data in urban scenes, based on normalised and dynamic flood monitoring and a flood knowledge map, can be supplied as part of a smart city flood strategy.

13.9 FUTURE SCOPE

Digital twin technology helps to create virtual replicas of cities or individual build-ings, infrastructure, and transportation systems to optimise performance. Some of the futuristic trustworthy applications that will rely on digital twin technology are energy efficient building through mapping the necessity of renewable energy sources, mak-ing transportation and the logistics system free of congestion and emissions, sup-port waste management through increasing recycling rates, and provide green space planning such as garden and park placement and plantation planning in the city for reducing the heatwave effect and improving the air quality.

REFERENCES

[1] Daniel, A. U., Sunmola, F. T., & Khoudian, P. (2017). "A systematic literature review on visibility in in sustainable supply chain ecosystems," in *Proceedings of the 2017 International Conference on Industrial Engineering and Operations Management (IEOM)*, Bristol, UK, July 24–25, 2017.

[2] Viitanen, J., & Kingston, R. (2014). Smart cities and green growth: Outsourcing demo-cratic and environmental resilience to the global technology sector. *Environment and Planning A: Economy and Space*, 46(4), 803–819. https://doi.org/10.1068/a46242

[3] Mehrotra, S., & Dhande, R. (2015). "Smart cities and smart homes: From realization to reality," in *2015 International Conference on Green Computing and Internet of Things (ICGCIoT)*, Greater Noida, India, 2015, pp. 1236–1239. https://doi.org/10.1109/ICGCIoT.2015.7380652

[4] Hou, W., Ning, Z., & Guo, L. (2018). Green survivable collaborative edge computing in smart cities. *IEEE Transactions on Industrial Informatics*, 14, 1594–1605.

[5] Deng, T., Zhang, K., & (Max) Shen, Z.-J. (2021). A systematic review of a digital twin city: A new pattern of urban governance toward smart cities. *Journal of Management Science and Engineering*, 6(2), 125–134, ISSN 2096-2320. https://doi.org/10.1016/j.jmse.2021.03.003

[6] Duarte, S., & Cruz-Machado, V. (2017, July). "Exploring linkages between lean and green supply chain and the industry 4.0," in *Proceedings of the International Conference on Management Science and Engineering Management*, Springer, Cham, July 2017, pp. 1242–1252.

[7] Zheng, Y., Yang, S., & Cheng, H. (2018). An application framework of digital twin and its case study. *Journal of Ambient Intelligence and Humanized Computing*, 10(3), 1141–1153.

[8] Liu, M., Fang, S., Dong, S., & Xu, C. (2021). Review of digital twin about concepts, technologies, and industrial applications. *Journal of Manufacturing Systems*, 58, 346–361. https://doi.org/10.1016/j.jmsy.2020.06.017

[9] Ang, X., Wang, H., Liu, G. et al. (2022). Industry application of digital twin: From concept to implementation. *The International Journal of Advanced Manufacturing Technology*, 121, 4289–4312. https://doi.org/10.1007/s00170-022-09632-z

[10] Srivastava, A., Gupta, M. S., & Kaur, G. (2019). "Green smart cities," in *Green and Smart Technologies for Smart Cities*. CRC Press, pp. 1–18.

[11] Ghobakhloo, M., & Fathi, M. (2019). Corporate survival in Industry 4.0 era: The enabling role of lean-digitized manufacturing. *Journal of Manufacturing Technology Management*.

[12] Deren, L., Wenbo, Y., & Zhenfeng, S. (2021). Smart city based on digital twins. *Computational Urban Science*, 1(1), 1–11.

[13] Manavalan, E., & Jayakrishna, K. (2019). A review of internet of things (IoT) embedded sustainable supply chain for Industry 4.0 requirements. *Computers & Industrial Engineering*, 127, 925–953.

[14] Fuller, A., Fan, Z., Day, C., & Barlow, C., (2020). Digital twin: Enabling technologies, challenges and open research. *IEEE Access*, 8, 108952–108971. doi: 10.1109/ACCESS.20

14 Challenges and Opportunities in Green Smart City Adoption

Sk Md Abidar Rahaman and Md Azharuddin
Aliah University, Kolkata, India

14.1 INTRODUCTION

On a global scale, governmental regulations and academic studies focus on the development of smart cities. The Internet of Things (IoT) has transformed society and transformed our way of life by enabling smart cities. However, IoT technologies are not without their challenges. They result in higher energy consumption, harmful pollutants, and e-waste production in smart cities. Green IoT is necessary since applications for smart cities must be eco-friendly. Eco-friendly environments are more sustainable for smart cities due to green IoT. The approaches and tactics for lowering the risks associated with pollution, traffic waste, resource utilization, and energy consumption, as well as for ensuring public safety, a high standard of living, and environmental sustainability, must all be addressed.

Numerous smart city applications are now possible because of the quick development of communication and sensor technology, which improves our standard of living [1]. The IoT is a term used to describe this connectedness between objects in the smart city [2, 3]. In smart cities, everything is connected to the IoT at all times, anywhere, and on any channel. Due to adaptive communication networks, processing, and analysis, IoT components are becoming smarter. Mobile phones, drones, radio frequency identification (RFID) tags, sensors, actuators, and cameras are all examples of IoT devices. They may all potentially interact with one another and cooperate to accomplish shared objectives [4]. Using IoT devices, real-time monitoring applications can be conducted through a variety of components and devices [5–12]. The IoT also makes it possible for software agents to share knowledge, make decisions together, and perform jobs in the best way possible. By using cutting-edge communication technologies, the IoT gathers and distributes huge volumes of data, which can be evaluated for wise decision-making [13, 14]. In order for the IoT to be widely adopted, cloud computing, big data storage, and high transmission bandwidth are required. Data analysis and transmission are energy-intensive tasks for IoT devices. Artificial intelligence (AI) is enabling smart cities to become smarter than ever before [15–17].

DOI: 10.1201/9781003388814-14

The IoT has a significant impact on the creation of smart cities in a number of ways. Green IoT is a crucial technology to reduce carbon emissions and power consumption in order to achieve the goals of smart cities and sustainability. Energy use rises as there are more IoT devices on the market. Such devices are being introduced to reduce resource and energy usage [18].

In the 1990s, the term "smart city" was initially used to highlight the benefits of information and communication technology (ICT) for network improvements and enhanced infrastructure. The increasing adoption of information technology also helps cities develop vital services for governance, delivery, health, and safety.

These are the main areas of attention for green smart cities:

- Utilizing IoT technologies for green ICT;
- ICT has a significant impact on the creation of sustainable smart cities by reducing energy consumption and CO_2 emissions;
- In smart cities, several methods and tactics are employed to minimize resource use, CO_2 emissions, and energy use;
- Methods for waste management to enhance smart cities;
- Innovative methods are employed to sustain smart cities.

14.2 MOTIVATION

Large cities are being burdened by the inflow of individuals from rural regions moving quickly into metropolitan areas. Local governments and city councils are having trouble meeting the basic demands of large populations. Smart cities are motivated by climate change; technological advances in renewable energy could meet the needs of their residents without harming the environment. Currently, technological innovation has played a key role in advancing economic and social development [19]. In order to optimize efficiency in all city management operations while utilizing limited energy, these sophisticated cities must leverage new technology to enhance their whole systems. Economies and social development are transmitted mainly through cities, which are the primary hubs for innovation. Innovation in science and technology is therefore playing an increasingly important role. Future city expansion will be driven primarily by advances in urban development. Due to their recent commercial success, the IoT and AI have emerged as burning research topics, and they have proven superior in several fields, including surveillance, automation of factories, asset management, and waste management. The present information systems may be intelligently upgraded by fusing the IoT and AI.

14.3 RELATED WORK

Technological revolutions can frequently result from global economic crises. For instance, the global economic downturn of 1857 sparked the first technical revolution, the one of 1929 sparked the second, and the one of 1987 sparked the information technology revolution [19]. With the landmark IBM study "Smart Planet," smart cities have emerged as a global hot subject since the 2008 financial crisis [20].

A smart city, or digital city, is created by leveraging next-generation information technology. Smart cities have yet to fully take hold in the modern world. ICT and the IoT are, however, firmly embedded in smart cities, as is evident from their transdisciplinary growth [21]. In smart cities, resources are utilized more efficiently, life quality is improved, and city management is optimized. Intelligent transportation, security, energy, buildings, education, health, and other areas may all benefit from the IoT and AI [22]. Numerous sustainability metrics, including those for social and economic sustainability, are included in the smart city framework, and IoT and AI technologies also support sustainability. By 2030, more than 60% of the global population will be living in cities [23, 24].

Globally, the movement of people into big cities has become commonplace. For instance, there are already more than 20 million people living in Beijing, Shanghai, and Shenzhen [25]. Urban resources and services are subject to several effects and problems due to the massive population increase. An examination of news articles and official records reveals that China has been implementing more smart city pilot projects in recent years.

Information technology must be used to support real-time management, forecasting, and monitoring for a city to be deemed "smart." The integration of IoT sensors and AI algorithms can replace traditional management methods. Smart cities are no longer limited to intelligence as technology has advanced and citizens are becoming more aware of them, so the term "smart" is now more inclusive of all aspects of user participation. The type of engagement includes decision-making in addition to public participation. Decisions must be made efficiently and quickly, frequently relying on real-time information and changes in citizen behavior. Top-down or bottom-up methods, community or technology leadership, and intentional design or natural development are all used in decision-making. In well-run cities, residents' judgments are influenced by a variety of tools that help them make regular decisions [26, 27].

Cloud computing technologies enable computers to resolve problems previously caused by lack of mathematical power. Due to the development of cloud computing, many areas, such as the study of smart cities, are paying attention to AI algorithms. The majority of the time, cloud servers are used to deliver and manage software, apps, and plugins. The market leader in public cloud computing for Infrastructure as a Service is Amazon Web Services (AWS). However, the top players in the public cloud computing sector for Platform as a Service are Oracle, AWS, Microsoft, and IBM. Smart cities can be categorized into three layers based on their functionality: perception, network, and applications. During the development of smart cities, Cohen suggests three stages: 1.0 technology-driven, 2.0 technology-enabled, and 3.0 citizen-created. Due to the various stages that various nations and localities are currently in, there are various levels of acceptance. It is the adoption level of an object that measures technological acceptability [28].

14.4   USING ICT IN SMART CITIES

Software, databases, smart sensors, cameras, and data centers in smart cities are the backbone of the IoT [29]. The IoT reduces expenses, reduces environmental and health effects, and conserves natural resources. The green IoT emphasizes sustainable development, use, and disposal. Increasing IoT energy efficiency and reducing CO_2 emissions are additional green IoT solutions that can be used to create a sustainable smart world with intelligent everything. Designing with green elements in mind is part of the green IoT; and IoT green design components include evolving energy efficiency, communication protocols, computing devices, and topologies for networking. By utilizing the IoT component, CO_2 emissions can be decreased and energy efficiency can be improved. In order to create an intelligent decision-making model, smart cities must first collect data from their surroundings [30].

To make smart cities more sustainable and hospitable, ICT is essential. ICT can interact with city services, lowering costs, using fewer resources, and producing less pollution. The standard of living may also be raised. ICT is therefore required for smart cities; it also enhances them by automating, streamlining, enabling IoT, identifying security threats, and scaling. Additionally, ICT technology helps slow down global climate change as its applications become more energy-efficient as a result of environmental consciousness. "Greening IoT" refers to cutting-edge technologies that enable users to collect, access, store, and manage various types of information by utilizing infrastructure and storage [31, 32].

Information gathering, access, management, and storage are all made possible by green ICT. The IoT is made more environmentally friendly by ICT, which also helps society by reducing the energy required for creating, producing, and distributing ICT hardware and devices. Researchers state how implementing green IoT principles could enhance life quality, the economy, and the environment. ICT has the numerous benefits of minimizing how negatively modern technology affects society, human health, and the environment. Through methods for sharing infrastructure, ICT may manage data center optimization for sustainability, resulting in lower CO_2 emissions and material disposal for e-waste [33–35].

Reducing the negative effects of the IoT on the environment (such as CO_2 emissions, NO_2 emissions, and other pollution) is a step toward greening the IoT. To help the environment, the energy consumption of IoT devices must be reduced. In addition, as IoT technology advances, greening ICT technologies contribute to economic development and environmental sustainability, making the world greener and smarter [36] (Figure 14.1).

14.5   SMART CITIES NEED INTELLIGENT DATA CENTERS

Data centers are repositories and technological tools which can be used for managing smart cities, storing collected data, and disseminating it. Many IoT devices require constant internet access in the smart city. Without the data center, it would be impossible to manage the data and turn it into information. Yet, due to the fact that such a center handles many types of data from various applications, it has high operating expenses, high energy consumption, and a significant CO_2 footprint. In addition to mobile devices, actuators, sensors, RFID, and so on, a wide range of technologies are now contributing to the generation of big data. Researchers have discussed a number of methods to increase the data center's energy efficiency, such as designing novel energy-efficient data center architectures and utilizing data center power models and renewable energy [37, 38].

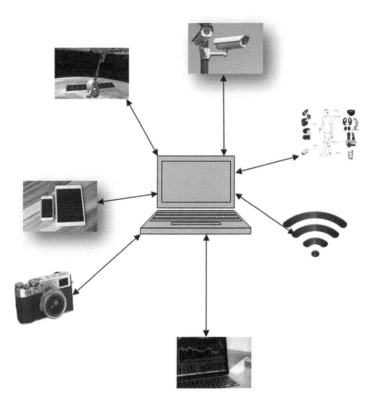

FIGURE 14.1   Different ICT applications in a smart city.

The future of data centers will be influenced by availability and sustainability. To manage the massive amounts of data produced by sensors positioned around the city, a data center with a significant processing capacity must be located in a smart city. To enhance the technological infrastructure and save costs, big data processing needs communication networks, virtualization systems, and storage access. Here, the intelligent data center will effectively and efficiently govern the intelligent cities. Consequently, smart data centers serve as a central part of smart cities, providing access security, passive sensing, environmental protection, and facilitating sustainable development for urban growth [39–41]. A smart data center will also be able to manage and coordinate the resources needed by smart cities. In addition, they will oversee other services, such as waste recycling, public transportation, and monitoring emissions and pollution. One example is monitoring and managing energy from renewable resources, safety, and health, among others. The development of new technology and architectural designs for managing smart cities will contribute to raising the standard of living for people [42, 43].

14.6   CLOUD COMPUTING FOR SMART CITIES

In order for smart cities to be successful, cloud computing must be incorporated into their physical infrastructure. Smart city implementation requires the cooperation of a distributed open-source network and a decentralized cloud. Application development for these cities requires cloud computing services, which is thus essential for managing the huge amounts of heterogeneous data generated by a variety of devices. The terms "smart cities" and "high quality of life" are used to describe the management of natural resources and economic growth. Smart city applications require many amenities, including police transportation, security, smart parking, public safety, electricity, and internet connectivity.

A variety of architectures, protocols, technologies, and algorithms were introduced in Deep et al.'s introduction to the IoT in cloud computing [44]. Zhu et al. [45] proposed a multi-method data distribution solution for inexpensive costs, sensor cloud (SC) users, and fast delivery. Data delivery via multiple methods involves four types of transportation: wireless sensor networks, cloudlets, clouds, and SC users. The primary concept behind green cloud computing is minimizing utility power [46].

Both public and private clouds require energy to process, switch, transmit, and store data [47]. There are several ways to use green cloud computing, including adopting energy-saving software and hardware, using virtual machine techniques, using various mechanisms for allocating resources in an energy-efficient manner, and performing related chores [48, 49].

To improve efficiency in power cost and server utilization with regard to green cloud computing, different correlations and similarities are examined [50]. A virtual machine (VM) scheduling technique, however, is crucial for green cloud computing, which reduces energy consumption [51, 52]. A machine algorithm is used in [51] for host load movement. Algorithm use consequently led to better power usage.

IoT and cloud computing integration is extremely beneficial for applications in healthcare, such as disease forecasting in smart cities [53]. Particle swarm optimization, a smart model for healthcare services in smart cities, has been presented in [54]. The suggested model resolves task scheduling, speeds up the use of medical resources, and reduces the amount of time needed to respond to medical requests. Combining AI, cloud computing, and the IoT has both financial benefits and drawbacks, as has been examined in [55].

Costs associated with CO_2 and radiation exposure were significantly decreased as a result of the authors' proposed genetic algorithm optimization for network planning in [56]. Methods for boosting data rates and lowering CO_2 emissions in cognitive wireless sensor networks (WSNs) were described in a study in [57]. In addition to the work of the authors in [56, 57], Chan et al. [58] presented a number of models to evaluate the use-phase power consumption and CO_2 emissions of wireless communications networks. A vehicular ad hoc network could reduce energy consumption [59].

Using theories, applications, and technological advancements, [60] examines the energy efficiency of mobile communication networks based on 5G. Also, Abrol and Jha [61] analyzed the impact of applying technologies that enable energy efficiency in next generation networks' (NGNs). Increased capacity, higher data rates, and improved NGN quality of service (QoS) require energy efficiency and a reduction in CO_2 emissions.

A utility-based adaptive duty cycle algorithm was also recommended [62] for reducing delay and improving energy efficiency. To cut down on waiting time and extend the time for reliable service, HTTP was used [63].

The goal of 5G is to utilize less energy, which leads to communication. By 2020, a green and healthy lifestyle will be achieved through all communication tools and objects. In order to increase communication dependability and QoS between machines and people, 5G technology is crucial. Additionally, 5G technology facilitates greater connection over a wider region, lowers latency, uses less energy, and offers faster data rates. The areas of robotic communication, eHealth, human–robot interaction, media, transportation, logistics, e-learning, e-governance, automotive, and so on are some of the services that 5G will provide for our society [64–67].

Energy harvesting methods and energy-efficient technologies have been applied in several ways, as mentioned in [68]. An energy reduction strategy was proposed by Wang et al. [69]. Orthogonal frequency division multiple access relay stations with sub-carriers were employed to optimize power efficiency. However, in order to increase communication network collaboration and energy efficiency, an energy-efficient incentive resource allocation strategy was used [70]. To increase energy efficiency and effectiveness, this method combines genetics and water droplets. Various research studies on energy harvesting, such as [71–73], concentrate on green energy. Green cognitive radio networks were studied in [73] for resource allocation strategies aimed at maximizing energy efficiency. Ge et al. [72] also described a cognitive radio network that uses multiple-input, single-output technologies to reduce the power of the information signal. A study of IEEE 802.11ah by Zheng et al. [73] led to the introduction of analysis of the smart grid's power consumption and performance. In terms of energy-efficiency measurements, the authors [74] offered various strategies for greening communication networks. The power consumption of network equipment has been calculated accurately and transparently [75]. Yang et al. [76] made a distinction between renewable and non-renewable energy. Additionally, Hoque et al. [77] looked at ways to improve the energy efficiency of portable electronic gadgets.

Through the internet, cloud computing offers limitless storage and delivery of computational services. The linking of various devices – including tablets, cameras, laptops, mobile phones, – shares data over the cloud. Several studies on the integration of cloud computing and the IoT have been conducted. The goal of cloud computing is to promote easily recyclable and reusable eco-friendly products. In order to increase energy consumption, decrease hazardous materials, and improve the capacity to recycle obsolete items, effective cloud computing is essential.

It may be simpler to store data in huge quantities, process that data, and analyze it in real time if cloud computing is available in smart cities. With cloud computing, computer resources can be shared on demand and expanded as needed, resulting in faster performance, sharper decision-making, and cost reductions. The IoT and cloud computing combination is crucial for healthcare applications like illness prediction in smart cities.

14.7 COMMUNICATION NETWORKS FOR SMART CITIES

WSNs were created as a result of wireless communication and sensing. WSNs have been used for a number of activities, including identifying fires [78–80], object monitoring [81–83], environmental monitoring [84–88], adapting to changing military requirements, maintaining machine health, and keeping an eye on industrial processes.

Local and global environmental variables can be monitored by sensor nodes (SNs), such as pollution, weather, and agricultural land. Additionally, SNs relay sensory data to the closest base station using wireless communication channels and ad hoc technologies. The inventors in [89, 90] proposed a sleep mode to promote green IoT and save sensor power over long periods of time. According to Khalil and Zaidi [91], the route of WSN energy conservation involves keeping the nearest SN active (transmitting and receiving data) while the other nodes remain in sleep mode and sense passively. Consequently, when an SN wants to transfer information, it wakes up all the SNs around its roots first.

As a result, all nodes will be put into sleep mode after data transmission is complete. The environment itself can provide energy to sensors in the form of sunlight, vibrations, kinetic energy, temperature differences, and so on [92–97]. Because energy harvesting is less expensive and has a longer battery life, combining WSNs with energy-gathering technologies is also important in the green world [98]. The selection of SNs [99], the overheads related to context-aware sensing [100], and the scheduling of sleep time [101] are some of the strategies that enable sensor networks to become green IoT devices. These techniques also help conserve energy and reduce communication delays between SNs.

A routing protocol developed by Mehmood and Song [102] promotes energy efficiency; and the energy efficiency of building hierarchical networks was also considered by Rekha et al. [103]. In [104], the authors propose a green WSN that improves system budgets, lowers relay SNs, increases energy efficiency, and extends network lifetimes to allow the greening of the IoT.

As part of their work on greening WSNs, the authors of [105] also investigated a cooperative energy-saving strategy. A multi-hop is used as a communication relay station in this cooperative strategy. Further, network resilience provisioning is explored along with energy consumption. Fog computing systems can use green WSNs [106]. This work was carried out in four steps, namely the development of: a hierarchical localization of sensor/actuator SNs, a system framework clustering of SNs, an optimization model to achieve a green IoT, and finally the discovery of the energy-efficient route. The proposed strategy proved to be adaptable, economical, and energy-efficient. Also, it applies to smart farm and smart city applications using the IoT.

In their study, Mahapatra et al. [107] presented ways to improve green WSN efficiency and lower CO_2 emissions by using different techniques. The required high power consumption and CO_2 emissions, both of which have a significant negative impact on the environment, are also decreased when WSN and cloud computing are combined [108]. A balanced tree-based WSN minimizes energy usage while maximizing the network lifespan of sensor nodes [109]. Araujo et al. [110] suggested cognitive WSNs as a way to significantly reduce power consumption. Three scenarios were used to show and assess their work, allowing for the creation of green procedures and power-saving measures for cognitive WSNs. The employment of sleep and active SNs to cut down on energy use, energy depletion, radio technology optimization, data compression mechanisms, power efficient routing schemes, and mixed transmission methods to boost lifespan reliability are possible techniques for green WSNs [111].

Smart cities are currently plagued by a number of problems, including traffic and pollution. To reduce the negative environmental impacts of sustainable development, more mobility is needed. The authors in [112] introduced linked and autonomous cars as well as smart mobility, and they also covered the problems with smart cities [113]. Among the benefits of mobility for improving smart city sustainability are increased public safety, decreased noise and pollution, decreased traffic, and decreased transfer costs. In addition, [114] covered how information exchanged via the IoT contributes to a network of sustainable value chains.

A genetic algorithm was suggested by the authors of [115] for improving IoT devices that use drones. In addition, Mozaffari et al. [116] investigated the optimum height for tiny drone cells to increase coverage area and decrease transmission power. The main function of IoT equipment is the processing it does on each machine. Data is collected, processed, analyzed, managed, stored, and sent to the cloud using drone-equipped IoT devices. Using drones for WSNs has been discussed in [117]. The drone sink, fixed-group leaders, and sensor nodes make up the drone and WSN structures. The results showed that the energy use decreased. Data collection methods using drone-based WSNs were covered [118].

The IoT may be made more environmentally friendly by using greener wireless communication technology. Green communications are those that are ecologically conscious, energy-efficient, and sustainable. Low CO_2 emissions, low radiation exposure, and minimal energy use are all considered to be characteristics of "green" communication networks.

The goal of 5G is to utilize less energy, which leads to communication that is green and beneficial for the environment. In order to increase communication dependability and QoS between machines and people, 5G technology is crucial. Additionally, 5G technology facilitates greater connection over a wider region, lowers latency, uses less energy, and offers faster data rates. Among the benefits of 5G, we find e-health, robotics communication, media, transportation and logistics, e-learning, and e-governance (Figure 14.2).

14.8   EFFICIENT ENERGY FOR SMART CITIES

Drones are crucial to the IoT's transition to sustainability. The power consumption of IoT devices has decreased since they offer effective energy use. IoT devices require significant transmission power for long-distance data transfer. In order to gather, process, and deliver data to a different device located elsewhere, the drone may approach IoT devices. Drone-assisted IoT devices are improved using genetic algorithms.

The methods employed might cut down on flight time, energy consumption, and data collection delays. In [119], the authors presented an algorithm for WSN data collection employing mobile agents and drones. Drones and mobile agents are employed to reduce SNs' energy usage and processing time. A mathematical model for IoT device energy efficiency was also created by Zorbas et al. [120]. The performance of the created model minimizes power usage while detecting events that have taken place on the ground in the coverage region. Sharma et al.'s [121] introduction of drone collaboration with WSNs for a better lifespan is another noteworthy development.

FIGURE 14.2 IoT applications in smart cities.

Energy-efficient components in new technologies have been discovered to improve energy efficiency and reduce the power required for drones [122]. Choi et al. [123] developed an energy-efficient relaying method based on traffic load and speed variables. The connected drone docking system was developed to perform several tasks with the help of drones and IoT devices to reduce waste, lower energy consumption, and ensure transmission security [124]. Using beacons and drones for IoT monitoring, security platforms, and emergency response in buildings was performed by Seo et al. [125]. The authors of [126] created a system for drone batteries that automatically replaces them after a certain amount of time. Drones were equipped with automated batteries so that there would be no need for manual battery changing.

For packet transmission, choosing the shortest path is crucial for energy conservation and good efficiency. On the basis of wind turbines, the Energy 4.0 fault diagnostic framework was proposed [127]. Intelligent route optimization is suggested to increase network usage and provide the shortest routing path in order to increase WSN efficiency [128]. Improved network performance results in a significant increase in traffic load and usage rate. Mahapatra et al. [129] investigated energy management in smart homes to develop environmentally friendly and sustainable smart cities. The authors also suggested neural network (NN)-based Q-learning for managing energy in Canadian households by reducing peak loads.

The most important step in creating apps for smart cities is big data analytics. With IoT gadgets connected to the existing infrastructure, smart cities can improve the quality of life. Thus, the authors in [130] developed a new protocol called QoS-IoT to improve energy efficiency and speed up the collection of massive amounts of sensor data.

TABLE 14.1

Methods and Strategies for Creating Energy-Efficient Smart Cities

Strategies	Features
Cloud computing and sensors	Reduce the use of computer resources that have negative consequences for the environment.
Switching of balanced tree nodes	Balancing the energy usage among SNs. Extending the lifetime of a network.
Hybrid transmission protocol	Conserve energy around the sinks. Increase network reliability and latency.
WSN that uses energy harvesting	Give WSN nodes an extended lifespan. Green world.
Duty cycling	Save energy.
Cluster technique	Conserve energy for green WSNs.
Intelligent transportation	Increasing network capacity. Decrease congestion. Boost security. Decrease the pollutants.
Self-healing depending on events and application profile.	Usage of energy. Provisioning for network resiliency.

The main function of IoT hardware is processing in each machine. Data is gathered, processed, analyzed, managed, stored, and transported to the cloud using IoT drones.

Table 14.1 outlines the methods and strategies used for creating energy-efficient smart cities.

14.9   HARMFUL POLLUTION REDUCTION IN SMART CITIES

Environmental, social, and economic benefits have risen sharply from monitoring air pollution.

For the purpose of monitoring pollution, smart sensors are used. For transferring data in real time, its transmission power is nonetheless constrained. Drones can carry SNs, which are capable of collecting and transmitting data in real time. Climate change has effects, one of which is air pollution. But drone technology is currently the most crucial for air pollution monitoring, so as to enhance the standard of living in smart cities. It is used in a variety of situations to track and forecast air pollution.

As a result, Villa et al.'s [131] solution for mounting gas SNs inside a hexacopter is described. The writers demonstrate how a built-in drone system could recognize emissions coming from certain sources. CO_2, NO_2, and other gas sensors are examined for tracking pollution emissions in a specific area, focusing on airflow behavior. Potential drone uses include monitoring crops remotely, detecting soil moisture, checking the quality of the water, keeping an eye on infrastructure, and deploying sensors remotely [132]. In order to control greenhouse gas emissions, it is also important to take greenhouse pollution into account. A solar-powered drone with CO_2 sensors and a WSN was introduced by Hamilton and Magdalene [133].

The authors of [134] suggested using drones to remotely monitor food quality and safety. A number of applications have been found to be suitable for dynamic and flexible deployment, with air pollution monitoring being one of them [135, 136]. The methods currently being used for drone monitoring applications were examined by the author in [137]. Additionally, the authors of [138] suggested drones with readily available sensors for tracking activities, but they disregarded the guidance system. A few researchers have recommended using a drone control system that is based on pollution to address this problem. This was built on the particle swarm optimization (PSO) approach and the chemotaxis metaheuristic, which monitor certain locations in the most contaminated zones [139]. The authors in [140] suggested using a Raspberry Pi to store and sense data on environmental pollution and a drone fitted out with a Pixhawk Autopilot to operate it.

A useful drone platform model was also created by the authors in [141] to monitor a variety of air contaminants. Additionally, Šmídl and Hofman [142] proposed the concept of autonomous drone navigation for pollution monitoring. Drone platforms are not recommended for air pollution control, according to these authors. They focused on monitoring for air pollution events and the profile of roadside air pollution. In southwest China, Zang et al. [143] used drones to investigate water contamination. Furthermore, [144] examined the projection of the carbon footprint in the ICT industries.

One effect of climate change is air pollution. Drone technology is now the most important technology for monitoring air pollution in smart cities. In several scenarios, it is used to track and forecast air pollution.

14.10 MANAGING WASTE IN SMART CITIES

Smarter and greener are the two main objectives of smart cities. This is why businesses and governments are looking for ways to increase the collection of waste through intelligent approaches and smart equipment such as smart sensors, cloud platforms, and the IoT. Through IoT-connected SNs, trashcan contents are tracked and transmitted to the cloud for processing and archiving. Through the optimization process, garbage collection may be effectively and dynamically managed, based on data gathered from workers. The effectiveness of garbage collection techniques is increased in real time by ensuring that workers collect dustbins when they are full, so reducing waste overflow. The linked sensors included in the container have made it possible for smart cities to install garbage monitoring and management systems.

Smarter, healthier, and greener cities can all be the result of a comprehensive system. The smart waste management (SWM) system therefore facilitates information processing and decision-making, ensures that employers adhere to policies, and enhances the provision of garbage collection services [145]. In public universities like Oradea University, the SWM technique was studied and greatly improved [146]. The Oradea University system was created to promote recycling, reduce pollution, and safeguard the environment. Additionally, the authors of [147] discussed the circular economy in Italy and ICT applications for smart management in Europe.

In order to increase efficiency and improve environmental protection, waste management is essential for sustainable smart cities. Numerous technologies exist to manage garbage, including the rate of recycling, route optimization, autonomous garbage pickup, and clean energy. When it comes to the automated collection of garbage, IoT gadgets like sensors create alerts if the container fills up and has to be emptied, or needs to be serviced, effectively managing the garbage. Smart in-vehicle monitoring speeds up waste removal and guarantees driver security. It also expedites the waste disposal method and ensures driver safety. Hence an efficient solution for waste management may be provided by the IoT.

The authors of [148] also emphasized the need for more automated garbage collection systems, as well as increased productivity and capacity. They assessed how well the technology interacted with the smart city's infrastructure. Here, the IoT made it possible to monitor and gather data in real time while also connecting to the automated garbage collection system's cloud. The IoT improves system performance by connecting devices, and processing and analyzing data in real time. The various kinds of waste in the containers may therefore be continuously monitored by the suggested system, which aids in the provision of information on the total amount of garbage collected.

The Italian legal framework for eco-friendly waste management in smart cities are presented by the authors of [149]. The waste management strategy should take the risk level and reduced trash quantity into account for sustainable growth in smart cities.

While also improving the environmental impact, productivity, safety, and efficiency of automated waste systems, modern technologies like AI and the IoT have significantly decreased the cost and complexity of these systems. Because of the potential health risks, disposed garbage is hazardous (Figure 14.3).

FIGURE 14.3 Green smart city.

14.11 SMART CITY SUSTAINABILITY

Sustainable, environmentally friendly smart cities require urban planning to ensure human survival. The maintenance of everyone's health and well-being is an important consideration. From human rights to garbage disposal, the areas are interconnected. Low obesity rates are followed by favorable mental health outcomes for people. Human health and welfare are directly impacted by the layout and design of sustainable green cities. Ecological resources are inspected and preserved, and the environment greatly benefits from smart networking and ecologically friendly environments. These technological applications aim to improve not just human health but also that of plants, animals, and trees. Energy-saving techniques are essential for a green, sustainable city. Efficient and environmentally friendly disposal methods aid in preventing the disastrous problem of greenhouse gas emissions.

Water and food also play a part in the creation of sustainable smart cities. Clean water accessibility is necessary for their growth. The interaction between the government, the population, the environment, ecosystems, infrastructure, and the use of resources is greatly influenced by modern technology. As a result, technological and social advancements are influenced by sustainable and green cities, which also refer to open spaces and cutting-edge agricultural resources. To maintain the city as smart and green as possible, renewable resources, a smaller ecological imprint, and less pollution are required. The IoT is essential for making smart cities more livable, resilient, environmentally friendly, and sustainable.

The research in [150] reveals an investigation into the role played by smart cities in creating sustainable urban environments. Green energy, renewable energy, energy efficiency, and water quality are the main areas of focus. For smart cities to be sustainable and eco-friendly, the green IoT is essential. With smart IoT technologies, large amounts of data can easily be analyzed, improving the safety, intelligence, and sustainability of smart cities significantly. By reducing pollution and making better use of the resources that are already available, the authors of [151] investigated how big data can improve living conditions.

The authors of [152] present streaming delay tolerance, bandwidth, and energy approaches for green smart cities. The authors of [153] also write about a green IoT ecosystem that is sustainable. The IoT can be used to improve the environment, living standards, and economy while reducing adverse effects on the environment. The authors proposed this strategy for environmentally friendly, smart city technology.

Smart cities are built on an intelligent foundation and a sophisticated system of pervasive networks and things [154]. The data that are collected from the cloud of smart cities are handled and reviewed properly, enabling real-time action transformation and decision-making from the information available.

Smart IoT approaches and technologies perform well in a large data analysis, considerably improving the safety, intelligence, and sustainability of smart cities.

A smart, sustainable city uses ICT to increase the effectiveness of urban services and operations. The management of energy resources, sustainability, sharing, and the applications of developing technology require additional discussion as more people want to live in urban and smart cities. Additionally, meeting needs is crucial, including increasing people's quality of life and resource management, expanding commercial possibilities, and having a positive influence on the environment.

14.12   OBSTACLES TO THE ADOPTION OF GREEN SMART CITIES

The various forms and major obstacles to the creation of smart cities are discussed below.

14.12.1   LACK OF INFRASTRUCTURE

Modern, highly developed infrastructure is required for smart cities, and every piece of machinery has to be online for monitoring. Associated IoT devices gather information from the physical environment to help decision-makers improve urban services for residents. Due to population expansion, there is an increased need for housing, healthcare facilities, educational institutions, and entertainment venues. All the residents of smart cities who will use the IoT infrastructure should have their infrastructure demands met by AI and the IoT. Sustainable development, which is closely related to the growth of infrastructure in developing countries, has a significant impact on AI. In addition, connecting to and using the old (existing) infrastructure for various reasons should be handled by AI and the IoT. IoT technology has greatly benefited the majority of the specific elements of smart city infrastructures and technologies. There are numerous economic prospects and significant development potential since IoT technologies and smart city infrastructures share many fundamental principles and ideas.

14.12.2   INSUFFICIENT FUNDS

Money and budget are important factors before beginning any project. To create, advance, and sustain smart city development, the government or local authorities need adequate funding. Lack of funding results in projects taking longer to complete, which raises project expenses once again. Projects should be prioritized using AI and IoT approaches based on their importance, need, timeframe, and other factors, and money should be distributed as efficiently as possible using the best optimization methods. The government might need to obtain more money for these smart city initiatives from private sector organizations in order to handle this problem successfully.

14.12.3   CYBER SECURITY AND DATA RISK

The interconnected networks of smart cities foster innovation, generate vast volumes of data with rich information, and link businesses, governments, and individuals. The information establishes a framework for managing cities that will increase their productivity and sustainability. A lot of issues and difficulties arise when sharing and keeping a lot of useful data. Private information about residents, official records, and data from all private entities might all be included in this. On the other hand, cyber vulnerability causes issues with data privacy and poses risks to technology used in smart cities. For instance, a leading US pipeline network operator suffered a cyber attack, forcing it to temporarily halt operations. To restore operations, the company paid the hackers a significant sum of money. New personal data privacy legislation in China adopted in 2021 requires that individuals separately approve the processing of sensitive data, such as biometrics, medical information, and location data. Furthermore, it prohibits the collection of personal data from individuals, such as biological and facial details. It could be difficult for the big IT companies to handle information without mismanagement and abuse as a result. Strong networking technology is essential for success due to the rising number of sensors and their data. Such a concept is quite unlikely to succeed without robust citywide coverage. Millions of people will live in smart cities, making it difficult for AI and the IoT to manage, analyze, and prevent cyber attacks. To address data security, trust, and user privacy in smart cities, an integrated approach is required. All parties participating in the process of creating a smart city, such as city managers, inhabitants, and the community at large, must share the challenge of cyber security and data threats, as well as the accompanying obligations. Data hazards and cyber security come in many forms, with a cyber assault being one of them. Effective action must be taken in order for smart cities to develop attack detection techniques.

14.12.4 SMART WASTE AND HYGIENE MANAGEMENT

Several manufacturing systems and municipal services are considering the implementation of these technologies in order to ensure flexibility and efficiency. For many nations, managing garbage is still a serious challenge. Waste management, from generation to disposal, is one of the most difficult challenges faced by governments around the world. Most food and other objects are now wrapped in plastic, necessitating a massive effort to handle all of the decay produced by people. Garbage collection must be finished within a certain amount of time because cities produce a lot of waste. Cities must also collect waste more efficiently. In some waste management companies, AI is employed to separate various types of waste (such as glassware, metals, paper, and plastic) without the need for human interference. The problems related to the collection, transportation, treatment, recycling, and disposal of waste should be addressed via AI and the IoT. The intelligent disposal method, for instance, was made possible by ant colony optimization technology. Smart city residents can receive high-quality services because the entire process can be centrally managed.

14.12.5 LACK OF PROFESSIONALS

IoT and AI adoption both call for highly qualified professionals. Organizations frequently misinterpret the advantages of these technologies because they lack the necessary expertise and understanding. The scarcity of experts with knowledge of various fields, including computer technology, is one of the major issues smart cities face. The use of AI and the IoT in the development of sustainable smart cities is hampered by a lack of qualified professionals. In the post-pandemic age, it will be especially crucial to recruit and adopt professionals. As a result of the pandemic, the economy of cities has suffered. Professional expertise aids in achieving objectives for the creation of smart cities.

14.12.6 MANAGING ENERGY DEMANDS

With the aid of cutting-edge computing methods, ordinary things are merged and transformed into smart devices, which then function as intelligent terminals for data transfer with other devices or a cloud server. As a result, keeping up with the smart city demands a lot of energy. The challenge of supplying a city with the energy it requires is difficult, and as more focus shifts to renewable energy sources, some cities have found it difficult to make the transition. The cost of and demand for energy, on the other hand, are rising over time. TVs, air conditioners, washing machines, smart phones, and computers contribute exponentially to domestic energy consumption. Higher energy demands are a result of both technological advancements and consumer behavior changes. As a result, energy producers are now looking to AI and the IoT for assistance in optimizing the distribution of energy. AI proposes new guidelines for activity scheduling. A number of issues must be addressed by improving the design, implementation, and manufacturing of energy infrastructure. Installing facilities producing solar power for residences is one way to address the issue of excessive energy consumption. Additionally, by adding auxiliary solar power systems, modern residences can contribute to increasing the value of fixed assets. After the sun sets, residences start to use power from the main grid as usual because solar panels do not produce electricity at night. The Mississippi Power Company and Tesla have declared their partnership in order to construct the first smart neighborhood equipped with a solar and battery unit.

14.12.7 MANAGING TRANSPORTATION

Transportation makes up an important portion of the modern economy, making up around 6–12% of the potential GDP of a nation. Even though transportation has significantly enhanced our lives, there are still a great deal of excessive issues. Due to inefficiencies, the transportation sector has grown to be the second-largest carbon emitter. Petroleum, diesel, and other products made from crude oil are significantly used in modern transportation technologies. Electric vehicles are a viable option for reducing emissions. An electric motor and a battery pack in electric cars provide the power required for propulsion. Electric vehicles can be charged if existing fuel stations are converted to hybrid models that offer both petroleum products and e-charging stations. It is crucial to plan a city to eliminate the need for everyday public transit that may fill in the gaps and shorten commute times for inhabitants in order to avoid traffic congestion. Smart cities are based on smart transportation. Without a dependable and effective transportation infrastructure, there can be no smart city. Because of this requirement, intelligent transportation systems are an essential part of any smart city idea. This has an impact on both the environment and intelligent transportation. A key feature of AI is machine learning, which is able to identify the root causes of traffic jams and accident hotspots as well as the best preventative measures. In addition to smart taxi dispatch methods for cars, there are smart bicycle dispatch systems, also known as shared bicycles. The algorithm proposes a logical scheduling strategy based on current traffic conditions to increase the effectiveness of shared bicycle use. Due to the fact that many users prefer riding bicycles to short distances of walking, bicycle sharing can address "last mile" problems for locals.

14.12.8 ENVIRONMENTAL RISKS

Climate change and environmental hazards are posing a growing threat to cities. Several European and Chinese nations suffered significant flooding in July 2021, resulting in millions of dollars in property damage and human fatalities. Disaster management systems, such as those that monitor the weather and warn people about preventative actions to lower pollution levels, should be extremely quick and flexible in smart cities. There is a continuing need to sustain economic growth because of population expansion, which puts environmental dangers in place. Larger population centers also cause a rise in trash, single-use plastics, and transportation-related pollutants, which is a major factor in environmental contamination. Several environmental risks are also brought on by the requirement for large-scale housing. The design process may include AI and the IoT to reduce environmental hazards. Automated drones have been employed in a number of industries, including air pollution monitoring, environmental risk detection and capture, and traffic control. Monitoring environmental dangers can be supported by air quality sensors on publicly available web networks. Urban settings boost economic viability and enable cities to address environmental problems.

14.12.9 Managing Public Health and Education

It can be difficult to provide healthcare services to everyone in a densely populated area. Without considering their prior health status, it is challenging to be able to offer all age groups essential medical care. AI should be able to determine the state of a patient's health based on their past medical history and current physical condition. Governments will be able to analyze data at anytime, anywhere in the nation, as a result of the digitalization of medical records on a centralized server. This enables medical professionals to carefully examine and treat patients. The use of a central server to manage medical records can be replaced by block-chain-based solutions. Operation rooms, diagnostic instruments, and other IoT-enabled medical equipment should help with health management. For instance, government established measures for efficient protocols for exchanging health data during the COVID-19 pandemic. The improper use of personal health data is the antithesis of the wise application of modern technologies. Data breaches or abuse are always caused by measures involving the gathering and use of data. The COVID-19 epidemic and records of challenging medical appointments are two examples of situations that may jeopardize patient rights.

Residents' top priority also includes education. AI should aid in the planning and creation of educational initiatives that meet the demands of business and academia. One way AI could be used is in the ongoing improvement of curricula, the evaluation of student abilities, and the analysis of the industrial employment needs for the skill development of students. The IoT is forcing universities to reconsider how they approach teaching and learning in the context of a global marketplace in this age of abundant data and an exponential increase in new knowledge generation. The IoT can be used to share information and use it in highly effective ways to improve student engagement and connections with teachers and fellow students.

14.12.10 Low Trust in the IoT and AI

Trust typically exists in interpersonal relationships. A lack of trust in AI and the IoT could slow the growth of smart cities. Building trust in AI models can assist in dispelling any misconceptions about the use of IoT and AI technologies. Research suggests that these models may help people make decisions and alter the social, political, and corporate contexts. Governmental organizations could work to raise awareness by disseminating information that supports the use of AI in sustainable practices, upholding openness, and releasing certain case studies that might increase public confidence.

14.13 OPPORTUNITIES

The use of IoT devices, including sensors, actuators, and wearables, in smart city technology has various benefits. Autonomous cars are a disruptive technology for smart cities because they may offer services made possible by wireless communication from vehicles to other vehicles and from vehicles to the internet. This will change how taxis have historically been owned and operated. For instance, using collaborative IoT devices and intelligent systems to improve traffic flow and decrease accidents would improve communication with autonomous vehicles.

Furthermore, depending on loading and unloading zones, autonomous cars may potentially attract passengers. Additionally, by enhancing traffic flow, public safety may be enhanced during evacuation preparations for natural disasters. Machine learning and IoT devices are required to increase efficiency in order to make our lives easier. In order to apply models to minimize trash in the near future through recycling and separation, smarter waste management uses IoT technology to evaluate our garbage disposal through the data obtained and how much waste is created. IoT technology is now essential to creating cities with happier, healthier, and cleaner residents, and enhancing health care and quality of life by keeping an eye on the environment, the air we breathe, and lowering stress levels. Therefore, there is much potential for the development of a population that is wiser, healthier, more environmentally conscious, and happier, which will result in a cleaner, greener world.

A nation's policies, laws, and regulations are based on the citizens who elect its government. The objective of participatory governance is to improve citizen involvement in policy-making, to build citizen–government interaction, and to instill social inclusion through democratic engagement. A smart city would finally benefit from good governance.

Businesses are expected to create innovative technology-based goods and services. Local players and innovative start-ups may be essential. Around the world, public–private partnerships have been the method of choice for creating smart and sustainable city initiatives. This is also applicable to India.

For the INR 7.5 trillion needed over 20 years, the Indian government's smart city plan has identified a number of potential funding sources, both conventional and non-conventional. To achieve smart and sustainable cities, additional innovative financial instruments must be encouraged. In the same way that National Bank for Agriculture and Rural Development (NABARD) has taken action in this area, corporate social responsibility (CSR) may also support supply chain management (SCM).

Urban life will be better because of the use of digital platforms and enhanced citizen awareness of infrastructure planning and so on. More chances for decent work will result from more thriving informal and formal labor marketplaces. Over time, this can reduce poverty thanks to improved digital skill matching with employment prospects, pay, and social protection.

The use of renewable energy sources like solar and wind energy could reduce CO_2 emissions.

The most important things are household dedication to water and energy conservation and alternative modes of transportation, including bicycles, trains, and electric buses.

Sustainable cities must encourage local food production and consumption as well as ethical consumerism in order to support more environmentally friendly local supply chains and sources.

To build a sustainable city, it is crucial to emphasize the value of recycling and lowering unnecessary and careless consumption.

Accelerating the adoption of the Jan Dhan-Aadhaar-Mobile, the switch to digital cash, the decline in transaction costs, and the rise in productivity of various businesses are all possible.

Prior ambitious programs like the Jawaharlal Nehru National Urban Renewal Mission and Providing Urban Amenities to Rural Areas will be expanded in scope and execution.

Numerous opportunities for more effective service delivery, digital inclusion, inclusive service delivery, and novel forms of decision-making are presented by smart cities. Data-driven innovation may help integrate urban systems into a "system of systems" that is more effective, sustainable, and resilient. Dynamic electricity pricing options and smart meters have the ability to fundamentally alter how businesses and consumers use energy. Air and noise pollution might be significantly reduced by using electric vehicles for transportation. Digital innovation may also help the circular economy, a paradigm that aims to boost financial and resource efficiency, through more precise management of consumption and production processes. Early warning systems for natural disasters like floods and other sorts might enhance preparedness, response, and recovery. Digital technologies may enable a more fluid and responsive type of municipal administration through e-government services and technology to simplify access to information through online forums, citizen monitoring, and public innovation labs. Especially for programs and infrastructure projects pursuing inclusive goals, innovative participatory budgeting may give residents a say in how public monies are spent. Cities may also use digitalization to strengthen their organizational and administrative capacities in order to cope with a variety of associated issues. Many local governments are reassessing how to use capacity in light of difficulties and the requirement to offer additional public services. There is also the possibility of promoting integrated contracts by incorporating cutting-edge collaboration technology. Smart cities of all sizes must enable an agile and adaptable form of municipal management.

14.14 CONCLUSION

In the 21st century, a wide range of technologies have advanced dramatically, enhancing life in smart cities, where the quality of life has lately been improved by IoT technology. However, developing new technology is energy-intensive and leads to unintended pollution and e-waste releases. Smart objects in smart cities are made intelligent so that they can complete their tasks on their own. In order to reduce e-waste and mitigate harmful emissions, these devices interact with one another and with people in a way that is both energy and bandwidth efficient. In building sustainable smart cities, the challenges and potential future opportunities have been studied.

REFERENCES

1. Atzori L, Iera A, Morabito G (2010). The internet of things: A survey. *Comput Netw* 54(15):2787–2805.
2. Minerva R, Biru A, Rotondi D (2015). Towards a definition of the Internet of Things (IoT). *IEEE Internet Initiative* 1(1):1–86.
3. Perera C, Zaslavsky A, Christen P, Georgakopoulos D (2014). Context aware computing for the internet of things: A survey. *IEEE Commun Surv Tutor* 16(1):414–454.
4. Gubbi J, Buyya R, Marusic S, Palaniswami M (2013). Internet of things (IoT): A vision, architectural elements, and future directions. *Future Gen Comput Syst* 29(7):1645–1660.
5. Tellez M, El-Tawab S, Heydari HM (2016). Improving the security of wireless sensor networks in an iot environmental monitoring system. In: *Systems and information engineering design symposium (SIEDS) IEEE*. IEEE, Conference Proceedings, pp. 72–77.
6. Shah J, Mishra B (2016). Iot enabled environmental monitoring system for smart cities. In: *Internet of things and applications (IOTA), International Conference on*. IEEE, Conference Proceedings, pp 383–388.
7. Chen X, Ma M, Liu A (2018). Dynamic power management and adaptive packet size selection for iot in e-healthcare. *Comput Electric Eng* 65:357–375.
8. Kong L, Khan MK, Wu F, Chen G, Zeng P (2017). Millimeterwave wireless communications for iot-cloud supported autonomous vehicles: Overview, design, and challenges. *IEEE Commun Mag* 55(1):62–68.
9. Popa D, Popa DD, Codescu M-M (2017). Reliabilty for a green internet of things. *Buletinul AGIR* nr 45–50.
10. Prasad SS, Kumar C (2013). A green and reliable internet of things. *Commun Netw* 5(1):44.
11. Pavithra D, Balakrishnan R (2015). Iot based monitoring and control system for home automation. In: *Communication Technologies (GCCT) Global Conference on*. IEEE, Conference Proceedings, pp 169–173.
12. Kodali RK, Jain V, Bose S, Boppana L (2016). Iot based smart security and home automation system. In: *Computing, Communication and Automation (ICCCA) International Conference on*. IEEE, Conference Proceedings, pp 1286–1289.
13. Gu M, Li X, Cao Y (2014). Optical storage arrays: A perspective for future big data storage. *Light Scie Appl* 3(5): e177.
14. Hashem IAT, Yaqoob I, Anuar NB, Mokhtar S, Gani A, Khan SU (2015). The rise of big data on cloud computing: Review and open research issues. *Inf Syst* 47:98–115.
15. Syed F, Gupta SK, Hamood Alsamhi S, Rashid M, Liu X (2020). A survey on recent optimal techniques for securing unmanned aerial vehicles applications. *Trans Emerg Telecommun Technol* e4133.
16. Alsamhi SH, Ansari MS, Zhao L, Van SN, Gupta SK, Alammari AA, Saber AH, Hebah MYAM, Alasali MAA, Aljabali HM (2019). Tethered balloon technology for green communication in smart cities and healthy environment. In: *First International Conference of Intelligent Computing and Engineering (ICOICE)*. IEEE, Conference Proceedings, pp 1–7.

17. Alsamhi SH, Ma O, Ansari MS, Almalki FA (2019). Survey on collaborative smart drones and internet of things for improving smartness of smart cities. *IEEE Access* 7:128125–128152.

18. Shuja J, Ahmad RW, Gani A, Ahmed AIA, Siddiqa A, Nisar K, Khan SU, Zomaya AY (2017). Greening emerging it technologies: Techniques and practices. *J Int Serv Appl* 8(1):9.

19. Wu X., Yang Z (2010). Smart City concept and future city development. *Urban Dev Res* 11: 60.

20. Palmisano SJ (2008). A smarter planet: The next leadership agenda. *IBM* 6:1–8.

21. Fernandez-Anez V. (2016). Stakeholders approach to smart cities: A survey on Smart City definitions. In *Proceedings of the International Conference on Smart Cities*. Springer: Berlin/Heidelberg, Germany, pp. 157–167.

22. Gracia TJH, García AC (2018). Sustainable smart cities. Creating spaces for technological, social and business development. *Boletín Científico Cienc. Económico Adm. ICEA* 6:3074.

23. Kim T, Ramos C, Mohammed S. (2017). Smart City and IoT. *Future Gener Comput Syst* 76:159–162.

24. UN DESA (2014). *World's Population Increasingly Urban with More than Half Living in Urban Areas*. United Nations Department of Economic and Social Affairs: New York, NY.

25. Yang D, Li P (2019). The economic effects of population agglomeration: An empirical study based on instrumental variables. *J Demogr* 3:28–37.

26. Haas K (2021). How to build a Smart City: Decision making and modeling. Skyfii|Omni data Intelligence Solutions for Physical Venues. Available online: https://skyfii.io/blog/how-to-build-a-smart-city-decision-making-and-modeling/ (accessed on 4 September 2021).

27. ANBOUND Research Center. How to design and operate smart cities? Available online: https://www.anbound.my/Section/ArticalView.php?Rnumber=18377&SectionID=1 (accessed on 1 October 2021).

28. Cohen B. The 3 generations of Smart Cities. Available online: https://www.fastcompany.com/3047795/the-3-generations-of-smart-cities (accessed on 17 July 2021).

29. Nandyala CS, Kim H-K (2016) Green iot agriculture and healthcareapplication (gaha). *Int J Smart Home* 10(4):289–300.

30. Gapchup A, Wani A, Wadghule A, Jadhav S (2017). Emerging trends of green iot for smart world. *Int J Innov Res Comput Commun Eng* 5(2):2139–2148.

31. Zanamwe N, Okunoye A (2013). Role of information and communication technologies (icts) in mitigating, adapting to and monitoring climate change in developing countries. In: *International Conference on ICT for Africa. Conference Proceedings*.

32. Mickoleit A (2010). *Greener and smarter: ICTs, the environment and climate change*. OECD Publishing, Report.

33. Murugesan S (2008). Harnessing green IT: Principles and practices. *IT professional* 10(1): 24–33.

34. Rani S, Talwar R, Malhotra J, Ahmed SH, Sarkar M, Song H (2015). A novel scheme for an energy efficient Internet of Things based on wireless sensor networks. *Sensors* 15(11):28603–28626.

35. Huang J, Meng Y, Gong X, Liu Y, Duan Q (2014). A novel deployment scheme for green internet of things. *IEEE Internet Things J* 1(2):196–205.

36. Radu L-D (2016). Determinants of green ICT adoption in organizations: A theoretical perspective. *Sustainability* 8(8):731.

37. Baccarelli E, Amendola D, Cordeschi N (2015). Minimumenergy bandwidth management for qos live migration of virtual machines. *Comput Netw* 93:1–22.

38. Amendola D, Cordeschi N, Baccarelli E (2016). Bandwidth management VMs live migration in wireless fog computing for 5G networks. In: *Cloud Networking (Cloudnet), 5th IEEE International Conference on*. IEEE, Conference Proceedings, pp 21–26.

39. Dayarathna M, Wen Y, Fan R (2016). Data center energy consumption modeling: A survey. *IEEE Commun Surv Tutor* 18(1):732–794.

40. Cordeschi N, Shojafar M, Amendola D, Baccarelli E (2015). Energy-efficient adaptive networked datacenters for the qos support of real-time applications. *J Supercomput* 71(2):448–478.

41. Shuja J, Bilal K, Madani SA, Othman M, Ranjan R, Balaji P, Khan SU (2016). Survey of techniques and architectures for designing energy-efficient data centers. *IEEE Syst J* 10(2):507–519.

42. Roy A, Datta A, Siddiquee J, Poddar B, Biswas B, Saha S, Sarkar P (2016). Energy-efficient data centers and smart temperature control system with iot sensing. In: *Information Technology, Electronics and Mobile Communication Conference (IEMCON), IEEE 7Th Annual*. IEEE, Conference Proceedings, pp 1–4.

43. Peoples C, Parr G, McClean S, Scotney B, Morrow P (2013). Performance evaluation of green data centre management supporting sustainable growth of the internet of things. *Simul Model Pract Theory* 34:221–242.

44. Deep B, Mathur I, Joshi N (2020). *An Approach toward More Accurate Forecasts of Air Pollution Levels through Fog Computing and IoT*. Springer: Berlin, pp 749–758.

45. Zhu C, Leung VC, Wang K, Yang LT, Zhang Y (2017). Multimethod data delivery for green sensor-cloud. *IEEE Commun Mag* 55(5):176–182.

46. Kumar, S., & Buyya, R. (2012). Green cloud computing and environmental sustainability. *Harnessing Green IT: Principles and Practices*, 315–339.

47. Chen F, Schneider J, Yang Y, Grundy J, He Q (2012). An energy consumption model and analysis tool for cloud computing environments. In: *First International Workshop on Green and Sustainable Software (GREENS)*. Conference Proceedings, pp 45–50.

48. Shaikh, F. K., Zeadally, S., & Exposito, E. (2015). Enabling technologies for green internet of things. *IEEE Systems Journal*, *11*(2), 983–994.

49. Liu X-F, Zhan Z-H, Zhang J (2017). An energy aware unified antcolony system for dynamic virtual machine placement in cloud computing. *Energies* 10(5):609.

50. Peoples C, Parr G, McClean S, Morrow P, Scotney B (2013). Energy aware scheduling across 'green' cloud data centres. In: *Integrated Network Management (IM 2013), IFIP/IEEE International Symposium on*. IEEE, Conference Proceedings, pp 876–879.

51. Lago DG, Madeira ER, Bittencourt LF (2011). Power-aware virtual machine scheduling on clouds using active cooling control and dvfs. In: *Proceedings of the 9th International Workshop on Middleware for Grids, Clouds and e-Science*. ACM, Conference Proceedings, p 2.

52. Cotes-Ruiz IT, Prado RP, García-Galan S, Muñoz Expósito JE, Ruiz-Reyes N (2017). Dynamic voltage frequency scaling simulator for real workflows energy-aware management in green cloud computing. *PloS One* 12(1):e0169803.

53. Abdelaziz A, Salama AS, Riad AM, Mahmoud AN (2019). *A Machine Learning Model for Predicting of Chronic Kidney Disease Based Internet of Things and Cloud Computing in Smart Cities*. Springer International Publishing: Cham, pp 93–114. [Online]. Available: https://doi.org/10.1007/978-3-030-01560-2_5

54. Abdelaziz A, Salama AS, Riad AM (2019). *A Swarm Intelligence Model for Enhancing Health Care Services in Smart Cities Applications*. Springer: Berlin, pp 71–91.

Adelin A, Owezarski P, Gayraud T (2010). On the impact of monitoring router energy consumption for greening the internet. In: *Grid Computing (GRID), 11th IEEE/ACM International Conference on*. IEEE, Conference Proceedings, pp 298–304.

Yang Y,Wang D, Pan D, Xu M (2016). Wind blows, traffic flows: Green internet routing under renewable energy. In: *Computer communications, IEEE INFOCOM-The 35th Annual IEEE international conference on*. IEEE, Conference Proceedings, pp 1–9.

Hoque MA, Siekkinen M, Nurminen JK (2014). Energy efficient multimedia streaming to mobile devices—A survey. *IEEE Commun Surv Tutor* 16(1):579–597.

Lloret J, Garcia M, Bri D, Sendra S (2009) A wireless sensor network deployment for rural and forest fire detection and verification. *Sensors* 9(11):8722–8747.

Aslan YE, Korpeoglu I, Ulusoy Z (2012). A framework for use of wireless sensor networks in forest fire detection and monitoring. *Computers Environ Urban Syst* 36(6): 614–625.

Viani F, Lizzi L, Rocca P, Benedetti M, Donelli M, Massa A (2008) Object tracking through rssi measurements in wireless sensor networks. *Electron Lett* 44(10):653–654.

Han G, Shen J, Liu L, Qian A, Shu L (2016). Tgm-cot: Energyefficient continuous object tracking scheme with two-layer grid model in wireless sensor networks. *Pers Ubiquit Comput* 20(3):349–359.

2. Han, G., Shen, J., Liu, L., & Shu, L. (2016). BRTCO: A novel boundary recognition and tracking algorithm for continuous objects in wireless sensor networks. *IEEE Systems Journal*, *12*(3), 2056–2065.

3. Wu F, Rüdiger C, Yuce MR (2017). Real-time performance of a self-powered environmental iot sensor network system. *Sensors* 17(2):282.

4. Prabhu, B., Balakumar, N., & Antony, A. (2017). Wireless sensor network based smart environment applications. *Wireless Sensor Network Based Smart Environment Applications (January 31, 2017)*. IJIRT, *3*(8).

5. Trasvina-Moreno CA, Blasco R, Marco L, Casas R, Trasvina-Castro A (2017). Unmanned aerial vehicle based wireless sensor network for marine-coastal environment monitoring. *Sensors* 17(3):460.

86. Sharma, D. (2017). Low cost experimental set up for real time temperature, humidity monitoring through wsn. *Int. J. Eng. Sci*, *7*(1), 4340–4342.

87. Almalki, F. A., Soufiene, B. O., Alsamhi, S. H., & Sakli, H. (2021). A low-cost platform for environmental smart farming monitoring system based on IoT and UAVs. *Sustainability*, *13*(11), 5908.

88. Prabhu, S. B., Balakumar, N., & Antony, A. J. (2017). Evolving constraints in military applications using wireless sensor networks. *International Journal of Innovative Research in Computer Science & Technology (IJIRCST) ISSN*, 2347-5552.

89. Ye W, Heidemann J, Estrin D (2002). An energy-efficient mac protocol for wireless sensor networks. In: *INFOCOM Twenty-First Annual Joint Conference of the IEEE Computer and Communications Societies*. Proceedings IEEE, vol 3. IEEE, Conference Proceedings, pp 1567–1576.

90. Anastasi G, Francesco MD, Conti M, Passarella A (2013). *How to Prolong the Lifetim of WSNs*. CRC Press, Boca Raton. book Section 6.

91. Khalil HB, Zaidi SJH (2012). Mnmu-ra: Most nearest most used routing algorithm greening the wireless sensor networks. *Wirel Sens Netw* 4(06):162.

92. Azevedo J, Santos F (2012). Energy harvesting from wind and water for autono wireless sensor nodes. *IET Circ Dev Syst* 6(6):413–420.

93. Eu ZA, Tan H-P, Seah WK (2011). Design and performance analysis of mac for wireless sensor networks powered by ambient energy harvesting. *Ad I* 9(3):300–323.

55. Mishra KN, Chakraborty C (2020). *A Novel Approac* *Quality of Life in Smart Cities Using Clouds and IoT-Ba* International Publishing: Cham, pp 19–35. Online]. Availab 978-3-030-18732-3_2

56. Koutitas G (2010). Green network planning of single freque *Broadcast* 56(4):541–550.

57. Naeem M, Pareek U, Lee DC, Anpalagan A (2013). Estimatio for resource allocation in green cooperative cognitive radio 13(4):4884–4905.

58. Chan CA, Gygax AF, Wong E, Leckie CA, Nirmalathas Methodologies for assessing the usephase power consumpt emissions of telecommunications network services. *Environ Sc*

59. Feng W, Alshaer H, Elmirghani JM (2010). Green information a nology: Energy efficiency in a motorway model. *IET Commun*

60. Mao G (2017). 15G green mobile communication networks. *China*

61. Abrol A, Jha RK (2016). Power optimization in 5G networks communication. *IEEE Access* 4:1355–1374.

62. Wang, J., Hu, C., & Liu, A. (2017). Comprehensive optimizatic tion and delay performance for green communication in inte *Information Systems, 2017.*

63. Liu A, Zhang Q, Li Z, Choi Y-J, Li J, Komuro N (2017). A green cation modeling for industrial internet of things. *Comput Electric*

64. Sahal R, Alsamhi SH, Breslin JG, Ali MI (2021). Industry 4.0 tow detection use case. *Sensors* 21(3): 694.

65. Alsamhi SH, Lee B, Guizani M, Kumar N, Qiao Y, Liu X (2021). tralized multi-drone to combat COVID-19 and future pandemics: posed solutions. *Trans Emerg Telecommun Technol*, e4255.

66. Sahal R, Alsamhi SH, Breslin JG, Brown KN, Ali MI (2021). Digita for automatic erratic operational data detection in industry 4.0. *App*

67. Alsamhi, S. H., & Lee, B. (2020). Blockchain-empowered multi-rc fight COVID-19 and future pandemics. *Ieee Access, 9,* 44173–4419

68. Anwar, M., Abdullah, A. H., Altameem, A., Qureshi, K. N., Masud, & Kharel, R. (2018). Green communication for wireless body are aware link efficient routing approach. *Sensors, 18*(10), 3237.

69. Wang, T., Ma, C., Sun, Y., Zhang, S., & Wu, Y. (2018). Energy eff resource allocation for opportunistic relay-aided ofdma downlink w ing. *Wireless Communications and Mobile Computing, 2018.*

70. Liu, Z. Y., Mao, P., Feng, L., & Liu, S. M. (2018). Energy-efficient i allocation scheme in cooperative communication system. *Wireless Co Mobile Computing, 2018.*

71. Yang Z, Jiang W, Li G (2018). Resource allocation for green cogniti efficiency maximization. *Wirel Commun Mob Comput* 2018.

72. Ge W, Zhu Z, Wang Z, Yuan Z (2018). An-aided transmit beamfo secured cognitive radio networks with swipt. *Wirel Commun Mob Con*

73. Zheng, Z., Cui, W., Qiao, L., & Guo, J. (2018). Performance and tion analysis of ieee802. 11ah for smart grid. *Wireless Communicati Computing, 2018,* 1–8.

74. Wang X, Vasilakos AV, Chen M, Liu Y, Kwon TT (2012). A survey networks: Opportunities and challenges. *Mob Netw Appl* 17(1):4–20.

94. Shaikh FK, Zeadally S (2016). Energy harvesting in wireless sensor networks: A comprehensive review. *Renew Sust Energ Rev* 55:1041–1054.

95. Almalki, F. A., Soufiene, B. O., Alsamhi, S. H., & Sakli, H. (2021). A low-cost platform for environmental smart farming monitoring system based on IoT and UAVs. *Sustainability*, *13*(11), 5908.

96. Busaileh, O., Hawbani, A., Wang, X., Liu, P., Zhao, L., & Al-Dubai, A. (2020). Tuft: Tree based heuristic data dissemination for mobile sink wireless sensor networks. *IEEE Transactions on Mobile Computing*, *21*(4), 1520–1536.

97. Hawbani A, Wang X, Zhao L, Al-Dubai A, Min G, Busaileh O (2020). Novel architecture and heuristic algorithms for softwaredefined wireless sensor networks. *IEEE/ACM Trans Netw* 28(6):2809–2822.

98. Jain PC (2015). Recent trends in energy harvesting for green wireless sensor networks. In: *International Conference on Signal Processing and Communication (ICSC) Conference Proceedings*, pp 40–45.

99. Abedin SF, Alam MGR, Haw R, Hong CS (2015). A system model for energy efficient green-iot network. In: *Information Networking (ICOIN) International Conference on.* IEEE, Conference Proceedings, pp 177–182.

100. Sun K, Ryoo I (2015). A study on medium access control scheme for energy efficiency in wireless smart sensor networks. In: *Information and Communication Technology Convergence (ICTC) International Conference on.* IEEE, Conference Proceedings, pp 623–625.

101. Uzoh PC, Li J, Cao Z, Kim J, Nadeem A, Han K (2015). Energy efficient sleep scheduling for wireless sensor networks. In: *International Conference on Algorithms and Architectures for Parallel Processing.* Springer, Conference Proceedings, pp 430–444.

102. Mehmood A, Song H (2015) Smart energy efficient hierarchical data gathering protocols for wireless sensor networks. *SmartCR* 5(5):425–462.

103. Rekha RV, Sekar JR (2016). An unified deployment framework for realization of green internet of things (giot). *Middle-East J Sci Res* 24(2):187–196.

104. Naranjo PGV, Shojafar M, Mostafaei H, Pooranian Z, Baccarelli E (2017). P-sep: A prolong stable election routing algorithm for energy-limited heterogeneous fog-supported wireless sensor networks. *J Supercomput* 73(2):733–755.

105. Yaacoub E, Kadri A, Abu-Dayya A (2012). Cooperative wireless sensor networks for green internet of things. In: *Proceedings of the 8h ACM Symposium on QoS and Security for Wireless and Mobile Networks.* ACM, Conference Proceedings, pp 79–80.

106. Castillo-Cara, M., Huaranga-Junco, E., Quispe-Montesinos, M., Orozco-Barbosa, L., & Antúnez, E. A. (2018). FROG: a robust and green wireless sensor node for fog computing platforms. *Journal of Sensors*, *2018*.

107. Mahapatra C, Sheng Z, Kamalinejad P, Leung VC, Mirabbasi S (2017). Optimal power control in green wireless sensor networks with wireless energy harvesting, wake-up radio and transmission control. *IEEE Access* 5:501–518.

108. Amirthavarshini LJ, Varshini R, Kavya S (2015). Wireless sensor networks in green cloud computing. *International Journal of Scientific & Engineering Research* 6(10):98–100. https://www.ijser.org/researchpaper/Wireless-Sensor-Networks-in-Green-Cloud-Computing.pdf

109. Khatri A, Kumar S, Kaiwartya O, Aslam N, Meena N, Abdullah AH (2018). Towards green computing in wireless sensor networks: Controlled mobility–aided balanced tree approach. *Int J Commun Syst* 31(7): e3463.

110. Araujo A, Romero E, Blesa J, Nieto-Taladriz O (2012). Cognitive wireless sensor networks framework for green communications design. In: *Proceedings of the 2nd International Conference on Advances in Cognitive Radio (COCORA'12).* Conference Proceedings, pp 34–34.

111. Rault T, Bouabdallah A, Challal Y (2014). Energy efficiency in wireless sensor networks: A top-down survey. *Comput Netw* 67:104–122.
112. Seuwou P, Banissi E, Ubakanma G (2020). *The Future of Mobility with Connected and Autonomous Vehicles in Smart Cities*. Springer: Berlin, pp 37–52.
113. Bencardino, M., & Greco, I. (2014). Smart communities. Social innovation at the service of the smart cities. *TeMA-Journal of Land Use, Mobility and Environment*.
114. Jraisat L (2020). *Information Sharing in Sustainable Value Chain Network (SVCN)— The Perspective of Transportation in Cities*. Springer: Berlin, pp 67–77.
115. Yoo S-J, Park J-H, Kim S-H, Shrestha A (2016). Flying path optimization in uav-assisted iot sensor networks. *ICT Express* 2(3):140–144.
116. Mozaffari M, Saad W, Bennis M, Debbah M (2015). Drone small cells in the clouds: Design, deployment and performance analysis. In: *Global Communications Conference (GLOBECOM)*. IEEE, Conference Proceedings, pp 1–6.
117. Cao H-R, Yang Z, Yue X-J, Liu Y-X (2017). An optimization method to improve the performance of unmanned aerial vehicle wireless sensor networks. *Int J Distrib Sensor Netw* 13(4):1550147717705614.
118. Cao H, Liu Y, Yue X, Zhu W (2017). Cloud-assisted uav data collection for multiple emerging events in distributed wsns. *Sensors* 17(8):1818.
119. Dong M, Ota K, Lin M, Tang Z, Du S, Zhu H (2014). UAV-assisted data gathering in wireless sensor networks. *J Supercomput* 70(3):1142–1155.
120. Zorbas D, Razafindralambo T, Guerriero F (2013). Energy efficient mobile target tracking using flying drones. *Procedia Comput Sci* 19:80–87.
121. Sharma, V., You, I., & Kumar, R. (2016). Energy efficient data dissemination in multi-UAV coordinated wireless sensor networks. *Mobile Information Systems, 2016*.
122. Uragun B (2011). Energy efficiency for unmanned aerial vehicles. In: *Machine Learning and Applications and Workshops (ICMLA), 10th International Conference on*, vol 2. IEEE, Conference Proceedings, pp 316–320.
123. Choi DH, Kim SH, Sung DK (2014). Energy-efficient maneuvering and communication of a single uav-based relay. *IEEE Trans Aerosp Electron Syst* 50(3):2320–2327.
124. Yu Y, Lee S, Lee J, Cho K, Park S (2016). Design and implementation of wired drone docking system for cost-effective security system in iot environment. In: *Consumer Electronics (ICCE) IEEE International Conference on*. IEEE, Conference Proceedings, pp 369–370.
125. Seo S-H, Choi J-I, Song J (2017). Secure utilization of beacons and UAVs in emergency response systems for building fire hazard. *Sensors* 17(10):2200.
126. Fujii K, Higuchi K, Rekimoto J (2013). Endless flyer: A continuous flying drone with battery replacement. In: *Ubiquitous Intelligence and Computing, IEEE 10th International Conference on and 10th International Conference on Autonomic and Trusted Computing (UIC/ATC)*. IEEE, Conference Proceedings, pp 216–223.
127. Sahal R (2021). Digital twins collaboration for automatic erratic operational data detection in industry 4.0. *Appl Sci* 11:15.
128. Luo Z, Zhong L, Zhang Y, Miao Y, Ding T (2017). An efficient intelligent algorithm based on wsns of the drug control system. *Tehniˇcki vjesnik* 24(1):273–282.
129. Mahapatra C, Moharana AK, Leung V (2017). Energy management in smart cities based on internet of things: Peak demand reduction and energy savings. *Sensors* 17(12):2812.
130. Rani, S., & Chauhdary, S. H. (2018). A novel framework and enhanced QoS big data protocol for smart city applications. *Sensors, 18*(11), 3980.
131. Villa TF, Salimi F, Morton K, Morawska L, Gonzalez F (2016). Development and validation of a UAV based system for air pollution measurements. *Sensors* 16(12):2202.

132. Wang J, Schluntz E, Otis B, Deyle T (2015). A new vision for smart objects and the internet of things: Mobile robots and long-range uhf rfid sensor tags. arXiv:1507.02373.

133. Hamilton A, Magdalene AHS (2017). Study of solar powered unmanned aerial vehicle to detect greenhouse gases by using wireless sensor network technology. *J Sci Eng Educ* (ISSN 2455-5061) 2: 1–11.

134. Almalki FA (2020). Utilizing drone for food quality and safety detection using wireless sensors. In: *IEEE 3rd International Conference on Information Communication and Signal Processing (ICICSP)*. IEEE, Conference Proceedings, pp 405–412.

135. Klimkowska A, Lee I, Choi K (2016). Possibilities of UAS for maritime monitoring. *ISPRS-International Archives of the Photogrammetry. Remote Sens Spatial Inform Sci* 885–891.

136. Villa TF, Gonzalez F, Miljievic B, Ristovski ZD, Morawska L (2016). An overview of small unmanned aerial vehicles for air quality measurements: Present applications and future prospectives. *Sensors* 16(7):1072.

137. Telesetsky A (2016). Navigating the legal landscape for environmental monitoring by unarmed aerial vehicles. *GeoWash J Energy Envtl L* 7:140.

138. Alvear O, Calafate CT, Hernández E, Cano J-C, Manzoni P (2015). Mobile pollution data sensing using UAVs. In: *Proceedings of the 13th International Conference on Advances in Mobile Computing and Multimedia*. ACM, Conference Proceedings, pp 393–397.

139. Alvear OA, Zema NR, Natalizio E, Calafate CT (2017). A chemotactic pollution-homing uav guidance system. In: *Wireless Communications and Mobile Computing Conference (IWCMC), 13th International*. IEEE, Conference Proceedings, pp 2115–2120.

140. Alvear, O., Zema, N. R., Natalizio, E., & Calafate, C. T. (2017). Using UAV-based systems to monitor air pollution in areas with poor accessibility. *Journal of advanced Transportation, 2017*.

141. Koo VC, Chan YK, Vetharatnam G, Chua MY, Lim CH, Lim C-S, Thum C, Lim TS, Bin Ahmad Z, Mahmood KA (2012). A new unmanned aerial vehicle synthetic aperture radar for environmental monitoring. *Prog Electromagn Res* 122:245–268.

142. Šmídl V, Hofman R (2013). Tracking of atmospheric release of pollution using unmanned aerial vehicles. *Atmos Environ* 67:425–436.

143. Zang W, Lin J, Wang Y, Tao H (2012). Investigating small-scale water pollution with uav remote sensing technology. In: *World automation congress (WAC)*. IEEE, Conference Proceedings, pp 1–4.

144. Bronk C, Lingamneni A, Palem K (2010). *Innovation for Sustainability in Information and Communication Technologies (ICT)*. James A. Baker III Institute for Public Policy Rice University.

145. Omar M, Termizi A, Zainal D, Wahap N, Ismail N, Ahmad N (2016). Implementation of Spatial Smart Waste Management System in Malaysia. In: *IOP Conference Series: Earth and Environmental Science*, vol 37. IOP Publishing, Conference Proceedings, p 012059.

146. Popescu DE, Bungau C, Prada M, Domuta C, Bungau S, Tit D (2016). Waste management strategy at a public university in smart city context. *J Environ Prot Ecol* 17(3):1011–1020.

147. Del Borghi A, Gallo M, Strazza C, Magrassi F, Castagna M (2014). Waste management in smart cities: The application of circular economy in Genoa (Italy). *Impresa Progetto Electronic J. Manage.* 4:1–13.

148. Popa CL, Carutasu G, Cotet CE, Carutasu NL, Dobrescu T (2017). Smart City platform development for an automated waste collection system. *Sustainability* 9(11):2064.

149. Pirlone F, Spadaro I (2014). Towards a waste management plan for smart cities. *WIT Trans Ecol Environ* 191:1279–1290.
150. Ismagiloiva E, Hughes L, Rana N, Dwivedi Y (2019). Role of smart cities in creating sustainable cities and communities: A systematic literature review. In: *International Working Conference on Transfer and Diffusion of IT*. Springer, Conference Proceedings, pp 311–324.
151. Maksimovic M (2017). The role of Green Internet of Things (G-IoT) and Big Data in making cities smarter, safer and more sustainable. *Int J Comput Digit Syst* 6(04):175–184.
152. Sodhro AH, Pirbhulal S, Luo Z, de Albuquerque VHC (2019) Towards an optimal resource management for iot based green and sustainable smart cities. *J Clean Prod* 220:1167–1179.
153. Tuysuz MF, Trestian R (2020). From serendipity to sustainable Green IoT: Technical, industrial and political perspective. *Comput Netw* 182:107469.
154. Maksimovic M (2018). *Greening the Future: Green Internet of Things (G-IoT) as a Key Technological Enabler of Sustainable Development*. Springer, Berlin, pp 283–313.

15 Quantum Digital Twin Technology for Health Care

Pradheep Kumar
BITS Pilani, Pilani, India

B. Amutha and J. Ramkumar
SRM Institute of Science and Technology, KTR Campus,
Chennai, India

15.1 INTRODUCTION

In today's world of health, several patients may suffer from vital organ failure. In several cases the unavailability of organ donors in a timely manner results in a catastrophe. Digital twins provide a promising solution in designing effective organs which may help humans to lead longer lives. The digital twin is a digital representation of the physical object which exists in reality. To accomplish this we need to mathematically model the organ based on its functionality. The functionality model is designed by identifying the vital parameters and their dependent physical parameters. Boundary values are set for the vital parameters and an objective function is created to maximise these values.

A basis is formed with the initial boundary values, which are converted to a quantum dataset. A neural fuzzy algorithm is used to decide the optimal bounds for the dependent factors. Based on the optimal rules generated, a finalised digital prototype is created which would help in the making of organs which could be used for the patient. Again this may vary based on the patient, whether they are a child, an adult, or a senior citizen. Organ design could be made more effective by quantum machine learning algorithms as the skin texture and other factors are controlled according to the optimal bounds. The quantum digital twin algorithm improves reliability by 33% and reduces processing time by 33%, compared to the conventional machine learning approaches.

The chapter is organised as follows. Section 15.2 reports on the literature; section 15.3 explains the proposed model and the salient features of the algorithm; section 15.4 explains the tool and analyses the results; section 15.5 concludes; and section 15.6 suggests future research directions.

DOI: 10.1201/9781003388814-15

15.2 LITERATURE REVIEW

Quantum gate issues and the computation's optimality from the data and control signal flow was explained by Arrighi et al. in [1]. The Bell state and the associated Hilbert space is here discussed. The quantum circuit transformation was explained by Zhou and Feng in [2]. The circuit analysis was performed using a Monte Carlo tree approach. The random sampling was done using the steps of selection, expansion, simulation, backpropagation, and decision. The random moves facilitate the gate optimisation. The transformation of circuits was also explained and highlighted. The technique of cut enumeration was discussed by Haner and Soeken in [3]. This helps in logic network optimisation by partitioning the network using consistent cuts. The truth tables of the circuits were normalised. The multiplication depth reduction was based on balancing the network. A quantum multi-level intermediate representation was discussed by Ittah et al. in [4]. The quantum classical co-optimisation was explained using dialects. The compiler representations for each operation were explained. A quantum programming stack model was also explained using different quantum programs. The quantum algorithm implementation was set out by Abhijith and Vuffray in [5]. Qubit programming explains the features of superposition and entanglement. The practical pattern matching approach was discussed by Iten et al. in [6]. Quantum circuit matching with different strategies like maximal matching, greedy matching, and disturbing gates was discussed. An optimality based approach was also discussed. Leveraging state sparsity was set out by Jaques and Haner in [7]. Quantum states highlighting the sparse states and measuring the information amount was explained. The sparse states with operations were also simulated. Harwood et al. in [8] explain a variational quantum eigen solver network. This work uses an adiabatic quantum computing approach. Bootstrapping was one of the approaches used. FlexSim in [9] explains the features of modelling a digital twin based on the physical and mathematical model. The boundary conditions illustrate the data sets with initial and final bounds. The virtual reality (VR) and artificial intelligence (AI) enabled datasets and their illustrations were also explained. The digital twin technology to improve personalised health care was explained by Harita in [10]. The AI based revolution techniques explaining the "slippery slope problem" were discussed.

The digital twin paradigm shift in a systematic derivable approach was discussed by Semararo et al. in [11]. The concept was explained using a "formal concept analysis" method. The application contexts and the associated life cycles were illustrated. An architecture model showing the functionalities of the associated components were also discussed. The quantum machine learning model using TensorFlow and the datasets were explained by Broughton et al. in [12]. The work explains the modelling of a quantum dataset and the creating of a quantum neural network to process the same. The qubits are then modelled and processed using quantum gates, which are then optimised based on the states using the principles of superposition and entanglement.

Mixed state and hybrid state analysis using superposition and entanglement was highlighted using graphical languages, as explained by Carette et al. in [13]. The quantum evolutions and their associated graphs were illustrated. Discard construction using isometries were also discussed. The 3D design printing of organs was set out by Catterton et al. in [14]. The 3D printing technique also shows laser etching. The tissue viability dimensions of the slices and mapping the process onto a digital twin model were explained. The objective function criterion for the critical and dependent parameters was also discussed. The design parameter optimisation is made based on the detailed analysis of the vital and dependent factors. The essence of quantum machine learning technologies explaining the kernel were explained by Huang et al. in [15]. The projected quantum kernels explain the data and code availability. The relationship between the dimension, geometric difference, and prediction performance was also analysed extensively using empirical measurements. The enabling technologies and challenges of the digital twin model were explained by Fuller et al. [16]. The model highlights creating a digital shadow and modifying the state-based constraints to amend the criteria for the same. These technologies have several applications which include smart cities, manufacturing, and healthcare, to name a few.

The taxonomy of digital twins was explained by Van Der Walk et al. in [17]. Multi-dimensional taxonomies based on the functionality and application were also explained. These categories were decided on according to the dimensions and characteristics. The conceptual elements, data link, and purpose were also analysed. Almudever in [18] shows that quantum computing has noise associated with the data. The author also explains the top-down and bottom-up approach. Enders and Hobbach in [19] discuss the dimensions of digital twin applications. They explain the creation time and the connection between the digital and physical twin. Fitzsimons and Yuen in [20] discuss the protocol compression and the history states of the protocol circuits. They also illustrate the recursive compression. The compression rate improvement based on a theorem was also explained. A prover merging policy connecting the states and compression was also discussed. The design, discovery, and technology which facilitate organ development were explained by Kaushik et al. in [21]. The biomimicry of organ structure function was also explained in this work. The hollow 3D and 2D modelling techniques and their associated boundary conditions were also discussed. A multiscale biological imaging solution was also explained. The tissue stiffness which also depends on the variation of blood pressure and other physio-chemical properties was discussed. The biochemical signalling strategies were explained in this work.

The conceptual framework to print an organ was provided by Rezende et al. in [22]. The processing of the printing and the associated features along with the virtual bio-organ mapping was also illustrated. The different quantum theoretical approaches has been explained by Aburated et al. [23]. The work makes a detailed survey approach by analysing, denoising, edge detection, image storage, and so on. The storage retrieval and compression techniques to provide appropriate recommendations were also explained.

15.3 PROPOSED WORK

The block diagram showing the various modules is illustrated in Figure 15.1.

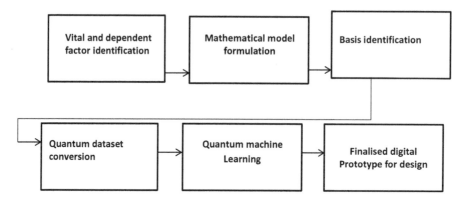

FIGURE 15.1   Block diagram of a quantum digital twin system.

15.3.1 VITAL AND DEPENDENT FACTOR IDENTIFICATION

This process involves first identifying the vital and dependent factors based on the functionality of the organ. The vital factors chosen are:

1. Type of patient;
2. Tissue parameters;
3. Functional parameters of the organ;
4. Drug allergies;
5. Current condition of the failed organ.

The dependent factors chosen are:

1. Thickness of organ;
2. Dimensions of the organ;
3. Cell morphology;
4. Tissue morphology;
5. Cell structure;
6. Tissue structure.

15.3.2 MATHEMATICAL MODEL FORMULATION

The mathematical model is formulated by prioritising the vital and dependent factors. Based on the priorities assigned, an objective function is created, which is a mathematical representation of the weightage of the factors involved in the digital organ design.

The priorities for the vital and dependent factors are indicated in Tables 15.1 and 15.2.

15.3.3 BASIS IDENTIFICATION

The basis consists of an initial dataset with values set to the lowest bound for normal functionality. For example, if we consider the case of the kidney, creatinine is the vital parameter which should be between 0.8 and 1.2. The initial value would be chosen as 0.8. If multiple parameters are involved based on the vital factors a vector is formulated which acts as the initial basis.

TABLE 15.1
Vital Factors and Their Associated Priorities

Vital Factor	Priority
Functional parameter of organ	5
Tissue parameters	4
Type of patient	3
Drug allergies	2
Current condition of failed organ	1

TABLE 15.2
Dependent Factors and Their Associated Priorities

Dependent Factor	Priority
Tissue structure	6
Cell structure	5
Tissue morphology	4
Cell morphology	3
Thickness of organ	2
Dimensions of the organ	1

15.3.4 QUANTUM DATASET FORMULATION

The quantum dataset consists of qubits. Each qubit would contain a tag, orientation ranging from 0 to 360, and the data. Each qubit can store 2^n vectors where n is the number of basis vectors. For each qubit the equivalent fuzzy qubit is computed. A qubit is given as illustrated in Equation 15.1, where a and b are complex numbers.

$$a\langle 1| + b|0\rangle \tag{15.1}$$

This could be modeled as a product of a vector and its conjugate transpose. Equations 15.2 and 15.3 illustrate the complex representation. Equation 15.4 gives the product of the complex representation of the numbers. Equation 15.5 is the magnitude of the qubit. The fuzzy qubits are first decided based on the angle orientation which could be mapped on a circle homeomorphism to a square, as shown in Table 15.3.

$$a = x + jy \tag{15.2}$$

$$b = p + jq \tag{15.3}$$

$$\begin{bmatrix} a & b \end{bmatrix} \times \begin{bmatrix} p \\ -q \end{bmatrix} = \left((a \times p) + (b \times q)\right) \tag{15.4}$$

$$M(q) = \sqrt{\left((a \times p) + (b \times q)\right)} \tag{15.5}$$

TABLE 15.3
Fuzzy Qubit Values

Fuzzy Qubit Name	Value
FQB1(1)	1+j1
FQB1(2)	0.66+j1
FQB1(3)	0.5+j1
FQB1(4)	0.33+j1
FQB1(5)	0+j1
FQB1(6)	0+j0.66
FQB1(7)	0+j0.5
FQB1(8)	0+j0.33
FQB2(1)	0+j0
FQB2(2)	0.33+j0
FQB2(3)	0.5+j0
FQB2(4)	0.66+j1
FQB2(5)	1+j0
FQB2(6)	1+j0.33
FQB2(7)	1+j0.5
FQB2(8)	1+j0.66

It can be observed from Table 15.3 that we obtain two sets of fuzzy qubit values for eight orientations: as FQB1(1), FQB1(2), FQB1(3), FQB1(4), FQB1(5), FQB1(6), FQB1(7), FQB1(8) and FQB2(1), FQB2(2), FQB2(3), FQB2(4), FQB2(5), FQB2(6), FQB2(7), FQB2(8). The corresponding fuzzy qubit rule set is formulated as illustrated in Table 15.12 and as shown in [24].

Each fuzzy qubit has a real and imaginary part which is termed FQBR1 and FQBI1. Similarly for the second fuzzy qubit it is FQBR2 and FQBI2. We then compute the resultant fuzzy qubit by adding the real and imaginary parts separately which is given by FRQR and FRQI.

It could be observed that we have 128 rules based on the combinations. The original fuzzy set illustrates a number of combinations of the real and imaginary parts of the fuzzy qubits, which have phase angles of 0, 45, 90, 135, 180, 225, 270, 315, and 360. Of these, the phase angles 0, 45, 90, 135, and 180 represent positive phase shifts and the phase angles 225, 270, 315, and 360 represent negative phase angles. The negative phase angles are a mirror image of the positive phase angles. The vital factor priorities were shown in Table 15.1. The magnitude of the fuzzy qubit was computed using Equation 15.5.

As we have five vital factors and two fuzzy qubit sets of values, we split the analysis into three parts: vital factor-1 analysis, vital factor-2 analysis, and vital factor-3 analysis. The vital factor-1 analysis of Type of Patient and Tissue parameters is shown in Table 15.13. The other vital factors, taken two at a time, are shown in Tables 15.6 and 15.7 respectively.

TABLE 15.4
Original Fuzzy Set

Rule No.	IF	FQB 1	AND	FQB 2	THEN	RFQB
1	IF	FQB 1(1)	AND	FQB2(1)	THEN	RFQB1
2	IF	FQB 1(1)	AND	FQB2(2)	THEN	RFQB2
3	IF	FQB 1(1)	AND	FQB2(3)	THEN	RFQB3
4	IF	FQB 1(1)	AND	FQB2(4)	THEN	RFQB4
5	IF	FQB 1(1)	AND	FQB2(5)	THEN	RFQB5
6	IF	FQB 1(1)	AND	FQB2(6)	THEN	RFQB6
7	IF	FQB 1(1)	AND	FQB2(7)	THEN	RFQB7
8	IF	FQB 1(1)	AND	FQB2(8)	THEN	RFQB8
9	IF	FQB 1(2)	AND	FQB2(1)	THEN	RFQB9
10	IF	FQB 1(2)	AND	FQB2(2)	THEN	RFQB10
11	IF	FQB 1(2)	AND	FQB2(3)	THEN	RFQB11
12	IF	FQB 1(2)	AND	FQB2(4)	THEN	RFQB12
13	IF	FQB 1(2)	AND	FQB2(5)	THEN	RFQB13
14	IF	FQB 1(2)	AND	FQB2(6)	THEN	RFQB14
15	IF	FQB 1(2)	AND	FQB2(7)	THEN	RFQB15
16	IF	FQB 1(2)	AND	FQB2(8)	THEN	RFQB16
17	IF	FQB 1(3)	AND	FQB2(1)	THEN	RFQB17
18	IF	FQB 1(3)	AND	FQB2(2)	THEN	RFQB18
19	IF	FQB 1(3)	AND	FQB2(3)	THEN	RFQB19
20	IF	FQB 1(3)	AND	FQB2(4)	THEN	RFQB20
21	IF	FQB 1(3)	AND	FQB2(5)	THEN	RFQB21
22	IF	FQB 1(3)	AND	FQB2(6)	THEN	RFQB22
23	IF	FQB 1(3)	AND	FQB2(7)	THEN	RFQB23
24	IF	FQB 1(3)	AND	FQB2(8)	THEN	RFQB24
25	IF	FQB 1(4)	AND	FQB2(1)	THEN	RFQB25
26	IF	FQB 1(4)	AND	FQB2(2)	THEN	RFQB26
27	IF	FQB 1(4)	AND	FQB2(3)	THEN	RFQB27
28	IF	FQB 1(4)	AND	FQB2(4)	THEN	RFQB28
29	IF	FQB 1(4)	AND	FQB2(5)	THEN	RFQB29
30	IF	FQB 1(4)	AND	FQB2(6)	THEN	RFQB30
31	IF	FQB 1(4)	AND	FQB2(7)	THEN	RFQB31
32	IF	FQB 1(4)	AND	FQB2(8)	THEN	RFQB32
33	IF	FQB 1(5)	AND	FQB2(1)	THEN	RFQB33
34	IF	FQB 1(5)	AND	FQB2(2)	THEN	RFQB34
35	IF	FQB 1(5)	AND	FQB2(3)	THEN	RFQB35
36	IF	FQB 1(5)	AND	FQB2(4)	THEN	RFQB36
37	IF	FQB 1(5)	AND	FQB2(5)	THEN	RFQB37
38	IF	FQB 1(5)	AND	FQB2(6)	THEN	RFQB38
39	IF	FQB 1(5)	AND	FQB2(7)	THEN	RFQB39
40	IF	FQB 1(5)	AND	FQB2(8)	THEN	RFQB40
41	IF	FQB 1(6)	AND	FQB2(1)	THEN	RFQB41
42	IF	FQB 1(6)	AND	FQB2(2)	THEN	RFQB42

(Continued)

TABLE 15.4 (Continued)
Original Fuzzy Set

Rule No.	IF	FQB 1	AND	FQB 2	THEN	RFQB
43	IF	FQB 1(6)	AND	FQB2(3)	THEN	RFQB43
44	IF	FQB 1(6)	AND	FQB2(4)	THEN	RFQB44
45	IF	FQB 1(6)	AND	FQB2(5)	THEN	RFQB45
46	IF	FQB 1(6)	AND	FQB2(6)	THEN	RFQB46
47	IF	FQB 1(6)	AND	FQB2(7)	THEN	RFQB47
48	IF	FQB 1(6)	AND	FQB2(8)	THEN	RFQB48
49	IF	FQB 1(7)	AND	FQB2(1)	THEN	RFQB49
50	IF	FQB 1(7)	AND	FQB2(2)	THEN	RFQB50
51	IF	FQB 1(7)	AND	FQB2(3)	THEN	RFQB51
52	IF	FQB 1(7)	AND	FQB2(4)	THEN	RFQB52
53	IF	FQB 1(7)	AND	FQB2(5)	THEN	RFQB53
54	IF	FQB 1(7)	AND	FQB2(6)	THEN	RFQB54
55	IF	FQB 1(7)	AND	FQB2(7)	THEN	RFQB55
56	IF	FQB 1(7)	AND	FQB2(8)	THEN	RFQB56
57	IF	FQB 1(8)	AND	FQB2(1)	THEN	RFQB57
58	IF	FQB 1(8)	AND	FQB2(2)	THEN	RFQB58
59	IF	FQB 1(8)	AND	FQB2(3)	THEN	RFQB59
60	IF	FQB 1(8)	AND	FQB2(4)	THEN	RFQB60
61	IF	FQB 1(8)	AND	FQB2(5)	THEN	RFQB61
62	IF	FQB 1(8)	AND	FQB2(6)	THEN	RFQB62
63	IF	FQB 1(8)	AND	FQB2(7)	THEN	RFQB63
64	IF	FQB 1(8)	AND	FQB2(8)	THEN	RFQB64
65	IF	FQB1(1)	AND	FQB2(1)	THEN	RFQB65
66	IF	FQB1(2)	AND	FQB2(1)	THEN	RFQB66
67	IF	FQB1(3)	AND	FQB2(1)	THEN	RFQB67
68	IF	FQB1(4)	AND	FQB2(1)	THEN	RFQB68
69	IF	FQB1(5)	AND	FQB2(1)	THEN	RFQB69
70	IF	FQB1(6)	AND	FQB2(1)	THEN	RFQB70
71	IF	FQB1(7)	AND	FQB2(1)	THEN	RFQB71
72	IF	FQB1(8)	AND	FQB2(1)	THEN	RFQB72
73	IF	FQB1(1)	AND	FQB2(2)	THEN	RFQB73
74	IF	FQB1(2)	AND	FQB2(2)	THEN	RFQB74
75	IF	FQB1(3)	AND	FQB2(2)	THEN	RFQB75
76	IF	FQB1(4)	AND	FQB2(2)	THEN	RFQB76
77	IF	FQB1(5)	AND	FQB2(2)	THEN	RFQB77
78	IF	FQB1(6)	AND	FQB2(2)	THEN	RFQB78
79	IF	FQB1(7)	AND	FQB2(2)	THEN	RFQB79
80	IF	FQB1(8)	AND	FQB2(2)	THEN	RFQB80
81	IF	FQB1(1)	AND	FQB2(3)	THEN	RFQB81
82	IF	FQB1(2)	AND	FQB2(3)	THEN	RFQB82
83	IF	FQB1(3)	AND	FQB2(3)	THEN	RFQB83
84	IF	FQB1(4)	AND	FQB2(3)	THEN	RFQB84
85	IF	FQB1(5)	AND	FQB2(3)	THEN	RFQB85

(Continued)

TABLE 15.4 (Continued)
Original Fuzzy Set

Rule No.	IF	FQB 1	AND	FQB 2	THEN	RFQB
86	IF	FQB1(6)	AND	FQB2(3)	THEN	RFQB86
87	IF	FQB1(7)	AND	FQB2(3)	THEN	RFQB87
88	IF	FQB1(8)	AND	FQB2(3)	THEN	RFQB88
89	IF	FQB1(1)	AND	FQB2(4)	THEN	RFQB89
90	IF	FQB1(2)	AND	FQB2(4)	THEN	RFQB90
91	IF	FQB1(3)	AND	FQB2(4)	THEN	RFQB91
92	IF	FQB1(4)	AND	FQB2(4)	THEN	RFQB92
93	IF	FQB1(5)	AND	FQB2(4)	THEN	RFQB93
94	IF	FQB1(6)	AND	FQB2(4)	THEN	RFQB94
95	IF	FQB1(7)	AND	FQB2(4)	THEN	RFQB95
96	IF	FQB1(8)	AND	FQB2(4)	THEN	RFQB96
97	IF	FQB1(1)	AND	FQB2(5)	THEN	RFQB97
98	IF	FQB1(2)	AND	FQB2(5)	THEN	RFQB98
99	IF	FQB1(3)	AND	FQB2(5)	THEN	RFQB99
100	IF	FQB1(4)	AND	FQB2(5)	THEN	RFQB100
101	IF	FQB1(5)	AND	FQB2(5)	THEN	RFQB101
102	IF	FQB1(6)	AND	FQB2(5)	THEN	RFQB102
103	IF	FQB1(7)	AND	FQB2(5)	THEN	RFQB103
104	IF	FQB1(8)	AND	FQB2(5)	THEN	RFQB104
105	IF	FQB1(1)	AND	FQB2(6)	THEN	RFQB105
106	IF	FQB1(2)	AND	FQB2(6)	THEN	RFQB106
107	IF	FQB1(3)	AND	FQB2(6)	THEN	RFQB107
108	IF	FQB1(4)	AND	FQB2(6)	THEN	RFQB108
109	IF	FQB1(5)	AND	FQB2(6)	THEN	RFQB109
110	IF	FQB1(6)	AND	FQB2(6)	THEN	RFQB110
111	IF	FQB1(7)	AND	FQB2(6)	THEN	RFQB111
112	IF	FQB1(8)	AND	FQB2(6)	THEN	RFQB112
113	IF	FQB1(1)	AND	FQB2(7)	THEN	RFQB113
114	IF	FQB1(2)	AND	FQB2(7)	THEN	RFQB114
115	IF	FQB1(3)	AND	FQB2(7)	THEN	RFQB115
116	IF	FQB1(4)	AND	FQB2(7)	THEN	RFQB116
117	IF	FQB1(5)	AND	FQB2(7)	THEN	RFQB117
118	IF	FQB1(6)	AND	FQB2(7)	THEN	RFQB118
119	IF	FQB1(7)	AND	FQB2(7)	THEN	RFQB119
120	IF	FQB1(8)	AND	FQB2(7)	THEN	RFQB120
121	IF	FQB1(1)	AND	FQB2(8)	THEN	RFQB121
122	IF	FQB1(2)	AND	FQB2(8)	THEN	RFQB122
123	IF	FQB1(3)	AND	FQB2(8)	THEN	RFQB123
124	IF	FQB1(4)	AND	FQB2(8)	THEN	RFQB124
125	IF	FQB1(5)	AND	FQB2(8)	THEN	RFQB125
126	IF	FQB1(6)	AND	FQB2(8)	THEN	RFQB126
127	IF	FQB1(7)	AND	FQB2(8)	THEN	RFQB127
128	IF	FQB1(8)	AND	FQB2(8)	THEN	RFQB128

TABLE 15.5
Vital Factor Analysis-1

S.No.	FQBR1	FQBI1	FQBR2	FQBI2	5*FQBR1	4*FQBI1	5*FQBR2	4*FQBI2	FRQR/9	FRQI/9	Magnitude
1	1	1	0	0	5	4	0	0	0.74	0.44	0.86
2	1	1	0.33	0	5	4	1.65	0	0.83	0.44	0.94
3	1	1	0.5	0	5	4	2.5	0	0.92	0.44	1.02
4	1	1	0.66	1	5	4	3.3	4	1.11	0.89	1.42
5	1	1	1	0	5	4	5	0	1.11	0.44	1.2
6	1	1	1	0.33	5	4	5	1.32	1.11	0.59	1.26
7	1	1	1	0.5	5	4	5	2	1.11	0.67	1.3
8	1	1	1	0.66	5	4	5	2.64	0.56	0.74	0.92
9	0.66	1	0	0	3.3	4	0	0	0.55	0.44	0.71
10	0.66	1	0.33	0	3.3	4	1.65	0	0.64	0.44	0.78
11	0.66	1	0.5	0	3.3	4	2.5	0	0.73	0.44	0.86
12	0.66	1	0.66	1	3.3	4	3.3	4	0.92	0.89	1.28
13	0.66	1	1	0	3.3	4	5	0	0.92	0.44	1.02
14	0.66	1	1	0.33	3.3	4	5	1.32	0.92	0.59	1.1
15	0.66	1	1	0.5	3.3	4	5	2	0.92	0.67	1.14
16	0.66	1	1	0.66	3.3	4	5	2.64	0.37	0.74	0.82
17	0.5	1	0	0	2.5	4	0	0	0.46	0.44	0.64
18	0.5	1	0.33	0	2.5	4	1.65	0	0.56	0.44	0.71
19	0.5	1	0.5	0	2.5	4	2.5	0	0.64	0.44	0.78
20	0.5	1	0.66	1	2.5	4	3.3	4	0.83	0.89	1.22
21	0.5	1	1	0	2.5	4	5	0	0.83	0.44	0.94
22	0.5	1	1	0.33	2.5	4	5	1.32	0.83	0.59	1.02
23	0.5	1	1	0.5	2.5	4	5	2	0.83	0.67	1.07
24	0.5	1	1	0.66	2.5	4	5	2.64	0.28	0.74	0.79

25	0.33	1	0	0	1.65	4	0	0	0.37	0.44	0.58
26	0.33	1	0.33	0	1.65	4	1.65	0	0.46	0.44	0.64
27	0.33	1	0.5	0	1.65	4	2.5	0	0.55	0.44	0.71
28	0.33	1	0.66	1	1.65	4	3.3	4	0.74	0.89	1.16
29	0.33	1	1	0	1.65	4	5	0	0.74	0.44	0.86
30	0.33	1	1	0.33	1.65	4	5	1.32	0.74	0.59	0.95
31	0.33	1	1	0.5	1.65	4	5	2	0.74	0.67	1
32	0.33	1	1	0.66	1.65	4	5	2.64	0.18	0.74	0.76
33	0	1	0	0	0	4	0	0	0.18	0.44	0.48
34	0	1	0.33	0	0	4	1.65	0	0.28	0.44	0.52
35	0	1	0.5	0	0	4	2.5	0	0.37	0.44	0.58
36	0	1	0.66	1	0	4	3.3	4	0.56	0.89	1.05
37	0	1	1	0	0	4	5	0	0.56	0.44	0.71
38	0	1	1	0.33	0	4	5	1.32	0.56	0.59	0.81
39	0	1	1	0.5	0	4	5	2	0.56	0.67	0.87
40	0	1	1	0.66	0	4	5	2.64	0	0.74	0.74
41	0	0.66	0	0	0	2.64	0	0	0.18	0.29	0.35
42	0	0.66	0.33	0	0	2.64	1.65	0	0.28	0.29	0.4
43	0	0.66	0.5	0	0	2.64	2.5	0	0.37	0.29	0.47
44	0	0.66	0.66	1	0	2.64	3.3	4	0.56	0.74	0.92
45	0	0.66	1	0	0	2.64	5	0	0.56	0.29	0.63
46	0	0.66	1	0.33	0	2.64	5	1.32	0.56	0.44	0.71
47	0	0.66	1	0.5	0	2.64	5	2	0.56	0.52	0.76
48	0	0.66	1	0.66	0	2.64	5	2.64	0	0.59	0.59
49	0	0.5	0	0	0	2	0	0	0.18	0.22	0.29
50	0	0.5	0.33	0	0	2	1.65	0	0.28	0.22	0.36
51	0	0.5	0.5	0	0	2	2.5	0	0.37	0.22	0.43
52	0	0.5	0.66	1	0	2	3.3	4	0.56	0.67	0.87

(Continued)

TABLE 15.5 (Continued)
Vital Factor Analysis-1

S.No.	FQBR1	FQBI1	FQBR2	FQBI2	5*FQBR1	4*FQBI1	5*FQBR2	4*FQBI2	FRQR/9	FRQI/9	Magnitude
53	0	0.5	1	0	0	2	5	0	0.56	0.22	0.6
54	0	0.5	1	0.33	0	2	5	1.32	0.56	0.37	0.67
55	0	0.5	1	0.5	0	2	5	2	0.56	0.44	0.71
56	0	0.5	1	0.66	0	2	5	2.64	0	0.52	0.52
57	0	0.33	0	0	0	1.32	0	0	0.18	0.15	0.23
58	0	0.33	0.33	0	0	1.32	1.65	0	0.28	0.15	0.31
59	0	0.33	0.5	0	0	1.32	2.5	0	0.37	0.15	0.39
60	0	0.33	0.66	1	0	1.32	3.3	4	0.56	0.59	0.81
61	0	0.33	1	0	0	1.32	5	0	0.56	0.15	0.57
62	0	0.33	1	0.33	0	1.32	5	1.32	0.56	0.29	0.63
63	0	0.33	1	0.5	0	1.32	5	2	0.56	0.37	0.67
64	0	0.33	1	0.66	0	1.32	5	2.64	0	0.44	0.44
65	1	1	0	0	5	4	0	0	0.56	0.44	0.71
66	0.66	1	0	0	3.3	4	0	0	0.37	0.44	0.58
67	0.5	1	0	0	2.5	4	0	0	0.28	0.44	0.52
68	0.33	1	0	0	1.65	4	0	0	0.18	0.44	0.48
69	0	1	0	0	0	4	0	0	0	0.44	0.44
70	0	0.66	0	0	0	2.64	0	0	0	0.29	0.29
71	0	0.5	0	0	0	2	0	0	0	0.22	0.22
72	0	0.33	0	0	0	1.32	0	0	0	0.15	0.23
73	1	1	0.33	0	5	4	1.65	0	0.74	0.44	0.86
74	0.66	1	0.33	0	3.3	4	1.65	0	0.55	0.44	0.71
75	0.5	1	0.33	0	2.5	4	1.65	0	0.46	0.44	0.64
76	0.33	1	0.33	0	1.65	4	1.65	0	0.37	0.44	0.58

77	0	1	0.33	0	0	4	1.65	0	0.18	0.44	0.48
78	0	0.66	0.33	0	0	2.64	1.65	0	0.18	0.29	0.35
79	0	0.5	0.33	0	0	2	1.65	0	0.18	0.22	0.29
80	0	0.33	0.33	0	0	1.32	1.65	0	0.28	0.15	0.31
81	1	1	0.5	0	5	4	2.5	0	0.83	0.44	0.94
82	0.66	1	0.5	0	3.3	4	2.5	0	0.64	0.44	0.78
83	0.5	1	0.5	0	2.5	4	2.5	0	0.56	0.44	0.71
84	0.33	1	0.5	0	1.65	4	2.5	0	0.46	0.44	0.64
85	0	1	0.5	0	0	4	2.5	0	0.28	0.44	0.52
86	0	0.66	0.5	0	0	2.64	2.5	0	0.28	0.29	0.4
87	0	0.5	0.5	0	0	2	2.5	0	0.28	0.22	0.36
88	0	0.33	0.5	0	0	1.32	2.5	0	0.37	0.15	0.39
89	1	1	0.66	1	5	4	3.3	4	0.92	0.89	1.28
90	0.66	1	0.66	1	3.3	4	3.3	4	0.73	0.89	1.15
91	0.5	1	0.66	1	2.5	4	3.3	4	0.64	0.89	1.1
92	0.33	1	0.66	1	1.65	4	3.3	4	0.55	0.89	1.05
93	0	1	0.66	1	0	4	3.3	4	0.37	0.89	0.96
94	0	0.66	0.66	1	0	2.64	3.3	4	0.37	0.74	0.82
95	0	0.5	0.66	1	0	2	3.3	4	0.37	0.67	0.76
96	0	0.33	0.66	1	0	1.32	3.3	4	0.56	0.59	0.81
97	1	1	1	0	5	4	5	0	1.11	0.44	1.2
98	0.66	1	1	0	3.3	4	5	0	0.92	0.44	1.02
99	0.5	1	1	0	2.5	4	5	0	0.83	0.44	0.94
100	0.33	1	1	0	1.65	4	5	0	0.74	0.44	0.86
101	0	1	1	0	0	4	5	0	0.56	0.44	0.71
102	0	0.66	1	0	0	2.64	5	0	0.56	0.29	0.63
103	0	0.5	1	0	0	2	5	0	0.56	0.22	0.6
104	0	0.33	1	0	0	1.32	5	0	0.56	0.15	0.57

(Continued)

TABLE 15.5 (Continued)
Vital Factor Analysis-1

S.No.	FQBR1	FQBI1	FQBR2	FQBI2	5*FQBR1	4*FQBI1	5*FQBR2	4*FQBI2	FRQR/9	FRQI/9	Magnitude
105	1	1	1	0.33	5	4	5	1.32	1.11	0.59	1.26
106	0.66	1	1	0.33	3.3	4	5	1.32	0.92	0.59	1.1
107	0.5	1	1	0.33	2.5	4	5	1.32	0.83	0.59	1.02
108	0.33	1	1	0.33	1.65	4	5	1.32	0.74	0.59	0.95
109	0	1	1	0.33	0	4	5	1.32	0.56	0.59	0.81
110	0	0.66	1	0.33	0	2.64	5	1.32	0.56	0.44	0.71
111	0	0.5	1	0.33	0	2	5	1.32	0.56	0.37	0.67
112	0	0.33	1	0.33	0	1.32	5	1.32	0.56	0.29	0.63
113	1	1	1	0.5	5	4	5	2	1.11	0.67	1.3
114	0.66	1	1	0.5	3.3	4	5	2	0.92	0.67	1.14
115	0.5	1	1	0.5	2.5	4	5	2	0.83	0.67	1.07
116	0.33	1	1	0.5	1.65	4	5	2	0.74	0.67	1
117	0	1	1	0.5	0	4	5	2	0.56	0.67	0.87
118	0	0.66	1	0.5	0	2.64	5	2	0.56	0.52	0.76
119	0	0.5	1	0.5	0	2	5	2	0.56	0.44	0.71
120	0	0.33	1	0.5	0	1.32	5	2	0.56	0.37	0.67
121	1	1	1	0.66	5	4	5	2.64	1.11	0.74	1.33
122	0.66	1	1	0.66	3.3	4	5	2.64	0.92	0.74	1.18
123	0.5	1	1	0.66	2.5	4	5	2.64	0.83	0.74	1.11
124	0.33	1	1	0.66	1.65	4	5	2.64	0.74	0.74	1.04
125	0	1	1	0.66	0	4	5	2.64	0.56	0.74	0.92
126	0	0.66	1	0.66	0	2.64	5	2.64	0.56	0.59	0.81
127	0	0.5	1	0.66	0	2	5	2.64	0.56	0.52	0.76
128	0	0.33	1	0.66	0	1.32	5	2.64	0	0.44	0.44

TABLE 15.6
Vital Factor Analysis-2

S.No.	QBR1	QBI1	QBR2	QBI2	3*QBR1	2*QBI1	3*QBR2	2*QBI2	RQR/5	RQI/5	Magnitude
1	1	1	0	0	3	2	0	0	0.8	0.4	0.89
2	1	1	0.33	0	3	2	0.99	0	0.9	0.4	0.98
3	1	1	0.5	0	3	2	1.5	0	1	0.4	1.07
4	1	1	0.66	1	3	2	1.98	2	1.2	0.8	1.44
5	1	1	1	0	3	2	3	0	1.2	0.4	1.26
6	1	1	1	0.33	3	2	3	0.66	1.2	0.53	1.31
7	1	1	1	0.5	3	2	3	1	1.2	0.6	1.34
8	1	1	1	0.66	3	2	3	1.32	0.6	0.66	0.89
9	0.66	1	0	0	1.98	2	0	0	0.59	0.4	0.72
10	0.66	1	0.33	0	1.98	2	0.99	0	0.7	0.4	0.8
11	0.66	1	0.5	0	1.98	2	1.5	0	0.79	0.4	0.89
12	0.66	1	0.66	1	1.98	2	1.98	2	1	0.8	1.28
13	0.66	1	1	0	1.98	2	3	0	1	0.4	1.07
14	0.66	1	1	0.33	1.98	2	3	0.66	1	0.53	1.13
15	0.66	1	1	0.5	1.98	2	3	1	1	0.6	1.16
16	0.66	1	1	0.66	1.98	2	3	1.32	0.4	0.66	0.77
17	0.5	1	0	0	1.5	2	0	0	0.5	0.4	0.64
18	0.5	1	0.33	0	1.5	2	0.99	0	0.6	0.4	0.72
19	0.5	1	0.5	0	1.5	2	1.5	0	0.7	0.4	0.8
20	0.5	1	0.66	1	1.5	2	1.98	2	0.9	0.8	1.2
21	0.5	1	1	0	1.5	2	3	0	0.9	0.4	0.98
22	0.5	1	1	0.33	1.5	2	3	0.66	0.9	0.53	1.05
23	0.5	1	1	0.5	1.5	2	3	1	0.9	0.6	1.08

(Continued)

TABLE 15.6 (Continued)
Vital Factor Analysis-2

S.No.	QBR1	QBI1	QBR2	QBI2	3*QBR1	2*QBI1	3*QBR2	2*QBI2	RQR/5	RQI/5	Magnitude
24	0.5	1	1	0.66	1.5	2	3	1.32	0.3	0.66	0.73
25	0.33	1	0	0	0.99	2	0	0	0.4	0.4	0.56
26	0.33	1	0.33	0	0.99	2	0.99	0	0.5	0.4	0.64
27	0.33	1	0.5	0	0.99	2	1.5	0	0.59	0.4	0.72
28	0.33	1	0.66	1	0.99	2	1.98	2	0.8	0.8	1.13
29	0.33	1	1	0	0.99	2	3	0	0.8	0.4	0.89
30	0.33	1	1	0.33	0.99	2	3	0.66	0.8	0.53	0.96
31	0.33	1	1	0.5	0.99	2	3	1	0.8	0.6	1
32	0.33	1	1	0.66	0.99	2	3	1.32	0.2	0.66	0.69
33	0	1	0	0	0	2	0	0	0.2	0.4	0.45
34	0	1	0.33	0	0	2	0.99	0	0.3	0.4	0.5
35	0	1	0.5	0	0	2	1.5	0	0.4	0.4	0.56
36	0	1	0.66	1	0	2	1.98	2	0.6	0.8	1
37	0	1	1	0	0	2	3	0	0.6	0.4	0.72
38	0	1	1	0.33	0	2	3	0.66	0.6	0.53	0.8
39	0	1	1	0.5	0	2	3	1	0.6	0.6	0.85
40	0	1	1	0.66	0	2	3	1.32	0	0.66	0.66
41	0	0.66	0	0	0	1.32	0	0	0.2	0.26	0.33
42	0	0.66	0.33	0	0	1.32	0.99	0	0.3	0.26	0.4
43	0	0.66	0.5	0	0	1.32	1.5	0	0.4	0.26	0.48
44	0	0.66	0.66	1	0	1.32	1.98	2	0.6	0.66	0.89
45	0	0.66	1	0	0	1.32	3	0	0.6	0.26	0.66
46	0	0.66	1	0.33	0	1.32	3	0.66	0.6	0.4	0.72
47	0	0.66	1	0.5	0	1.32	3	1	0.6	0.46	0.76

48	0	0.66	1	0.66	0	1.32	3	1.32	0	0.53	0.53
49	0	0.5	0	0	0	1	0	0	0.2	0.2	0.28
50	0	0.5	0.33	0	0	1	0.99	0	0.3	0.2	0.36
51	0	0.5	0.5	0	0	1	1.5	2	0.4	0.2	0.44
52	0	0.5	0.66	1	0	1	1.98	0	0.6	0.6	0.85
53	0	0.5	1	0	0	1	3	0.66	0.6	0.2	0.63
54	0	0.5	1	0.33	0	1	3	1	0.6	0.33	0.69
55	0	0.5	1	0.5	0	1	3	1.32	0.6	0.4	0.72
56	0	0.5	1	0.66	0	1	3	0	0	0.46	0.46
57	0	0.33	0	0	0	0.66	0	0	0.2	0.13	0.24
58	0	0.33	0.33	0	0	0.66	0.99	0	0.3	0.13	0.33
59	0	0.33	0.5	0	0	0.66	1.5	2	0.4	0.13	0.42
60	0	0.33	0.66	1	0	0.66	1.98	0	0.6	0.53	0.8
61	0	0.33	1	0	0	0.66	3	0.66	0.6	0.13	0.61
62	0	0.33	1	0.33	0	0.66	3	1	0.6	0.26	0.66
63	0	0.33	1	0.5	0	0.66	3	1.32	0.6	0.33	0.69
64	0	0.33	1	0.66	0	0.66	3	0	0	0.4	0.4
65	1	1	0	0	3	2	0	0	0.6	0.4	0.72
66	0.66	1	0	0	1.98	2	0	0	0.4	0.4	0.56
67	0.5	1	0	0	1.5	2	0	0	0.3	0.4	0.5
68	0.33	1	0	0	0.99	2	0	0	0.2	0.4	0.45
69	0	1	0	0	0	2	0	0	0	0.4	0.4
70	0	0.66	0	0	0	1.32	0	0	0	0.26	0.26
71	0	0.5	0	0	0	1	0	0	0	0.2	0.2
72	0	0.33	0	0	0	0.66	0	0	0.2	0.13	0.24
73	1	1	0.33	0	3	2	0.99	0	0.8	0.4	0.89
74	0.66	1	0.33	0	1.98	2	0.99	0	0.59	0.4	0.72
75	0.5	1	0.33	0	1.5	2	0.99	0	0.5	0.4	0.64

(Continued)

TABLE 15.6 (Continued)
Vital Factor Analysis-2

S.No.	QBR1	QBI1	QBR2	QBI2	3*QBR1	2*QBI1	3*QBR2	2*QBI2	RQR/5	RQI/5	Magnitude
76	0.33	1	0.33	0	0.99	2	0.99	0	0.4	0.4	0.56
77	0	1	0.33	0	0	2	0.99	0	0.2	0.4	0.45
78	0	0.66	0.33	0	0	1.32	0.99	0	0.2	0.26	0.33
79	0	0.5	0.33	0	0	1	0.99	0	0.2	0.2	0.28
80	0	0.33	0.33	0	0	0.66	0.99	0	0.3	0.13	0.33
81	1	1	0.5	0	3	2	1.5	0	0.9	0.4	0.98
82	0.66	1	0.5	0	1.98	2	1.5	0	0.7	0.4	0.8
83	0.5	1	0.5	0	1.5	2	1.5	0	0.6	0.4	0.72
84	0.33	1	0.5	0	0.99	2	1.5	0	0.5	0.4	0.64
85	0	1	0.5	0	0	2	1.5	0	0.3	0.4	0.5
86	0	0.66	0.5	0	0	1.32	1.5	0	0.3	0.26	0.4
87	0	0.5	0.5	0	0	1	1.5	0	0.3	0.2	0.36
88	0	0.33	0.5	0	0	0.66	1.5	0	0.4	0.13	0.42
89	1	1	0.66	1	3	2	1.98	2	1	0.8	1.28
90	0.66	1	0.66	1	1.98	2	1.98	2	0.79	0.8	1.13
91	0.5	1	0.66	1	1.5	2	1.98	2	0.7	0.8	1.06
92	0.33	1	0.66	1	0.99	2	1.98	2	0.59	0.8	1
93	0	1	0.66	1	0	2	1.98	2	0.4	0.8	0.89
94	0	0.66	0.66	1	0	1.32	1.98	2	0.4	0.66	0.77
95	0	0.5	0.66	1	0	1	1.98	2	0.4	0.6	0.72
96	0	0.33	0.66	1	0	0.66	1.98	2	0.6	0.53	0.8
97	1	1	1	0	3	2	3	0	1.2	0.4	1.26
98	0.66	1	1	0	1.98	2	3	0	1	0.4	1.07
99	0.5	1	1	0	1.5	2	3	0	0.9	0.4	0.98
100	0.33	1	1	0	0.99	2	3	0	0.8	0.4	0.89

101	0	1	1	0	0	2	3	0	0.6	0.4	0.72
102	0	0.66	1	0	0	1.32	3	0	0.6	0.26	0.66
103	0	0.5	1	0	0	1	3	0	0.6	0.2	0.63
104	0	0.33	1	0	0	0.66	3	0	0.6	0.13	0.61
105	1	1	1	0.33	3	2	3	0.66	1.2	0.53	1.31
106	0.66	1	1	0.33	1.98	2	3	0.66	1	0.53	1.13
107	0.5	1	1	0.33	1.5	2	3	0.66	0.9	0.53	1.05
108	0.33	1	1	0.33	0.99	2	3	0.66	0.8	0.53	0.96
109	0	1	1	0.33	0	2	3	0.66	0.6	0.53	0.8
110	0	0.66	1	0.33	0	1.32	3	0.66	0.6	0.4	0.72
111	0	0.5	1	0.33	0	1	3	0.66	0.6	0.33	0.69
112	0	0.33	1	0.33	0	0.66	3	0.66	0.6	0.26	0.66
113	1	1	1	0.5	3	2	3	1	1.2	0.6	1.34
114	0.66	1	1	0.5	1.98	2	3	1	1	0.6	1.16
115	0.5	1	1	0.5	1.5	2	3	1	0.9	0.6	1.08
116	0.33	1	1	0.5	0.99	2	3	1	0.8	0.6	1
117	0	1	1	0.5	0	2	3	1	0.6	0.6	0.85
118	0	0.66	1	0.5	0	1.32	3	1	0.6	0.46	0.76
119	0	0.5	1	0.5	0	1	3	1	0.6	0.4	0.72
120	0	0.33	1	0.5	3	0.66	3	1	0.6	0.33	0.69
121	1	1	1	0.66	1.98	2	3	1.32	1.2	0.66	1.37
122	0.66	1	1	0.66	1.5	2	3	1.32	1	0.66	1.2
123	0.5	1	1	0.66	0.99	2	3	1.32	0.9	0.66	1.12
124	0.33	1	1	0.66	0	2	3	1.32	0.8	0.66	1.04
125	0	1	1	0.66	0	2	3	1.32	0.6	0.66	0.89
126	0	0.66	1	0.66	0	1.32	3	1.32	0.6	0.53	0.8
127	0	0.5	1	0.66	0	1	3	1.32	0.6	0.46	0.76
128	0	0.33	1	0.66	0	0.66	3	1.32	0	0.4	0.4

TABLE 15.7
Vital Factor Analysis-3

S.No.	QBR1	QBI1	QBR2	QBI2	2*QBR1	1*QBI1	2*QBR2	1*QBI2	RQR/3	RQI/3	Magnitude
1	1	1	0	0	2	1	0	0	0.89	0.33	0.95
2	1	1	0.33	0	2	1	0.66	0	1	0.33	1.05
3	1	1	0.5	0	2	1	1	0	1.11	0.33	1.16
4	1	1	0.66	1	2	1	1.32	1	1.33	0.67	1.49
5	1	1	1	0	2	1	2	0	1.33	0.33	1.37
6	1	1	1	0.33	2	1	2	0.33	1.33	0.44	1.41
7	1	1	1	0.5	2	1	2	0.5	1.33	0.5	1.42
8	1	1	1	0.66	2	1	2	0.66	0.67	0.55	0.87
9	0.66	1	0	0	1.32	1	0	0	0.66	0.33	0.74
10	0.66	1	0.33	0	1.32	1	0.66	0	0.77	0.33	0.84
11	0.66	1	0.5	0	1.32	1	1	0	0.88	0.33	0.94
12	0.66	1	0.66	1	1.32	1	1.32	1	1.11	0.67	1.29
13	0.66	1	1	0	1.32	1	2	0	1.11	0.33	1.16
14	0.66	1	1	0.33	1.32	1	2	0.33	1.11	0.44	1.19
15	0.66	1	1	0.5	1.32	1	2	0.5	1.11	0.5	1.21
16	0.66	1	1	0.66	1.32	1	2	0.66	0.44	0.55	0.71
17	0.5	1	0	0	1	1	0	0	0.55	0.33	0.65
18	0.5	1	0.33	0	1	1	0.66	0	0.67	0.33	0.75
19	0.5	1	0.5	0	1	1	1	0	0.77	0.33	0.84
20	0.5	1	0.66	1	1	1	1.32	1	1	0.67	1.2
21	0.5	1	1	0	1	1	2	0	1	0.33	1.05
22	0.5	1	1	0.33	1	1	2	0.33	1	0.44	1.09
23	0.5	1	1	0.5	1	1	2	0.5	1	0.5	1.12
24	0.5	1	1	0.66	1	1	2	0.66	0.33	0.55	0.65

25	0.33	0	1	0	0.66	1	0	0	0.44	0.33	0.55
26	0.33	0.33	1	0	0.66	1	0.66	0	0.55	0.33	0.65
27	0.33	0.5	1	0	0.66	1	1	0	0.66	0.33	0.74
28	0.33	0.66	1	1	0.66	1	1.32	1	0.89	0.67	1.11
29	0.33	1	1	0	0.66	1	2	0	0.89	0.33	0.95
30	0.33	1	1	0.33	0.66	1	2	0.33	0.89	0.44	0.99
31	0.33	1	1	0.5	0.66	1	2	0.5	0.89	0.5	1.02
32	0.33	1	1	0.66	0.66	1	2	0.66	0.22	0.55	0.6
33	0	0	1	0	0	1	0	0	0.22	0.33	0.4
34	0	0.33	1	0	0	1	0.66	0	0.33	0.33	0.47
35	0	0.5	1	0	0	1	1	0	0.44	0.33	0.55
36	0	0.66	1	1	0	1	1.32	1	0.67	0.67	0.94
37	0	1	1	0	0	1	2	0	0.67	0.33	0.75
38	0	1	1	0.33	0	1	2	0.33	0.67	0.44	0.8
39	0	1	1	0.5	0	1	2	0.5	0.67	0.5	0.83
40	0	1	1	0.66	0	1	2	0.66	0	0.55	0.55
41	0	0	0.66	0	0	0.66	0	0	0.22	0.22	0.31
42	0	0.33	0.66	0	0	0.66	0.66	0	0.33	0.22	0.4
43	0	0.5	0.66	0	0	0.66	1	0	0.44	0.22	0.49
44	0	0.66	0.66	1	0	0.66	1.32	1	0.67	0.55	0.87
45	0	1	0.66	0	0	0.66	2	0	0.67	0.22	0.7
46	0	1	0.66	0.33	0	0.66	2	0.33	0.67	0.33	0.74
47	0	1	0.66	0.5	0	0.66	2	0.5	0.67	0.39	0.77
48	0	1	0.66	0.66	0	0.66	2	0.66	0	0.44	0.44
49	0	0	0.5	0	0	0.5	0	0	0.22	0.17	0.28
50	0	0.33	0.5	0	0	0.5	0.66	0	0.33	0.17	0.37
51	0	0.5	0.5	0	0	0.5	1	0	0.44	0.17	0.47
52	0	0.66	0.5	1	0	0.5	1.32	1	0.67	0.5	0.83

(Continued)

TABLE 15.7 (Continued)
Vital Factor Analysis-3

S.No.	QBR1	QBI1	QBR2	QBI2	2*QBR1	1*QBI1	2*QBR2	1*QBI2	RQR/3	RQI/3	Magnitude
53	0	0.5	1	0	0	0.5	2	0	0.67	0.17	0.69
54	0	0.5	1	0.33	0	0.5	2	0.33	0.67	0.28	0.72
55	0	0.5	1	0.5	0	0.5	2	0.5	0.67	0.33	0.75
56	0	0.5	1	0.66	0	0.5	2	0.66	0	0.39	0.39
57	0	0.33	0	0	0	0.33	0	0	0.22	0.11	0.25
58	0	0.33	0.33	0	0	0.33	0.66	0	0.33	0.11	0.35
59	0	0.33	0.5	0	0	0.33	1	0	0.44	0.11	0.45
60	0	0.33	0.66	1	0	0.33	1.32	1	0.67	0.44	0.8
61	0	0.33	1	0	0	0.33	2	0	0.67	0.11	0.68
62	0	0.33	1	0.33	0	0.33	2	0.33	0.67	0.22	0.7
63	0	0.33	1	0.5	0	0.33	2	0.5	0.67	0.28	0.72
64	0	0.33	1	0.66	0	0.33	2	0.66	0	0.33	0.33
65	1	1	0	0	2	1	0	0	0.67	0.33	0.75
66	0.66	1	0	0	1.32	1	0	0	0.44	0.33	0.55
67	0.5	1	0	0	1	1	0	0	0.33	0.33	0.47
68	0.33	1	0	0	0.66	1	0	0	0.22	0.33	0.4
69	0	1	0	0	0	1	0	0	0	0.33	0.33
70	0	0.66	0	0	0	0.66	0	0	0	0.22	0.22
71	0	0.5	0	0	0	0.5	0	0	0	0.17	0.17
72	0	0.33	0	0	0	0.33	0	0	0.22	0.11	0.25
73	1	1	0.33	0	2	1	0.66	0	0.89	0.33	0.95
74	0.66	1	0.33	0	1.32	1	0.66	0	0.66	0.33	0.74
75	0.5	1	0.33	0	1	1	0.66	0	0.55	0.33	0.65
76	0.33	1	0.33	0	0.66	1	0.66	0	0.44	0.33	0.55
77	0	1	0.33	0	0	1	0.66	0	0.22	0.33	0.4

78	0	0.66	0.33	0	0	0.66	0.66	0	0.66	0.22	0.22	0.31
79	0	0.5	0.33	0	0	0.5	0.66	0	0.5	0.22	0.17	0.28
80	0	0.33	0.33	0	0	0.33	0.66	0	0.33	0.33	0.11	0.35
81	1	1	0.5	0	2	1	1	0	1	1	0.33	1.05
82	0.66	1	0.5	0	1.32	1	1	0	1	0.77	0.33	0.84
83	0.5	1	0.5	0	1	1	1	0	1	0.67	0.33	0.75
84	0.33	1	0.5	0	0.66	1	1	0	1	0.55	0.33	0.65
85	0	0.66	0.5	0	0	0.66	1	0	0.66	0.33	0.33	0.47
86	0	0.5	0.5	0	0	0.5	1	0	0.5	0.33	0.22	0.4
87	0	0.33	0.5	0	0	0.33	1	0	0.33	0.33	0.17	0.37
88	0	1	0.5	0	0	1	1	0	1	0.44	0.11	0.45
89	1	1	0.66	1	2	1	1.32	1	1	1.11	0.67	1.29
90	0.66	1	0.66	1	1.32	1	1.32	1	1	0.88	0.67	1.1
91	0.5	1	0.66	1	1	1	1.32	1	1	0.77	0.67	1.02
92	0.33	1	0.66	1	0.66	1	1.32	1	1	0.66	0.67	0.94
93	0	0.66	0.66	1	0	0.66	1.32	1	0.66	0.44	0.67	0.8
94	0	0.5	0.66	1	0	0.5	1.32	1	0.5	0.44	0.55	0.71
95	0	0.33	0.66	1	0	0.33	1.32	1	0.33	0.44	0.5	0.67
96	0	1	0.66	1	0	1	1.32	1	1	0.67	0.44	0.8
97	1	1	1	0	2	1	2	0	1	1.33	0.33	1.37
98	0.66	1	1	0	1.32	1	2	0	1	1.11	0.33	1.16
99	0.5	1	1	0	1	1	2	0	1	1	0.33	1.05
100	0.33	1	1	0	0.66	1	2	0	0.89	0.89	0.33	0.95
101	0	0.66	1	0	0	0.66	2	0	0.67	0.67	0.33	0.75
102	0	0.5	1	0	0	0.5	2	0	0.67	0.67	0.22	0.7
103	0	0.33	1	0	0	0.33	2	0	0.67	0.67	0.17	0.69
104	0	0.33	1	0	0	0.33	2	0	0.67	0.67	0.11	0.68

(Continued)

TABLE 15.7 (Continued)
Vital Factor Analysis-3

S.No.	QBR1	QBI1	QBR2	QBI2	2*QBR1	1*QBI1	2*QBR2	1*QBI2	RQR/3	RQI/3	Magnitude
105	1	1	1	0.33	2	1	2	0.33	1.33	0.44	1.41
106	0.66	1	1	0.33	1.32	1	2	0.33	1.11	0.44	1.19
107	0.5	1	1	0.33	1	1	2	0.33	1	0.44	1.09
108	0.33	1	1	0.33	0.66	1	2	0.33	0.89	0.44	0.99
109	0	1	1	0.33	0	1	2	0.33	0.67	0.44	0.8
110	0	0.66	1	0.33	0	0.66	2	0.33	0.67	0.33	0.74
111	0	0.5	1	0.33	0	0.5	2	0.33	0.67	0.28	0.72
112	0	0.33	1	0.33	0	0.33	2	0.33	0.67	0.22	0.7
113	1	1	1	0.5	2	1	2	0.5	1.33	0.5	1.42
114	0.66	1	1	0.5	1.32	1	2	0.5	1.11	0.5	1.21
115	0.5	1	1	0.5	1	1	2	0.5	1	0.5	1.12
116	0.33	1	1	0.5	0.66	1	2	0.5	0.89	0.5	1.02
117	0	1	1	0.5	0	1	2	0.5	0.67	0.5	0.83
118	0	0.66	1	0.5	0	0.66	2	0.5	0.67	0.39	0.77
119	0	0.5	1	0.5	0	0.5	2	0.5	0.67	0.33	0.75
120	0	0.33	1	0.5	0	0.33	2	0.5	0.67	0.28	0.72
121	1	1	1	0.66	2	1	2	0.66	1.33	0.55	1.44
122	0.66	1	1	0.66	1.32	1	2	0.66	1.11	0.55	1.24
123	0.5	1	1	0.66	1	1	2	0.66	1	0.55	1.14
124	0.33	1	1	0.66	0.66	1	2	0.66	0.89	0.55	1.05
125	0	1	1	0.66	0	1	2	0.66	0.67	0.55	0.87
126	0	0.66	1	0.66	0	0.66	2	0.66	0.67	0.44	0.8
127	0	0.5	1	0.66	0	0.5	2	0.66	0.67	0.39	0.77
128	0	0.33	1	0.66	0	0.33	2	0.66	0	0.33	0.33

Each fuzzy qubit of the real and imaginary part is multiplied by the corresponding priority and the resultant weighted average of the fuzzy qubit real and imaginary parts is computed. Table 15.13 gives the vital factor-1 analysis as explained in [23] and [24].

The weightage for each vital factor was decided based on the criticality of the parameter. The digital twin model is designed first.

15.3.5 DIGITAL TWIN MODEL FORMULATION

Mathematically the digital twin model is derived by designing a physical model of the object under consideration. Once this is done a mathematical model is formulated, which could be illustrated as a mathematical equation with specific boundary conditions. The latter would involve an initial and final value. This forms the basis or the initial dataset to be trained iteratively for optimality.

The optimality and number of iterations depends on the quantum machine learning algorithm and the activation function chosen. In this work the fuzzy model provides a set of rules and the learning function is decided by a neural based optimisation which eliminates rules whose strength is less than the threshold, which depends on the number of quantum datasets formulated and the number of rules which decide the qubit mapping phase.

15.3.6 WEIGHTAGE OF VITAL FACTORS

The weightage is decided according to the criticality of the parameter which makes the decision choice. The parameters are first classified as critical and non-critical. Based on this analysis the optimal bounds are chosen, which are the basis for creating an objective function. The other constraints which are dependent on the criticality are dependent factors. This is a linear programming model and the optimality solution is obtained. In some cases we have an alternative optimal solution which would lead to a degeneracy issue.

When an alternative optimal solution occurs we use tie resolution factors to modify the dataset obtained in the iterations of the basis.

The objective function is modelled with the equation

$$\text{Obj} = \text{MaxorMin}\left(\text{Vitfac}(1) + \text{Vitfac}(2) + \text{Vitfac}(3)\right) \tag{15.6}$$

where MaxorMin maximises or minimises the attributes Vitfac(1), Vitfac(2), and Vitfac(3) which are the vital factors for the digital twin model.

The constraints are modelled with the equation

$$\text{Cons} = \text{MaxorMin}\left(\text{DF}(1) + \text{DF}(2) + \text{DF}(3)\right) \tag{15.7}$$

where MaxorMin maximises or minimises the attributes DF(1), DF(2), and DF(3), which are the dependent factors for the digital twin model.

The vital factors and their analysis based on the weights were shown in Table 15.6. This explains the initial boundary condition set for the basis, which was optimised according to the threshold computed. Table 15.6 gives the vital factor-2 analysis as explained in [24]. The next set of basis weights was shown in Table 15.7. The weights of the corresponding vital factors are shown there along with the vital factor-3 analysis, as explained in [24].

The dependent factors and their corresponding priorities are shown in Table 15.8. We have six dependent factors. The analysis of the vital factors result in three magnitudes. Hence we take three dependent factors at a time and compute the weighted average. Each magnitude of the vital qubit value is multiplied by the dependent factor's priority and the weighted average is computed which is then taken as the partial result. This is shown in Tables 15.6 and 15.7 respectively. The combined results of both these partial results are computed by assigning a priority of 2 and 1 to the partial result 1 and 2 respectively; the weighted average is computed for the same and illustrated in Table 15.8.

The threshold for this rule set is computed by using the Equation 15.6:

$$T = \left(\frac{\left(\sum_{i=1}^{i=n} (\mathrm{Rno} \times \mathrm{Magnitude}) \right)}{\sum_{i=1}^{i=n} \mathrm{Rno}} \right) \tag{15.8}$$

The rule strength is computed using Equation 15.7:

$$\mathrm{RS} = \left((\mathrm{Rno} \times \mathrm{Magnitude}) - T \right) \tag{15.9}$$

Rules with RS < 1 are neglected; it may be observed that out of 128 rules only 30 rules are used for the modified rule set. The dependent factor analysis for the first set of three output values of vital factors computed in Tables 15.13 and 15.6 is shown in Table 15.8.

The dependent factors explaining the constraints are shown in Table 15.8 and relate to the partial results of the analysis. The table also shows the error factors which exceed the threshold and their iterative analysis to reach an optimal value. Table 15.8 gives the dependent factor-1 analysis as mentioned and explained in [24].

The second set of dependent factors is shown in Table 15.9. The error associated beyond the threshold has been analysed and the optimality for the combined basis has been processed.

The dependent factor analysis for the next set of three output values of the vital factors computed in Tables 15.6 and 15.7 is shown in Table 15.9, which shows the dependent factor-2 analysis as explained in [24]. The combined analysis of the partial results assessing the total error and maximising the objective function is shown in Table 15.10. The objective function's individual maximised value for each attribute explains the criticality in choosing the appropriate bounds for the vital parameters.

The combined analysis for the partial results obtained in Tables 15.8 and 15.9 is shown in Table 15.10, which reveals the combined result analysis as explained in [24].

The threshold was computed for the modified basis and the rule set; a zero error rate was chosen. The chosen rules attempt to maximise the individual attributes'

TABLE 15.8
Dependent Factor-1 Analysis

R.No.	IVF1	IVF2	IVF3	Part Res-1 = ((6*IVF1)+ (5*IVF2)+(4*IVF3))/15
1	0.86	0.89	0.95	0.9
2	0.94	0.98	1.05	0.99
3	1.02	1.07	1.16	1.08
4	1.42	1.44	1.49	1.45
5	1.2	1.26	1.37	1.27
6	1.26	1.31	1.41	1.32
7	1.3	1.34	1.42	1.35
8	0.92	0.89	0.87	0.9
9	0.71	0.72	0.74	0.72
10	0.78	0.8	0.84	0.81
11	0.86	0.89	0.94	0.89
12	1.28	1.28	1.29	1.28
13	1.02	1.07	1.16	1.08
14	1.1	1.13	1.19	1.13
15	1.14	1.16	1.21	1.17
16	0.82	0.77	0.71	0.78
17	0.64	0.64	0.65	0.64
18	0.71	0.72	0.75	0.72
19	0.78	0.8	0.84	0.81
20	1.22	1.2	1.2	1.21
21	0.94	0.98	1.05	0.99
22	1.02	1.05	1.09	1.05
23	1.07	1.08	1.12	1.09
24	0.79	0.73	0.65	0.73
25	0.58	0.56	0.55	0.57
26	0.64	0.64	0.65	0.64
27	0.71	0.72	0.74	0.72
28	1.16	1.13	1.11	1.13
29	0.86	0.89	0.95	0.9
30	0.95	0.96	0.99	0.96
31	1	1	1.02	1
32	0.76	0.69	0.6	0.69
33	0.48	0.45	0.4	0.45
34	0.52	0.5	0.47	0.5
35	0.58	0.56	0.55	0.57
36	1.05	1	0.94	1
37	0.71	0.72	0.75	0.72
38	0.81	0.8	0.8	0.81
39	0.87	0.85	0.83	0.85
40	0.74	0.66	0.55	0.66
41	0.35	0.33	0.31	0.33

(Continued)

TABLE 15.8 (Continued)
Dependent Factor-1 Analysis

R.No.	IVF1	IVF2	IVF3	Part Res-1 = ((6*IVF1)+ (5*IVF2)+(4*IVF3))/15
42	0.4	0.4	0.4	0.4
43	0.47	0.48	0.49	0.48
44	0.92	0.89	0.87	0.9
45	0.63	0.66	0.7	0.66
46	0.71	0.72	0.74	0.72
47	0.76	0.76	0.77	0.76
48	0.59	0.53	0.44	0.53
49	0.29	0.28	0.28	0.28
50	0.36	0.36	0.37	0.36
51	0.43	0.44	0.47	0.44
52	0.87	0.85	0.83	0.85
53	0.6	0.63	0.69	0.63
54	0.67	0.69	0.72	0.69
55	0.71	0.72	0.75	0.72
56	0.52	0.46	0.39	0.46
57	0.23	0.24	0.25	0.24
58	0.31	0.33	0.35	0.33
59	0.39	0.42	0.45	0.42
60	0.81	0.8	0.8	0.81
61	0.57	0.61	0.68	0.61
62	0.63	0.66	0.7	0.66
63	0.67	0.69	0.72	0.69
64	0.44	0.4	0.33	0.4
65	0.71	0.72	0.75	0.72
66	0.58	0.56	0.55	0.57
67	0.52	0.5	0.47	0.5
68	0.48	0.45	0.4	0.45
69	0.44	0.4	0.33	0.4
70	0.29	0.26	0.22	0.26
71	0.22	0.2	0.17	0.2
72	0.23	0.24	0.25	0.24
73	0.86	0.89	0.95	0.9
74	0.71	0.72	0.74	0.72
75	0.64	0.64	0.65	0.64
76	0.58	0.56	0.55	0.57
77	0.48	0.45	0.4	0.45
78	0.35	0.33	0.31	0.33
79	0.29	0.28	0.28	0.28
80	0.31	0.33	0.35	0.33
81	0.94	0.98	1.05	0.99
82	0.78	0.8	0.84	0.81
83	0.71	0.72	0.75	0.72
84	0.64	0.64	0.65	0.64

(Continued)

TABLE 15.8 (Continued)
Dependent Factor-1 Analysis

R.No.	IVF1	IVF2	IVF3	Part Res-1 = ((6*IVF1)+ (5*IVF2)+(4*IVF3))/15
85	0.52	0.5	0.47	0.5
86	0.4	0.4	0.4	0.4
87	0.36	0.36	0.37	0.36
88	0.39	0.42	0.45	0.42
89	1.28	1.28	1.29	1.28
90	1.15	1.13	1.1	1.13
91	1.1	1.06	1.02	1.06
92	1.05	1	0.94	1
93	0.96	0.89	0.8	0.9
94	0.82	0.77	0.71	0.78
95	0.76	0.72	0.67	0.72
96	0.81	0.8	0.8	0.81
97	1.2	1.26	1.37	1.27
98	1.02	1.07	1.16	1.08
99	0.94	0.98	1.05	0.99
100	0.86	0.89	0.95	0.9
101	0.71	0.72	0.75	0.72
102	0.63	0.66	0.7	0.66
103	0.6	0.63	0.69	0.63
104	0.57	0.61	0.68	0.61
105	1.26	1.31	1.41	1.32
106	1.1	1.13	1.19	1.13
107	1.02	1.05	1.09	1.05
108	0.95	0.96	0.99	0.96
109	0.81	0.8	0.8	0.81
110	0.71	0.72	0.74	0.72
111	0.67	0.69	0.72	0.69
112	0.63	0.66	0.7	0.66
113	1.3	1.34	1.42	1.35
114	1.14	1.16	1.21	1.17
115	1.07	1.08	1.12	1.09
116	1	1	1.02	1
117	0.87	0.85	0.83	0.85
118	0.76	0.76	0.77	0.76
119	0.71	0.72	0.75	0.72
120	0.67	0.69	0.72	0.69
121	1.33	1.37	1.44	1.38
122	1.18	1.2	1.24	1.2
123	1.11	1.12	1.14	1.12
124	1.04	1.04	1.05	1.04
125	0.92	0.89	0.87	0.9
126	0.81	0.8	0.8	0.8
127	0.76	0.76	0.77	0.76
128	0.44	0.4	0.33	0.4

TABLE 15.9
Dependent Factor-2 Analysis

R.No.	IVF1	IVF2	IVF3	Part Res-2 = ((3*IVF1)+ (2*IVF2)+(1*IVF3))/6
1	0.86	0.89	0.95	0.89
2	0.94	0.98	1.05	0.98
3	1.02	1.07	1.16	1.06
4	1.42	1.44	1.49	1.44
5	1.2	1.26	1.37	1.25
6	1.26	1.31	1.41	1.3
7	1.3	1.34	1.42	1.33
8	0.92	0.89	0.87	0.9
9	0.71	0.72	0.74	0.72
10	0.78	0.8	0.84	0.8
11	0.86	0.89	0.94	0.88
12	1.28	1.28	1.29	1.28
13	1.02	1.07	1.16	1.06
14	1.1	1.13	1.19	1.12
15	1.14	1.16	1.21	1.16
16	0.82	0.77	0.71	0.79
17	0.64	0.64	0.65	0.64
18	0.71	0.72	0.75	0.72
19	0.78	0.8	0.84	0.8
20	1.22	1.2	1.2	1.21
21	0.94	0.98	1.05	0.98
22	1.02	1.05	1.09	1.04
23	1.07	1.08	1.12	1.08
24	0.79	0.73	0.65	0.74
25	0.58	0.56	0.55	0.57
26	0.64	0.64	0.65	0.64
27	0.71	0.72	0.74	0.72
28	1.16	1.13	1.11	1.14
29	0.86	0.89	0.95	0.89
30	0.95	0.96	0.99	0.96
31	1	1	1.02	1
32	0.76	0.69	0.6	0.71
33	0.48	0.45	0.4	0.46
34	0.52	0.5	0.47	0.51
35	0.58	0.56	0.55	0.57
36	1.05	1	0.94	1.01
37	0.71	0.72	0.75	0.72
38	0.81	0.8	0.8	0.81
39	0.87	0.85	0.83	0.86
40	0.74	0.66	0.55	0.68
41	0.35	0.33	0.31	0.33
42	0.4	0.4	0.4	0.4

(Continued)

TABLE 15.9 (Continued)
Dependent Factor-2 Analysis

R.No.	IVF1	IVF2	IVF3	Part Res-2 = ((3*IVF1)+ (2*IVF2)+(1*IVF3))/6
43	0.47	0.48	0.49	0.48
44	0.92	0.89	0.87	0.9
45	0.63	0.66	0.7	0.65
46	0.71	0.72	0.74	0.72
47	0.76	0.76	0.77	0.76
48	0.59	0.53	0.44	0.54
49	0.29	0.28	0.28	0.28
50	0.36	0.36	0.37	0.36
51	0.43	0.44	0.47	0.44
52	0.87	0.85	0.83	0.86
53	0.6	0.63	0.69	0.62
54	0.67	0.69	0.72	0.68
55	0.71	0.72	0.75	0.72
56	0.52	0.46	0.39	0.48
57	0.23	0.24	0.25	0.24
58	0.31	0.33	0.35	0.32
59	0.39	0.42	0.45	0.41
60	0.81	0.8	0.8	0.81
61	0.57	0.61	0.68	0.6
62	0.63	0.66	0.7	0.65
63	0.67	0.69	0.72	0.68
64	0.44	0.4	0.33	0.41
65	0.71	0.72	0.75	0.72
66	0.58	0.56	0.55	0.57
67	0.52	0.5	0.47	0.51
68	0.48	0.45	0.4	0.46
69	0.44	0.4	0.33	0.41
70	0.29	0.26	0.22	0.27
71	0.22	0.2	0.17	0.21
72	0.23	0.24	0.25	0.24
73	0.86	0.89	0.95	0.89
74	0.71	0.72	0.74	0.72
75	0.64	0.64	0.65	0.64
76	0.58	0.56	0.55	0.57
77	0.48	0.45	0.4	0.46
78	0.35	0.33	0.31	0.33
79	0.29	0.28	0.28	0.28
80	0.31	0.33	0.35	0.32
81	0.94	0.98	1.05	0.98
82	0.78	0.8	0.84	0.8
83	0.71	0.72	0.75	0.72
84	0.64	0.64	0.65	0.64
85	0.52	0.5	0.47	0.51

(Continued)

TABLE 15.9 (Continued)
Dependent Factor-2 Analysis

R.No.	IVF1	IVF2	IVF3	Part Res-2 = ((3*IVF1)+ (2*IVF2)+(1*IVF3))/6
86	0.4	0.4	0.4	0.4
87	0.36	0.36	0.37	0.36
88	0.39	0.42	0.45	0.41
89	1.28	1.28	1.29	1.28
90	1.15	1.13	1.1	1.14
91	1.1	1.06	1.02	1.07
92	1.05	1	0.94	1.01
93	0.96	0.89	0.8	0.91
94	0.82	0.77	0.71	0.79
95	0.76	0.72	0.67	0.73
96	0.81	0.8	0.8	0.81
97	1.2	1.26	1.37	1.25
98	1.02	1.07	1.16	1.06
99	0.94	0.98	1.05	0.98
100	0.86	0.89	0.95	0.89
101	0.71	0.72	0.75	0.72
102	0.63	0.66	0.7	0.65
103	0.6	0.63	0.69	0.62
104	0.57	0.61	0.68	0.6
105	1.26	1.31	1.41	1.3
106	1.1	1.13	1.19	1.12
107	1.02	1.05	1.09	1.04
108	0.95	0.96	0.99	0.96
109	0.81	0.8	0.8	0.81
110	0.71	0.72	0.74	0.72
111	0.67	0.69	0.72	0.68
112	0.63	0.66	0.7	0.65
113	1.3	1.34	1.42	1.33
114	1.14	1.16	1.21	1.16
115	1.07	1.08	1.12	1.08
116	1	1	1.02	1
117	0.87	0.85	0.83	0.86
118	0.76	0.76	0.77	0.76
119	0.71	0.72	0.75	0.72
120	0.67	0.69	0.72	0.68
121	1.33	1.37	1.44	1.36
122	1.18	1.2	1.24	1.2
123	1.11	1.12	1.14	1.12
124	1.04	1.04	1.05	1.04
125	0.92	0.89	0.87	0.9
126	0.81	0.8	0.8	0.8
127	0.76	0.76	0.77	0.76
128	0.44	0.4	0.33	0.41

TABLE 15.10
Combined Results Analysis

Rule No.	Part Res-1 = ((6*IVF1)+ (5*IVF2)+(4*IVF3))/15	Part Res-2 = ((3*IVF1)+ (2*IVF2)+(1*IVF3))/6	Combined Res=((2*Part Res-1)+(1*Part Res-2))/3	CR-0.01
1	0.9	0.89	0.89	0.88
2	0.99	0.98	0.98	0.97
3	1.08	1.06	1.07	1.06
4	1.45	1.44	1.45	1.44
5	1.27	1.25	1.26	1.25
6	1.32	1.3	1.31	1.3
7	1.35	1.33	1.34	1.33
8	0.9	0.9	0.9	0.89
9	0.72	0.72	0.72	0.71
10	0.81	0.8	0.8	0.79
11	0.89	0.88	0.89	0.88
12	1.28	1.28	1.28	1.27
13	1.08	1.06	1.07	1.06
14	1.13	1.12	1.13	1.12
15	1.17	1.16	1.16	1.15
16	0.78	0.79	0.78	0.77
17	0.64	0.64	0.64	0.63
18	0.72	0.72	0.72	0.71
19	0.81	0.8	0.8	0.79
20	1.21	1.21	1.21	1.2
21	0.99	0.98	0.98	0.97
22	1.05	1.04	1.05	1.04
23	1.09	1.08	1.08	1.07
24	0.73	0.74	0.74	0.73
25	0.57	0.57	0.57	0.56
26	0.64	0.64	0.64	0.63
27	0.72	0.72	0.72	0.71
28	1.13	1.14	1.14	1.13
29	0.9	0.89	0.89	0.88
30	0.96	0.96	0.96	0.95
31	1	1	1	0.99
32	0.69	0.71	0.7	0.69
33	0.45	0.46	0.45	0.44
34	0.5	0.51	0.5	0.49
35	0.57	0.57	0.57	0.56
36	1	1.01	1.01	1
37	0.72	0.72	0.72	0.71
38	0.81	0.81	0.81	0.8
39	0.85	0.86	0.85	0.84
40	0.66	0.68	0.67	0.66
41	0.33	0.33	0.33	0.32
42	0.4	0.4	0.4	0.39
43	0.48	0.48	0.48	0.47

(Continued)

TABLE 15.10 (Continued)
Combined Results Analysis

Rule No.	Part Res-1 = ((6*IVF1)+ (5*IVF2)+(4*IVF3))/15	Part Res-2 = ((3*IVF1)+ (2*IVF2)+(1*IVF3))/6	Combined Res=((2*Part Res-1)+(1*Part Res-2))/3	CR-0.01
44	0.9	0.9	0.9	0.89
45	0.66	0.65	0.65	0.64
46	0.72	0.72	0.72	0.71
47	0.76	0.76	0.76	0.75
48	0.53	0.54	0.53	0.52
49	0.28	0.28	0.28	0.27
50	0.36	0.36	0.36	0.35
51	0.44	0.44	0.44	0.43
52	0.85	0.86	0.85	0.84
53	0.63	0.62	0.63	0.62
54	0.69	0.68	0.69	0.68
55	0.72	0.72	0.72	0.71
56	0.46	0.48	0.47	0.46
57	0.24	0.24	0.24	0.23
58	0.33	0.32	0.33	0.32
59	0.42	0.41	0.42	0.41
60	0.81	0.81	0.81	0.8
61	0.61	0.6	0.61	0.6
62	0.66	0.65	0.65	0.64
63	0.69	0.68	0.69	0.68
64	0.4	0.41	0.4	0.39
65	0.72	0.72	0.72	0.71
66	0.57	0.57	0.57	0.56
67	0.5	0.51	0.5	0.49
68	0.45	0.46	0.45	0.44
69	0.4	0.41	0.4	0.39
70	0.26	0.27	0.27	0.26
71	0.2	0.21	0.2	0.19
72	0.24	0.24	0.24	0.23
73	0.9	0.89	0.89	0.88
74	0.72	0.72	0.72	0.71
75	0.64	0.64	0.64	0.63
76	0.57	0.57	0.57	0.56
77	0.45	0.46	0.45	0.44
78	0.33	0.33	0.33	0.32
79	0.28	0.28	0.28	0.27
80	0.33	0.32	0.33	0.32
81	0.99	0.98	0.98	0.97
82	0.81	0.8	0.8	0.79
83	0.72	0.72	0.72	0.71
84	0.64	0.64	0.64	0.63
85	0.5	0.51	0.5	0.49
86	0.4	0.4	0.4	0.39
87	0.36	0.36	0.36	0.35

(Continued)

TABLE 15.10 (Continued)
Combined Results Analysis

Rule No.	Part Res-1 = ((6*IVF1)+(5*IVF2)+(4*IVF3))/15	Part Res-2 = ((3*IVF1)+(2*IVF2)+(1*IVF3))/6	Combined Res=((2*Part Res-1)+(1*Part Res-2))/3	CR-0.01
88	0.42	0.41	0.42	0.41
89	1.28	1.28	1.28	1.27
90	1.13	1.14	1.13	1.12
91	1.06	1.07	1.07	1.06
92	1	1.01	1	0.99
93	0.9	0.91	0.9	0.89
94	0.78	0.79	0.78	0.77
95	0.72	0.73	0.72	0.71
96	0.81	0.81	0.81	0.8
97	1.27	1.25	1.26	1.25
98	1.08	1.06	1.07	1.06
99	0.99	0.98	0.98	0.97
100	0.9	0.89	0.89	0.88
101	0.72	0.72	0.72	0.71
102	0.66	0.65	0.65	0.64
103	0.63	0.62	0.63	0.62
104	0.61	0.6	0.61	0.6
105	1.32	1.3	1.31	1.3
106	1.13	1.12	1.13	1.12
107	1.05	1.04	1.05	1.04
108	0.96	0.96	0.96	0.95
109	0.81	0.81	0.81	0.8
110	0.72	0.72	0.72	0.71
111	0.69	0.68	0.69	0.68
112	0.66	0.65	0.65	0.64
113	1.35	1.33	1.34	1.33
114	1.17	1.16	1.16	1.15
115	1.09	1.08	1.08	1.07
116	1	1	1	0.99
117	0.85	0.86	0.85	0.84
118	0.76	0.76	0.76	0.75
119	0.72	0.72	0.72	0.71
120	0.69	0.68	0.69	0.68
121	1.38	1.36	1.37	1.36
122	1.2	1.2	1.2	1.19
123	1.12	1.12	1.12	1.11
124	1.04	1.04	1.04	1.03
125	0.9	0.9	0.9	0.89
126	0.8	0.8	0.8	0.79
127	0.76	0.76	0.76	0.75
128	0.4	0.41	0.4	0.39
8256			99.28	
		Threshold	0.01	

weights to reach the desired upper bound of the vital parameter. The chosen rule set is illustrated in Table 15.11.

The chosen rules with rule strength >=1 is shown in Table 15.11, which displays the chosen rules as explained in [24]. Based on the chosen rule set analysis in Table 15.10, we could have six groups for the dependent factors. The ranges for the modified rule strengths would be 0.8–1.04, 1.06–1.07, 1.11–1.13, 1.15–1.25, 1.27–1.30, and 1.33–1.44. The corresponding mid-values would be 0.92, 1.07, 1.12, 1.20, 1.29, and 1.39 respectively.

According to this analysis the organ could be designed based on the appropriate rule strength. An algorithm for this is given in the next section.

TABLE 15.11
Chosen Rules

Rule No.	Part Res-1 = ((6*IVF1)+ (5*IVF2)+(4*IVF3))/15	Part Res-2 = ((3*IVF1)+ (2*IVF2)+(1*IVF3))/6	Combined Res=((2*Part Res-1)+(1*Part Res-2))/3	CR-0.01
1	1.08	1.06	1.07	1.06
2	1.45	1.44	1.45	1.44
3	1.27	1.25	1.26	1.25
4	1.32	1.3	1.31	1.3
5	1.35	1.33	1.34	1.33
6	1.28	1.28	1.28	1.27
7	1.08	1.06	1.07	1.06
8	1.13	1.12	1.13	1.12
9	1.17	1.16	1.16	1.15
10	1.21	1.21	1.21	1.2
11	1.05	1.04	1.05	1.04
12	1.09	1.08	1.08	1.07
13	1.13	1.14	1.14	1.13
14	1	1.01	1.01	1
15	1.28	1.28	1.28	1.27
16	1.13	1.14	1.13	1.12
17	1.06	1.07	1.07	1.06
18	0.81	0.81	0.81	0.8
19	1.27	1.25	1.26	1.25
20	1.08	1.06	1.07	1.06
21	1.32	1.3	1.31	1.3
22	1.13	1.12	1.13	1.12
23	1.05	1.04	1.05	1.04
24	1.35	1.33	1.34	1.33
25	1.17	1.16	1.16	1.15
26	1.09	1.08	1.08	1.07
27	1.38	1.36	1.37	1.36
28	1.2	1.2	1.2	1.19
29	1.12	1.12	1.12	1.11
30	1.04	1.04	1.04	1.03

15.3.7 ALGORITHM

- Identify the vital parameters;
- Identify the dependent parameters;
- Design an objective function with the vital and dependent parameters;
- Initialise a basis for the dataset;
- Obtain the equivalent quantum dataset;
- Perform the neuro-fuzzy analysis;
- Modify the objective function as per the appropriate rule set and mid-value;
- End the procedure.

15.4 SIMULATION RESULTS

The algorithm was tested using the FlexSim tool [11] with about 5000 datasets and the performance was assessed based on reliability. The output of the model simulated, illustrating the different components and their associated functionality, is shown in Figure 15.2. This has a source and a sink component. Other components like the processor for tasks assigned and processing algorithm components like Basic FR1 and Basic FR2 show the algorithm's functionality. The results were compared with the conventional machine learning algorithms.

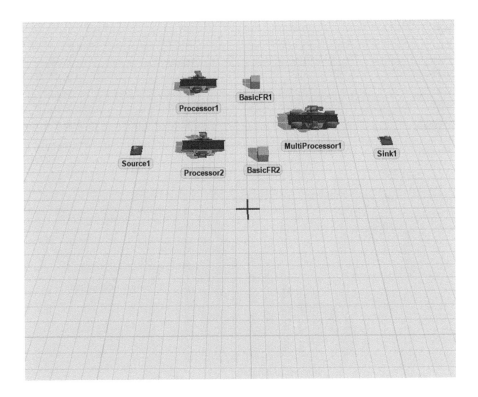

FIGURE 15.2 Output simulation using flexsim.

15.4.1 RELIABILITY

Reliability is measured as the difference between the theoretical dependent factor and the dependent factor obtained by the fuzzy analysis. The results of the simulation are shown in Table 15.12.

It may be observed from Table 15.12 and Figures 15.3 and 15.4 that the average improvement in reliability of the quantum neuro-fuzzy approach is 33%, compared to the conventional neuro-fuzzy approach, and has a maximum reliability for 5000 datasets.

TABLE 15.12

Comparison of Reliability (Machine Learning vs Quantum Digital Twin Approach)

| Number of Datasets | Reliability | | Improvement (%) |
	Machine Learning	Quantum Digital Twin	
100	8.25	11.45	38.79
200	11.45	15.67	36.86
400	14.65	19.45	32.76
700	23.45	32.34	37.91
1000	29.45	38.76	31.61
1500	34.24	45.67	33.38
2500	39.56	53.56	35.39
3000	43.56	57.65	32.35
3500	51.56	67.89	31.67
4000	71.56	92.56	29.35
4500	75.34	96.54	28.14
5000	79.65	98.67	23.88
Average	**40.23**	**52.52**	**32.67**

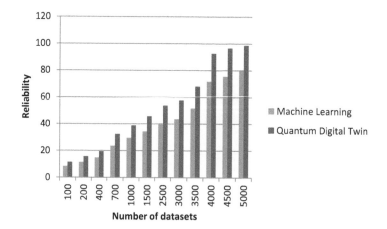

FIGURE 15.3 Plot comparing the reliability of the approaches.

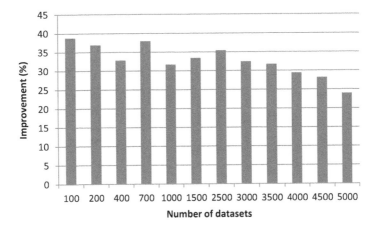

FIGURE 15.4 Plot showing improvement in reliability for the quantum digital twin approach.

15.4.2 PROCESSING TIME

It is the time duration to process a quantum dataset to provide valid inferences. The results of the simulation are given in Table 15.13.

It may be observed from Table 15.13 and Figures 15.5 and 15.6 that the average reduction in processing time for the quantum neuro-fuzzy approach is 33%, compared to the conventional neuro-fuzzy approach, and has a maximum processing time for 5000 datasets.

TABLE 15.13

Comparison of Processing Times (Machine Learning vs Quantum Digital Twin)

Number of Datasets	Processing Time		Reduction (%)
	Machine Learning	Quantum Digital Twin	
100	15.45	9.45	38.83
200	21.45	13.56	36.78
400	32.45	21.45	33.90
700	47.65	31.56	33.77
1000	56.54	35.65	36.95
1500	61.56	41.45	32.67
2500	73.56	49.56	32.63
3000	79.53	53.45	32.79
3500	81.45	55.65	31.68
4000	89.45	61.45	31.30
4500	95.76	69.56	27.36
5000	98.78	75.65	23.42
Average	**62.80**	**43.20**	**32.67**

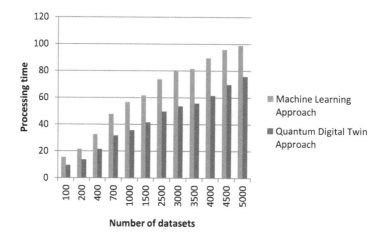

FIGURE 15.5   Plot comparing the processing times of the approaches.

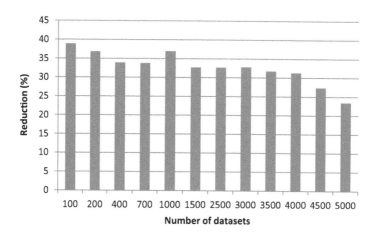

FIGURE 15.6   Plot showing reduction in processing time for the quantum digital twin approach.

15.5   CONCLUSION

The quantum digital twin approach improves the design of organs for the health care industry with significant improvement in reliability and reduced processing time. The design of the organs based on this approach is highly effective as it is performed by rigorous iterations to match the objective of the designed organ.

15.6   FUTURE WORK

The process could be extended to design the organs based on the analysis carried out by the digital twin approach. 3D printing could be used to design the desired organ which matches the analysis to effectively obtain the object organ.

REFERENCES

[1] Pablo Arrighi, Christopher Cedzich, and Marrin Coste. *Addressable Quantum Gates*, pp. 1–41, ACM Reference, 2023.

[2] Xiangzhen Zhou and Yuan Feng. "Quantum Circuit Transformation: A Monte Carlo Tree Search Framework", pp. 59:1–59:27, ACM, 2022.

[3] Thomas Haner and Mathias Soeken. *Lowering the T-depth of Quantum Circuits via Logic Network Optimisation*, pp. 1–15, ACM, 2022.

[4] David Ittah, Thomas Haner and Torsten Hoefler. "QIRO: A Static Single Assignment-Based Quantum Program Representation for Optimisation", pp. 14:1–14:32, ACM, 2022.

[5] Abhijith, William Casper and Marc Vuffray. "Quantum Algorithm Implementation for Beginners", pp. 18:1–18:92, ACM, 2022.

[6] Raban Iten, David Sutter and Stefan Worner. *Exact and Practical pattern for Quantum Circuit Optimisation*, pp. 12–53, ACM, 2022.

[7] Samuel Jaques and Thomas Haner. "Leveraging State Sparsity for More Efficient Quantum Simulations", pp. 15:1–15:17, *ACM Transactions on Quantum Computing*, 2022.

[8] Stuart Harwood, Panagiotis Barkpoutsos, Tanvi Gujarathi and Sarah Mostame. "Improving the Variation Quantum Eigensolver Using Variational Adiabatic Quantum Computing", pp. 1:1–1:20, ACM, 2022.

[9] www.flexsim.com. "End-to-End Process Improvement with Flexsim Simulation Modelling Software", white paper, pp. 1–13, 2021.

[10] Harita Garg. "Digital Twin Technology: Revolutionary to Improve Personalised Healthcare", *Science Progress and Research*, pp. 32–34, 2021.

[11] Concreta Semeraro, Mario Lezoche, Herve Panetto and Michele Dassisti. "Digital Twin Paradigm: A systematic Literature Review", Elsevier, pp. 1–42, 2021.

[12] Broughton Michael, Guillaume Verdon, Trevor Mc Court, Antonio J. Martinez, Jae Hyeon Yoo and Sergei V. Isakov. "TensorFlow Quantum: A Software Framework for Quantum Machine Learning", pp. 1–56, 2021.

[13] Titoujan Carette, Emanuel Jeandel and Renaud Vilmart. "Completeness of Graphical Languages for Mixed State Quantum Mechanics", pp 17:1–17:28 *ACM Transaction on Quantum Computing*, 2021.

[14] Megan A. Catterton, Alexander G. Bail and Rebecca R. Pompano. "Rapid Fabrication by Digital Light Processing 3D Printing a Slipchip with Movable Ports for Local Delivery to Ex Vivo Organ Cultures", pp. 1–15, *Micromachines* 2021.

[15] Hsin-Yuan Huang, Michael Broughton, Masoud Mohseni, Ryan Babbush, Sergio Boixo, Hartmut Neven and Jarrod R. Mc Clean. "Power of Data in Quantum Machine Learning", *Nature Communications*, pp. 1–9, 2021.

[16] Aidan Fuller, Charles Day and Chris Barlow. *Digital Twin: Enabling Technologies, Challenges and Research*, pp. 108952–108971, IEEE, 2020.

[17] Hendrik Van Der Walk, Hendrik Habe, Frederick Moller, Michael Arbter, Jan-Luca Henning and Boris Otto. "A Taxonomy of Digital Twins", *Americas Conference on Information Systems*, pp. 1–11, 2020.

[18] Carmen Almudever. *Designing and Benchmarking Full-Stack Quantum Systems*, pp. 1–12, ACM, 2020.

[19] Martin Robert Enders and Nadja Hobbach. "Dimensions of Digital Twin Applications – A Literature Review", *Twenty Fifth Americas Conference on Information Systems*, 2019.

[20] Joseph Fitzsimons and Henry Yuen. *Quantum Proof Systems for Iterated Exponential Time and Beyond*, pp. 473–481, ACM, 2019.

[21] Gaurav Kaushik, Jerden Leijten and Ali Khademhosseini. *Concise Review: Organ Engineering: Design, Technology and Integration*, pp. 51–60, STEM CELLS, 2017.

[22] Rodrigo A. Rezende, Vladimir Kasyanov, Vladimir Mironov and Jorge Vicente Lopes da Silva. "Organ Printing as an Information Technology", *ScienceDirect*, pp. 32–38, 2015.

[23] Nour Aburated, Faisal Shah Khan and, Harish Bhaskar. "Advances in the Quantum Theoretical Approach to Image Processing Algorithms", *ACM*, 2017.

[24] https://github.com/pradheep2812/Book-Chapter-Quantum-digital-twin/

16 The Future of Sustainable Green Smart Cities

C. M. Naga Sudha
Anna University MIT Campus, Chennai, India

J. Jesu Vedha Nayahi
Anna University Regional Campus, Tirunelveli, India

S. Saravanan
Anna University, Chennai, India

N. Renugadevi
Indian Institute of Information Technology, Tiruchirappalli, India

16.1 INTRODUCTION

Look after the land and the land will look after you, destroy the land and it will destroy you.

In the digital era, the requirements of people are satisfied instantaneously, which is known as "anytime, anywhere, and anything." The digital revolution started in the 1980s when information became available to people through advancements in technology, which made villages into modern cities. A tremendous focus on the digital revolution has brought about "automation" as the most highlighted word. Initially, digitization gave birth to four interesting branches: artificial intelligence (AI), big data, the Internet of Things (IoT), and cloud computing. These four technologies form the umbrella of the "digital age." With the combination of these factors, digitization has stepped into the modernization of cities, which are named "smart cities." This is considered an important role in the development of IoT application sectors. Smart cities are defined in various ways. Modern infrastructure fuels a high quality of life with the help of information and communication technology (ICT) when investment in traditional transport, social capital, and human lifestyle are increased. One of the most accepted definitions of "smart city" is a city which connects information-technology infrastructure, business infrastructure, physical infrastructure, and social infrastructure to support the collective intelligence of the city. These smart cities help in the selection of various criteria which will help in optimization and facilitate

DOI: 10.1201/9781003388814-16

urban development for the improvement of the lifestyle of the citizens. This can be achieved through certain rules and regulations which adhere to the implementation of ICT alongside the proper management. These rules enable technology exchange between private companies and public organizations (governmental ones) which have made their impact on concepts such as "e-government" [1]. This has induced positive effects within public administration in various contexts concerning public behavior.

Even though digitalization has made remarkable changes in society, sustainability is still a very important issue to be considered. Therefore, smart and sustainable cities are emerging as a hot topic globally, particularly due to their impact on climatic change. However, only with sustainable technological development natural resources can be preserved and so save future generations from reversing environmental effects.

Smart cities involve six components: smart living, smart citizens, smart economy, smart mobility, smart governance, and a smart environment, which are collectively known as user-oriented. The technology-oriented approach uses IoT to control the environment. Both user-oriented and technology-oriented approaches move towards a common goal to provide services to modern cities as shown in Figure 16.1 [2].

A smart city involves humanitarian, technical, and legal aspects. In the change from traditional cities to smart cities, users become key stakeholders. Technology leads as a dynamic enabler where businesses partner each other. Production is carried out only according to demands. Smart products are developed with improved life cycles. Smart transport provides services with advanced planning and efficiency. Sustainable smart cities improve the quality of human life without affecting natural resources globally. However, this whole transformation of people from modern cities to smart cities takes a long span of time.

Our research focuses on the significant paradigm shift toward smart cities and the higher-order human activities. Day-to-day activities have become more or less digital when the smart era arrived. Hence, when smart activities are introduced into the activities of a city, people acquire a sophisticated lifestyle. Therefore, these issues are the motivation behind our research work.

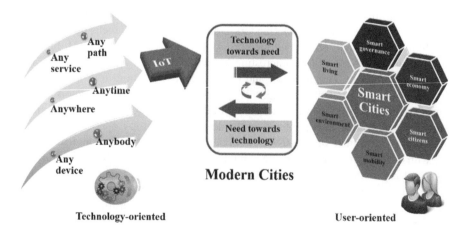

FIGURE 16.1 Interrelationship between the IoT and smart cities.

The organization of this chapter is as follows. As smart city development must follow some standard models, organizations have stepped into designing such models for smart cities – which is discussed in Section 16.2. Sustainable smart cities are developed with the help of three pillars – the economy, environment, and society – which are discussed in detail in Section 16.2. The technologies involved in sustainable smart city development and the major role of countries in initiating the startups are discussed in Section 16.3. The future of sustainable green IT research is discussed in Section 16.7.

16.2 MODELS OF SMART CITIES

Smart city models of varying scales have been proposed by various business companies which promote the emergence of smart cities in developing countries. These models are developed in such a way that some attributes and features are common even when various approaches have been applied – certain concerns are unique for every model. A few models of smart cities are discussed below.

The Smart Cities Wheel Model. This was developed by Boyd Cohen and is shown in Figure 16.2 [3], which highlights the six dimensional key factors through which the cities are identified or categorized. Key dimensions include smart environment, smart governance, smart living, smart people, smart mobility, and smart economy.

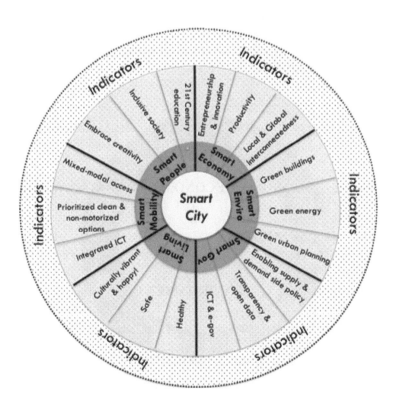

FIGURE 16.2   The smart cities wheel model proposed by Boyd Cohen.

A smart environment is achieved through the construction of green buildings, green urban planning, and green energy. Open data transparency, e-governance, supply-and-demand-side policy, and transparency are included under smart governance. Smart living focuses on improving the life standards of citizens in health, happiness, and security. Smart mobility is achieved through the availability of ICT and transportation systems which encourage clean and mixed-mode access and non-motorized options. Education in the 21st century has increased creativity among people which facilitates an inclusive society. Another important criterion insisted on by Boyd Cohen is "Go lean," which means that the smart city must identify the targets to be achieved easily in order to attain long term goals.

IBM Smarter Cities. These have developed smarter solutions for more than 2000 cities globally. IBM has a tripartite view of the smart city: people, infrastructure, and operations. These have laid a foundation for three basic services: infrastructure services, human services, city planning and management services. Human services outline healthcare, education, and social programs. Infrastructure services outline energy water and transportation. Three pillars of the smart city are shown in Figure 16.3 [4, 5]. The human-centric smarter city by IBM places the citizens at its center and expands the city boundaries in order to cooperate with citizen groups and universities for developing smarter cities. Interactions between the city's components are represented in the IBM model, which are absent in the wheel model.

Three characteristics identified by IBM are: leveraging information to make decisions better; proactiveness; and coordinating processes and resources. An important fact to be highlighted is the need to develop innovative solutions which help in satisfying the needs of the citizens in order to make the city smart. Once the basic needs are satisfied, citizens desire a luxury lifestyle. These needs are the basic key

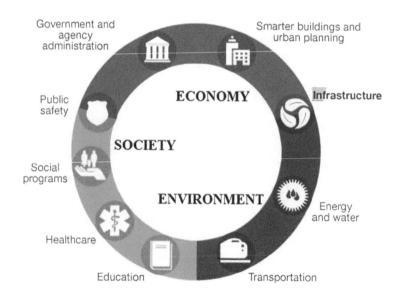

FIGURE 16.3   IBM pillars of the smart city.

considerations which are developed and applied in cities of varying scale. With the fact that the customer is the king of the competitive world, frameworks are being designed to emphasize the need for a human-centric environment and to increase citizen participation in governance. When all these considerations have been outlined, Maslow defines the frameworks with the terms: physiological, safety, belongingness and love, self-esteem, self-actualization, and self-transcendence.

Hitachi's Smart City. This has defined the smart city as "a city which is conscious environmentally and uses information technology resources efficiently." In simple terms, the smart city has been developed to satisfy the needs and values of its inhabitants along with the application of IT for improving energy and to safeguard the environment. This helps to maintain the "well-balanced relationship between people and Earth." Hitachi defines the smart city in three layers: the infrastructure layer, urban services layer, and urban life-style layer, which are included in providing services to the layers above them. The infrastructure layer comprises water supply, sewage and telecommunication generation, and distribution of electricity. The urban services layer consists of elevators and escalators. Distributed renewable energy (RE), which is now centralized, is considered a global trend. Horizontal mobility includes transportation such as energy vehicles (EVs) and hybrid electric vehicles (HEVs), whereas escalators and elevators with speed and high capacity are considered as vertical mobility [6]. Further, three key features which are identified to characterize the smart city are:

- Integration between the urban infrastructures makes the city smart. Also, the addition of some knowledge and information processing capabilities can be utilized for smart systems.
- Fusing of information and control. The main perspective of information usage is in its operational control.
- Smart cities are to be equipped with their own sensory systems which are enabled through sensor networks and smart meters.

16.2.1 Sustainable Smart Cities

Sustainability is a major factor which envisions future aspects such as environmental, social, and economic considerations that are equitable in improving the lifestyle of citizens and providing more gains to city actors. However, nowadays major cities have emerged as epicenters with unsustainable development which will affect natural resources and make trouble for future generations. Unsustainable development is caused through various components such as waste and pollution generation which will adversely impact ecological conditions and the core utilization of natural resources.

Recently, sustainability along with ICT has become more focused in urban and academic circles where debates concern addressing the challenges faced by rapid urbanization and unsustainable cities. Sustainability is a holistic approach which is the most effective paradigm to enable cities with restructuring demands and urban dynamics. It will alert people to the adverse effects caused, when urban system capacities are stretched too far. The urban system process organizes human services, ecosystem infrastructure, and administration which are under pressure due to the challenges of urbanization and sustainability. Urban growth has raised various problems concerning the economic, social, environmental, and social sustainability of cities. Exceptional urban thinking and new planning on urban development are required to overcome these challenges which can be associated with operational, infrastructural, and functional processes.

Smart cities are focused on digitalization with smart and innovative solutions depending on applications, models, and services. To achieve sustainability measures, ICT has taken the central spotlight and has ubiquity computing integrated with technologies that are applied in urban domains. ICT has also increased the contribution towards designing and developing sustainable smart cities through various sectors such as architecture, systems, infrastructure, applications, and data analytics [7]. The planning of sustainable smart cities requires sophisticated methods and innovative solutions which can lay the foundation for the application of data science and computer science through ICT. Also, the sensing of data and the processing of information are rapidly embedded in smart city applications where wireless networks are progressing at a fast rate. ICT enabled sectors have promised to make smart cities sustainable in terms of the environmental aspects, growth of social mobility, and in economic restructuring facilities. These are readily available in addressing a long range of challenges faced by contemporary cities.

Nowadays, new technologies and applications have been developed in order to help the urgent need to address unsustainability challenges. Hence, simulation models have been developed for the built environment, which is a part of sustainable smart cities. Powerful urban intelligence functions have also been designed for the smart and sustainable environment. An increased awareness of sustainability coupled with ICT innovation has resulted in a unique way to rethink designing the style of cities and in understanding new ways to address urban challenges.

According to the Brundtland report [9], it is a well-known fact that sustainable development can be made without affecting the needs of future generations. Sustainable development does not mean that development has to be made at a steady pace. It must aim for development in which the technology and environment are able to handle the effects of human activities. However, sustainable development will be a challenging task in countries which face poor finances, where the maintenance cost for dynamic growth and natural safety is considered. Therefore, sustainable development is expected by investing more in technology and which will help in the overall development of natural resources. Sustainable development is a long-term process which can sometimes be an exhaustive procedure. Also, it is considered to be a continuous process in which resources, technology trends, and governance are exploited harmoniously for existing and future needs [9]. Smart cities are termed as "deployers" that are involved in enhancing the quality of people's lives without disrupting natural resources. With this definition, researchers have described smart cities as using data and information technology for providing efficient services, optimizing the infrastructure, increasing collaboration among different actors in economic sectors, and motivating the business models in both (private and public) sectors [10]. Sustainable development is also defined as a condition which balances both urban development and environmental protection with equal weights on services, shelter, employment, social infrastructure, income, and transportation [11]. Furthermore, even in cases where the conditions of current production are favourable, the smart city must preserve the overtime conditions on reproduction [12].

Developing sustainable smart cities not only focuses on environmental factors but also on other dimensions such as the economic and social. Therefore, areas of smart cities must be properly mapped in suitable dimensions on sustainability measures. According to Hiremath et al., smart cities are generally sketched on a four dimensional umbrella of factors: the social, economic, environmental, and institutional [11]. Smart living, smart mobility, and smart people are considered under social factors. The economic category compromises smart energy, smart people, and smart living. The institutional category compromises smart government and smart people. It is clear that similar areas of smart cities are mapped in two or more dimensions of sustainability. For instance, smart living is influenced by social and economic factors as smart buildings help in reducing the amount of energy consumed.

The sustainability of smart cities is the main agenda for many smart communities in which sustainability is attained through the integration of ICT. Citizen-centric systems are implemented to measure social behavior, happiness of local areas, communal satisfactory impressions, and personal satisfaction, and need to be deployed successfully [13]. These attributes will retain sustainability via ICT-enabled smart city services which are also necessary in the design stage. In recent years, unified standards have been developed for environmental monitoring and accessibility which are followed by different standards than previously. Unified standards help in resolving issues in cross-platform development and provide increased usability. With respect to ICT, sustainable smart cities are defined as when investment in social capital and transport is increased, modern infrastructure helps in sustainable economic growth through participatory governance [14]. Two general categories of smart city disclosure are:

1. An ICT and innovation-centric approach;
2. A general population arranged approach.

Urban communities begin from the methodologies which describe the productivity and progressive foundations, such as transport, vitality and water. Further, urban communities are processed and passed to those who are concentrating on framework designs. Activities such as request driven and supply driven activities use top-down and bottom-up approaches.

16.2.2 Key Components of Sustainable Smart Cities

According to the European Union, sustainable smart cities are developed based on three major categories – the economy, environment, and society – which are mostly referred to as pillars of smart sustainable cities, as shown in Figure 16.4 [15].

16.2.2.1 Economy
The economy has attained an important position in the ranking of the country's position globally. A smart economy strategy can be fulfilled when the physical infrastructure of the smart city is focused more. Therefore, the combination of ICT, innovation,

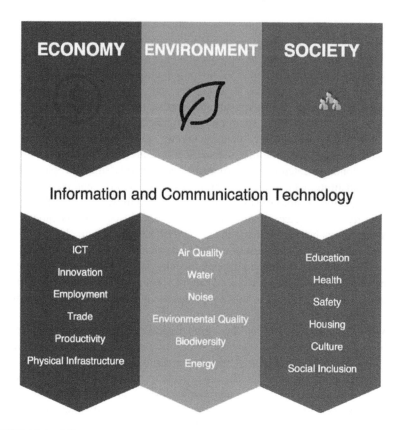

ECONOMY   ENVIRONMENT   SOCIETY

Information and Communication Technology

ICT	Air Quality	Education
Innovation	Water	Health
Employment	Noise	Safety
Trade	Environmental Quality	Housing
Productivity	Biodiversity	Culture
Physical Infrastructure	Energy	Social Inclusion

FIGURE 16.4   Pillars of smart sustainable cities.

employment, trade, and productivity is linked together with the physical infrastructure. When a core of the physical infrastructure is highlighted, smart transportation is concentrated more, which helps in all aspects of survival. However, dynamic changes in transportation show that most cities becoming unsustainable due to the threats caused to the environment and the economic and social aspects of future lives. Therefore, there arises a need for the collaboration of international, national, and regional stakeholders to make a change towards a sustainable transportation system. An initiative undertaken by the Netherlands and Germany for smart transport in modern cities, highlighting the role of women using bicycles as transport, has made a huge impact when compared to Kenya and Uganda [16].

When smart transportation is considered, it eventually leads to the concept of smart parking. This is the day-to-day progress in the urban scenario which is handled more smartly through IoT connected devices. Presently, various applications are developed for smart parking, namely a parking space finder application [2], which recommends a wide sensor architecture consisting of microphones, video cameras, and motion detectors used for finding the availability of parking slots. Cameras capture and generate a huge amount of data which consumes more energy and bandwidth. The processing and publishing of this information is a heavy task. High energy consumption and limited bandwidth are constraint in sensor networks. Responsive roadways and the Massachusetts Institute of Technology (MIT) has developed an intelligent transportation system which uses wireless magnetic sensors [17]. When the framework needs sensors to be attached on the pavement to distinguish vehicles, it is difficult for the parking areas to be managed as there is a large chance of damaging the sensors. A new protocol with the combination of ultrasonic sensors and magnetometers has been designed to detect vehicles in parking lots accurately. This is similar to Sipark PMA which provides solutions for multistorey parking lots integrated with cloud-based intelligent parking services deployed using the IoT.

One of the major functionalities which can be highlighted under civil infrastructure is smart homes. These can be expanded through the data analyzed by wireless sensors. The main objective of a smart home is to automate the home appliances. Sensors are attached for monitoring parameters and the data obtained are sent forward for analysis through IoT platforms. By using this information, residents can control their smart homes and connected objects (door locks, appliance control, temperature settings) in realtime and remotely as shown in Figure 16.5 [18]. Similarly, smart civil infrastructure can be developed in which continuous assessments are monitored. With IoT development, more researchers are focused on new paradigms concerning the cognitive IoT (CIoT) which integrates cognitive applications with IoT frameworks that are also termed as cognitive dynamic systems. Recently, a new methodology known as "sense now, retrieve now" has been implemented in combination with wireless sensors which are self-powered along with RFID-based data interrogative methods. This is based on measuring the sub-microwatt signals produced through mechanical strains by self-powered sensors which help in sending RFID-scanned data to the cloud through the IoT

Smart solutions in the smart economy perspective are provided through different actions that promote flexibility in productivity and transformation capacity. The goals of economic growth are achieved mainly with the help of innovation ecosystems or

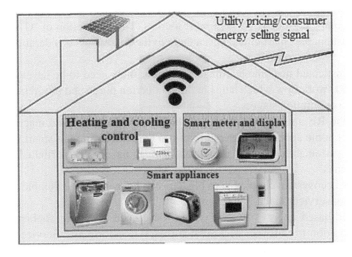

FIGURE 16.5   Illustration of IoT-induced smart home.

mechanisms generated using different human, economic, and technological infrastructures. In entrepreneurial environments, innovative infrastructeures have led to the advent of local authority, research institute, and business collaboration. New policies are framed in order to foster innovative environments. These policies can encourage expertise in knowledge and production. Therefore, key goals for developing economic productivity, commercial networks, and supporting businesses are encouraged more. Also, smart economy initiatives provide labor market flexibility to focus on reducing unemployment or in advancing technology which increases flexibility in work.

16.2.2.2   Environmental

What we are doing to the forests of the world is but a mirror reflection of what we are doing to ourselves and to one another.

Mahatma Gandhi

Among the three pillars of sustainable smart cities, the smart environment has been focused on more as it affects the lives of people in present as well as future generations. Renewable Energy (RE) resources are increased worldwide which helps in reducing the impact of global warming and climatic change [19]. RE is the major research field which has developed in recent times to provide the solutions for resource depletion and to protect the environment. RE integrated with AI has contributed more to the environment in terms of both accessibility and efficiency. It helps in managing the consumption and production of energy which has attained an important role in the changing environment and market. AI is integrated at the microeconomic level which can reduce the variability risk through predictive analysis, pattern identification, reduction of storage cost, and with the well-established connection between the grid and users to increase reliability and efficiency. The European Union plays a major role in the RE industry globally. These environmental concerns are transposing specific rules at theUnion level as well as at the member state level. AI implications along with its application is shown in Figure 16.6 [20]. AI is not only applied to mitigate depletion and degradation of natural resources but also to successfully compete globally for natural resources. The integration of AI with RE to develop sustainable smart cities must adhere to the following considerations:

- Sophisticated machine learning (ML) algorithms need to be developed for the RE work flow which helps in the production phase and progresses up to the consumption stage.
- AI and RE are coupled at all possible levels (the micro and macro) from smart home applications to platforms on the optimization of smart grids.
- The sustainability of the design and managing of the distributed energy systems.
- An AI ecosystem coupled with big data can be used for the flexible integration of services and applications.
- Cloud-based energy systems can be developed for more sophistication. In addition to Software as a Service (SaaS), Platform as a Service (PaaS), and Infrastructure as a Service (IaaS), Energy Infrastructure as a Service (EIaaS) should be considered.

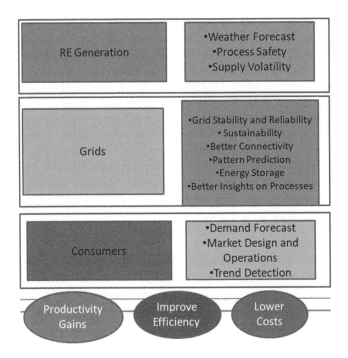

FIGURE 16.6   Energy system: Interactions based on AI application.

- Smart monitoring services with personalization can be provided at the local level.
- Global energy networks integrated with AI can be developed specifically for on-demand consumption.

RE still faces persistent barriers in the implementation phase. These issues are not related to technology aspects but in the making of the regulations and policies which should be explored in future research. The integration of AI with recent technologies has generated enormous energy maps which are applied in energy modeling along with planning purposes. RE sources can effectively replace fossil fuels to meet the energy demand efficiently and in a sustainable manner. Harvesting RE is more concerned with establishing the infrastructure at suitable places. AI tools are successfully tested through fuzzy logic tools for energy farm identification to achieve optimal output. Moreover, the energy sector is gaining focus in which AI can help in raising the adoption rate of RE sources. This mitigates the impact of climatic change which affects the livelihood of people. Researchers have highlighted the technical application of AI which can render better sustainable environments. To develop such tools, the education sector must be encouraged to generate more creative minds [21].

On the other hand, garbage management is one of the major smart city services that are implemented through the IoT. Air quality prediction is an important application for the smart city in which the urban IoT has the features of sensing and guiding air quality in fitness trails, parks, and other major crowded areas. Nitrous oxide and carbon monoxide are harmful gases which need to be monitored in urban areas. Do-it-yourself sensors help in measuring the gases. Also, Air Quality Eggs can find the safest and healthiest route on outdoor activities using the readings of gas levels [22]. An Air Quality Egg is a WiFi-enabled device that uses sensors to record changes in the levels of specified air contaminants. Each Egg can detect at least one air contaminant: NO_2, CO_2, CO, O_3, SO_2, H_2S, particulates, and volatile organic compounds (VOCs). The World Health Organization reports in Europe have insisted that particulate issues reduce each individual's lifetime by an average of one year. Belgrade city has attached sensors on the top of buses during 2011–2012 for measuring the quality of air. These data are shared to main points. Sensors in the Eko Bus framework help in measuring temperature and humidity and levels of nitrous oxide and carbon monoxide. These data are updated through an online mode for the inhabitants to view air quality information using an application.

16.2.2.3 Society

The sustainability of smart cities mainly depends on how technology has developed without affecting Mother Nature and people's lives (society). It's our responsibility to safeguard the environment for future generations. Green Smart City Smart Healthcare and Smart Education, which fall under the third pillar (i.e. society) of sustainable smart cities, are as follows.

16.2.3 GREEN SMART CITY

IoT and ML techniques have developed various applications such as waste management systems for households and industries which help sustainable smart cities as shown in Figure 16.7 [23].

16.2.4 SMART HEALTHCARE

A multi-agent system has been developed for observing users' activity on a daily basis for their well-being. Well-being is the pleasant state where users feel satisfaction of their spiritual and bodily requirements. This depends on various dimensions: the psychological, behavioral, and physical aspects. At the physical and psychological levels, the comfort of humans can be measured by sensors such as thermal, acoustic, visual, and respiratory. These will react to the environment according to the maximum or minimum threshold levels for the physical value of the sensors (light, temperature, voice). For the behavioral aspect, human capacity for action is measured as the environmental features vary in time. It is to be noted that humans respond to the systems according to their own environmental changes. The physiological aspects of humans can be measured in energy terms. The person must be conscious about his or her environment in order to satisfy certain psychological needs. An AI system learns about the activity context which includes the actions of users. On an average, 78.6 actions are performed by the system, whereas the user performs very few actions (e.g. 0.9 a day). As humans cannot perform exactly the same actions, simulations are repeated 20 times over a 50-day period. With these simulation results, the next step in learning the real reason underlying the human action can be predicted,

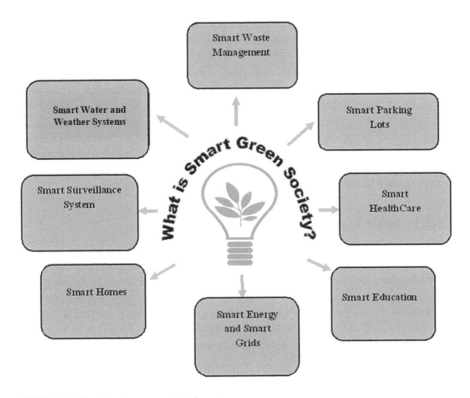

FIGURE 16.7   Smart green society layout.

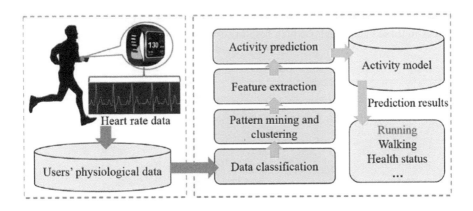

FIGURE 16.8   Running prediction based on heart rate.

which is not possible only with the actions of devices [24]. Moreover, monitoring the heart rate of the user traces their health condition, as shown in Figure 16.8 [28].

However, these tools cannot express human needs since they can only inform on human actions on devices in various contexts. It becomes more important to observe the requirements of humans and make the environment act accordingly. These responses from the environment will help in maintaining human well-being which will evolve in real time. However, these responses cannot be correctly provided to different humans, as actions differ from person to person. Self-adaptation to the changes is possible by sensors and actuators embedded with intelligence. A smart city ecosystem aims at improving citizen comfort in accordance to the environment. These strategies can be made without the interactions of the sensors, in which the "adaptive" and "bottom-up" approach system are developed. A cyber physical system is an example of this type of approach in which the information on inputs is provided through the sensors which receive the output through the effectors which are operated depending on the comfort of the users. The system which manages all comfort types is known as a cyber physical system [25].

16.2.5 SMART EDUCATION

"Education is the fundamental human right for all citizens and it must be granted in universal access with equal quality," states UNESCO. In order to achieve this goal, more effective ways have to be found for developing and underdeveloped countries. A computational intelligence model based on data mining and data science has been proposed to monitor a student's profile which helps in decision making using key performance indicators along with strategic planning for all students. Educational research and practices which are followed in the computational intelligence model are as shown in Figure 16.9 [26].

16.3 SUSTAINABLE DEVELOPMENT IN SMART CITIES

Sustainable development has been initiated in smart cities with the help of emerging technologies such as AI, ML, Deep Learning (DL), and Big Data. Some of the milestones in these technologies which help in sustainable development are as follows.

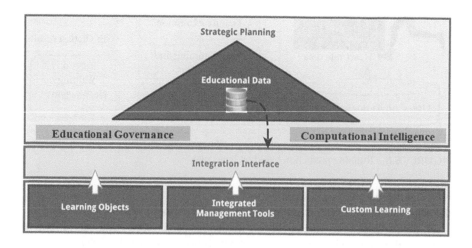

FIGURE 16.9 Education conceptual models.

16.3.1 Artificial Intelligence in Smart Cities

AI is defined as the brain of an intelligent city. It is the science which makes the computers for emulating human behavior in mind processes and helps in designing the intelligent agents. John McCarthy, a computer scientist, in 1979 defined AI as "science and engineering which makes intelligent machines" [27]. AI supports citizens in fulfilling their needs in smart cities, such as through public transport, supply of electricity, waste and paper management, healthcare, security, and digitization. AI can control all coordinated devices and thereby shows the active traffic of the whole system, which helps in the visualization of the whole network to solve possible problems. AI has an important role in the production, electronic control, and distribution of electricity. These have boosted the concept of smart grids. Intelligent robots have been programmed using the integration of AI. For instance, ZEN Robotics has built an intelligent system known as SITA Finland which can monitor waste management through its own predictions for any given task. AI plays an equal role as doctors in diagnosis, analysis, and treatment strategies. It is also applied in sitemonitoring, such as at prisons and government offices, with the help of neural networks. Digitization of data is made possible through AI which increases predictive capacity and system intelligence. The main functionality of a smart city is provided through the IoT, which is a key enabler. Different standards are followed by vendors for the manufacturing of IoT devices. AI-based, semantic IoT (AI-SioT) hybrid service architecture is modeled to emphasize AI technology which has more flexibility among different services as shown in Figure 16.10 [28].

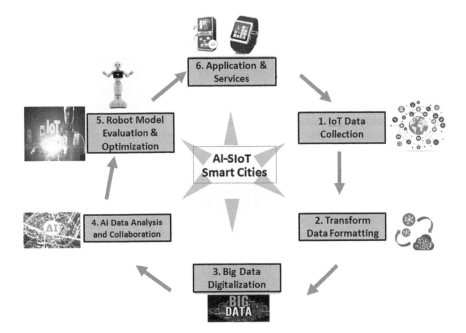

FIGURE 16.10 AI-SIoT hybrid service architecture.

The USA's infrastructure is highly dependent on the industrial control information, communication and society (ICS) approach to develop a smarter country. Recently, ICS components which are interconnected (a network) have become as the industrial IOT (IIOT). The IIOT is defined as the heart of the smart city and is involved in controlling infrastructure components such as electric grids, water networks, CCTV security networks, and transportation systems. Reliability measures can only be attained with the continuous functioning of ICS components. However, the IIOT is not secure from cyber attacks. Therefore, the future of sustainable smart cities is at risk regarding security aspects. Local governments are not focusing on the security risks possible in ICS components, as the public and administrators are not aware of the assets and components at risk. AI techniques have assisted in developing an automated tool which can evaluate cyber risks in critical infrastructure. AI-based tools can identify adversarial attacks (attack trees) which help specialists and security novices in identifying attack pathways [29].

16.3.2   Machine Learning in Smart Cities

Previously, the sending of raw data to the cloud for data processing was common practice. But these schemes have proved to be less effective in the deployment of smart cities. Therefore, a new trend toward the decentralization of data analytics has arisen which focuses on fog layer and IoT devices. An ML model can be made to work when datasets have the same features as those of the training datasets. When the distribution model is changed using a different feature, the ML model doesn't work. There is a need for compatibility when a trained localization model is transferred from one platform to a new platform without receiving signal strength indicator values. To achieve this kind of transformation, transfer learning has a role. Another important need for smart city deployment is the integration of semantic technologies. The integration of systems is built between citizens and social media. Virtual object interaction is added in smart city services along with deep reinforcement learning algorithms with the consideration that a physical object has virtual representation which can learn and act accordingly in an autonomous manner. Finally, the interaction with humans is the most important need, which is now fulfilled partially by mobile devices and wearable sensors [30].

16.3.3   Deep Learning in Smart Cities

Microgrids are gaining more attention in smart city development. DL is defined as a process in which natural data in raw form are processed in a limited way through ML methods. This is a type of ML which has emerged along with the representation of learning methods. DL can automatically discover data representation along with classification and natural data detection. DL has contributed more to AI applications which work on ML algorithms [31].

DL works on deep neural networks, which are similar to artificial neural networks, to process information and signals through hidden layers which are present between the input and output layers in order to determine weighted layers. These hidden layers help in providing a non-linear transformation of data which can model

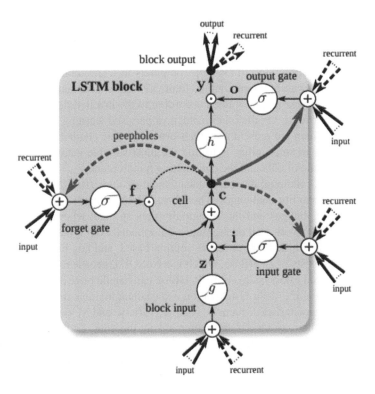

FIGURE 16.11 A long short-term memory block.

the complex relationship found in multi-layered structures. DL methods are mostly applied to time-dependent data which are more complex, noisy, and high-dimensional. One of the major applications of DL in detecting air quality in the smart city is implemented through long short-term memory (LSTM), which was introduced by Hochreiter and Schmidhuber. LSTM uses a recurrent neural network which has long term dependencies in learning. In this model, each ordinary neuron of a hidden layer is replaced by a memory block as shown in Figure 16.11 [32]. An LSTM block consists of a forget gate, input gate, and opposite gate which help in regulating the flow of information to and from the cell. An LSTM is frequently applied in speech recognition, machine translation, and neural language models.

16.3.4 Big Data in Smart Cities

Hadoop, which is a horizontally scalable platform, is more feasible for highlighting the scalability requirements among the big data analytics frameworks. It supports in processing huge volumes of big data, and is known as "fog computing" or "edge computing." This technology provides computing along with the storage facilities among cloud data centers and end devices. Also, it helps in detecting anomalous events on cloud storage and in providing robust platforms for computational fitting and data storage in the smart city. Data mining and ML are the most important technologies involved in data-centric applications such as smart cities. Data mining helps with the extraction of information provided by big data. ML is an important application of AI, which has numerous algorithms such as supervised and unsupervised algorithms. These algorithms help computers to learn about the environment and to increase their ability to predict the future as well as to respond to present conditions. Supervised learning is trained according to the labelled data which are able to predict future events. On the other hand, unsupervised learning is processed with unlabeled data which are present in hidden structures. The challenge of data mining and ML methods in smart cities is to find the most suitable dataset which exactly matches the volume of data and the data mining requirements.

Unfortunately, relational database management systems (RDBMSs) are not able to work with scalability challenges. Therefore, RDBMSs can't work with the processing power challenges and the commodity data storage, whereas notonly SQL databases (NoSQL) work well on big-data requirements. Four types of NoSQL databases are document-based, graph-based, column-based, and key-based. The Apache Hadoop suite consists of an HBase, which is a NoSQL database running on top of a Hadoop Distributed File System (HDFS). HBase can handle huge datasets with more rows and columns [33]. Data visualization is a challenging task in big data because of their "3V" characteristics, namely volume, variety, and velocity. Visualization helps decision makers in identifying patterns from the data. As it is flooded with a high volume of data, it is not feasible in big data technologies. However, technological developments have paved the way to make visualization possible through Google maps, virtual reality, augmented reality, and mixed reality. Some of the data visualization examples under smart cities are point of interest and traffic management [34].

16.4 BLOCKCHAIN IN SMART CITIES

Various applications are developed for the benefit of smart cities, though security is the main concern which has yet to be addressed. Therefore, blockchain technologies have emerged recently. Some of the blockchain applications in the smart city are as follows.

16.4.1 BLOCKCHAIN BASED SMART ECONOMY

The sharing economy is defined as the process of allocating under-utilized assets to the population through which supply and demand interact to provide products and services. The sharing economy greatly helps in expanding the scope of resource allocation. A representation of the links between sharing economic conditions and smart cities is depicted in Figure 16.12 [35]. The smart city is handled through three main role drivers: societal drivers, technology enablers, and economic drivers. Also, the growth of the smart city benefits from the improved utilization of urban resources such as food, money, services, goods, and transportation. The main goal of the sharing business is to reduce transaction costs and to improve the utilization of resources. The supply–demand life cycle benefits from a smart economy. The supply-side includes offering spare rooms in apartments and the short-term rentals of

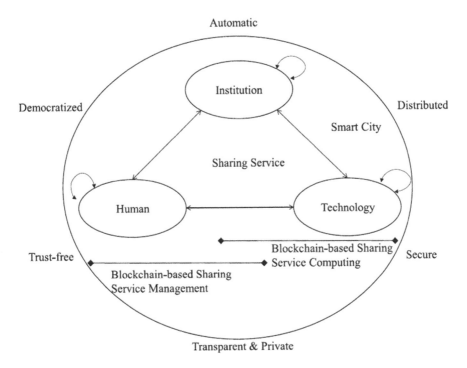

FIGURE 16.12   Links between the sharing economy and smart cities.

vehicles. The demand-side includes benefits provided from renting goods with lower transactional costs. The sharing economy is driven by digital connectivity in which citizens, utilities, and objects in smart cities are linked via ubiquitous computing to share information regarding idle resources.

16.4.2   DIGITAL CITIZEN PARTICIPATION

As the world revolves around innovative digital technologies, citizens are now termed "digital citizens." The notion of smart cities must be broadened beyond the technological aspects to incorporate the approach which invests in human and social growth along with environmental capital. Nowadays, digital citizen participation is discussed and holds a main role in sustainable smart cities, as shown in Figure 16.13 [36].

16.5   GLOBAL PERSPECTIVE OF SMART CITIES

The perspective of the sustainable smart city has been initiated globally. Various countries such as in the European Union, China, and Taiwan have stepped into the smart city mission through specific criteria framed according to each country. Some of the specialized features included in the regulations in order to make sustainable smart cities are discussed below.

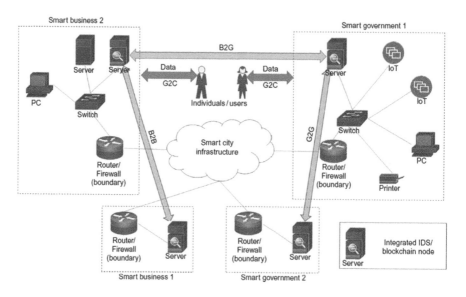

FIGURE 16.13 E-government security and privacy framework.

16.5.1 EUROPEAN UNION

The European Union has its assessment framework used for the ranking of smart cities, which depends on sustainability measures. These rankings are provided according to hierarchical clustering and principal component analysis, coupled with 32 indicators are divided into four components by which rank scores are calculated. Amongst European Union cities, Nordic capitals and Berlin lead the rankings, whereas Bucharest and Sofia are the lowest. Almost all the city rankings are correlated with the size of the city and population, which have a correlation with GDP, the wealth indicator of a country. Geographical divisions help to analyze where cities stand among others, hence policy makers are able to identify areas to be improved.

16.5.2 WUHAN CITY

According to the theoretical framework, four layers are considered – the digital city, open city, smart city, and intelligent city – and which are included in people-centric service intelligence. These layers are embedded with an infrastructure layer, data layer, and service layer respectively. The digital city includes soft infrastructure such as the people observation system (POS), professional processing system, cloud infrastructure, ICT, hard infrastructure (e.g. earth observation system orEOS), logistics systems, building information modelling, transport and global positioning, and communication systems. With the help of POS, EOS, and ICT, the dynamics of the earth and human dynamics are performed by infrastructure layers. A real-time geographical information system (GIS) and space-time are important for providing space-time analytic function fundamentals, and for organizing, digitalizing, and achieving smart cities.

The second layer (open city) is defined as a data layer in which the service intelligence level depends on openness in private and public data. Privacy and openness are considered as two sides in service intelligence. Handling these two sides is a very crucial problem. Private devices such as smart phones, eye-motion devices, and wearable glasses can sense individual schedules and social interactions. Various applications are involved in collecting the user's data with their permission during installation. These provide powerful sensing on technical systems for generating social, consumption, and personal data. Using these data, governance systems are built using geospatial information systems to relate business and people to locations. Technologies like virtual platforms and data clouds facilitate interoperability and data integration for supporting data as a service. An intelligent city is mainly dependent on a third layer which is termed a "service layer." The foundations of such technologies which are used by individuals and groups include AI, spatial behavior, and knowledge engineering modules. These are summarized as the need for detection, optimization, context sensing, and everything as a service. With the help of these layers, the smart practices in Wuhan city are listed in Table 16.1 [37].

TABLE 16.1
Wuhan Smart Practices

Projects	Functionalities
Administrative service center	Built on the IoT and cloud platforms
Smart-decision system (e-governance)	Provides integration between the intelligent services for e-governance
Digital City (Wuhan) Management Public Service Platform	Increases interaction between the government and citizens
Smart Fisheries Pilot Project	Improves smart agriculture through IoT platforms
Wireless Guanggu Project	Achieves the goal of full Wi-Fi in the main function zones
Non-parking charge system (urban bridge and road)	Tracks stolen vehicles and combat vehicles and supports smart transportation
Intelligent parking lot management system	Supports electronic toll collection (ETC) vehicle owners in finding a parking lot for convenient spacing
Vehicle network public service system	Provides services such as IaaS, PaaS, and SaaS for car-related information
"Wings travel"	Provides traffic information on the third ring road of Wuhan city
Geospatial cloudplatform	Provides 3D digital map for supporting municipal and GIS
Sewage treatment and operation management platform	Supports sewage water treatment using supervisory control and data acquisition (SCADA)
Intelligent pension platform	Supports old-aged people with 60 services such as human care, housekeeping service, and electrical maintenance
Smart campus	Provides smart education in the form of teaching content information services and home education through the internet

16.6 GLOBAL DEVELOPMENT IN AUTONOMOUS VEHICLES

Autonomous self-driving cars are now spotlighted globally. Various companies such as Mercedes-Benz, Tesla, BMW, and Toyota have stepped into the production of autonomous cars. The features of such cars are [38–40]:

- *Mercedes-Benz* team has introduced an autonomous research vehicle to support various aspects of smart mobility concepts and luxury which are to be experienced by users. The new design of the autonomous car includes four rotating lounge chairs which swing 30 degrees outwards on opening the doors. Also, these cars have the ability to exchange information between vehicles and passengers with the help of six display screens, visual and acoustic communication between the surroundings. They use light emitting diodes and lasers at the front and rear ends for visual communication with pedestrians and other vehicles. They have 3D video cameras and radars set at different ranges, namely low, mid-range, and high, for enabling autonomous operations. They have lane-assistance, brake automation, and are self-steering.

- *Tesla Autopilot* assists the driving system which enables adaptive cruise control, self-parking, and automatic lane changing functions. The Tesla Autopilot 8.0 which was released in August 2016 has a primary sensor with a camera. The next version 9.0 provides on-ramp and off-ramp features. Hardware 1, the 2014 autopilot, was an initial version of Tesla which had a monochrome camera and rear view camera for enabling human usage rather than for automation purposes. Hardware 2, the 2016 enhanced autopilot, had forward radar and cameras, forward-looking side cameras and reverse-looking cameras, and rearview cameras in which 12 sonars were included. The Test Autopilot crashed due to its inability to detect objects when speed exceeded 50 mph.

- *BMW Vision iNEXT* is designed in such a way to offer users entirely automatic and highly autonomous services. It was developed in collaboration with Intel, BMW and Mobileye. It was then deployed in November 2020. The amazing fact of BMW Vision iNEXT is that the driver can select the option to drive or to be driven (autonomously). It provides interactive services, entertainment mounted on the seats in front, panoramic flooding roofs, and a bench seat at the rear. The driver's area has two displays which are operated in two modes, namely boost and ease, which are more dependent on the involvement of drivers. BMW Vision iNEXT has planned to develop batteries, multiple sensors, and chip-based LiDAR integrated with Shy Tech which consists of an intelligent personal assistant, intelligent materials, and intelligent beams.

- *Toyota Lexus LS600HL*was designed in 2018 and has features such as a high-end navigation system, an adaptive front lighting system, radar cruise control, an intuitive park assist system, a pre-collision detection system, blind spot monitoring, and a lane-keeping assist system. Four-range LiDARs were developed by LUMINAR which are used to increase visibility, and short range LiDAR sensors are fixed on the rear end which are used to detect unusual activities on roads. A set of cameras with LiDARs are used for weatherproof panels.

16.7   FUTURE OF SUSTAINABLE GREEN IT RESEARCH

"Future cities" denotes the conceptualization of how cities will look and operate, systems that orchestrate and how they relate to stakeholders. The concepts of smart, eco, and resilient are the main focus regarding future cities. With the increased innovative technology and communications, future cities are coupled with smart concepts. Future cities have to be combined with social, economic, and environmental aspects along with digital, networked, and resilient technologies that are supported by governance systems. Digital literacy and digital culture play a key role in achieving new paradigms of sustainable future cities.

Smart cities have enabled recent technologies such as the IoT, digital twins, and 5G. Advanced analytics techniques have numerous open research issues related to smart cities. Medium k-nearest neighbors (KNN), fine KNN, You Only Look Once (YOLO) v4, YOLOv5, decision tree, and a convolutional neural network (CNN) were applied for autonomous driving related classification and traffic congestion. Current black-box technologies which depend on training data cannot help in decision-making in smart cities. Thus, Explainable AI (XAI) algorithms are used for applications such as traffic congestion, intrusion detection systems, self-driving cars, and computer vision problems. Therefore, it is mandatory to monitor and evaluate the behavior of smart city applications before ML and DL learning models are applied.

One of the most important applications of sustainable development is the health sector. Wearables have become more attractive among the public today for tracking all vital signs. Fitbit and Apple Watch have now been developed to an advanced level for recording data as well as assessing bodily conditions using sensors. These kinds of monitoring are defined as "intelligent associations." Networked contact lenses are used for analyzing the patient's tear fluid. These data are collected and then sent to an insulin pump for analyzing the blood sugar [43].

16.8   XAI FOR BLOCKCHAIN

Blockchain technology deals with an immutable ledger where all the transactions and asset management are carried out. Also, blockchain is used for tracking physical and virtual assets. Thus, it provides all the information in a secure environment. Smart contracts are defined as predetermined programs that are triggered in the blockchain network based on the conditions provided. It automates the agreements within the network without any interruptions. Blockchain technology is one among the essential solutions for cloud-based data. The advantages of blockchain technology have two main drawbacks: cross-layer implementation in cloud computing environments and the need for a control mechanism in automation tasks.

Blockchain prevents intruders from modifying data in the system. AI methodologies have security challenges for smart city problems, especially during the integration of blockchain with AI models which guarantees security and data privacy. These combinations will also have some security challenges.

16.9 APPLICATIONS OF XAI FOR BLOCKCHAIN

1. *Customer Profile Assessment.* The integration of blockchain with XAI affects the banking and finance sectors in a massive way. XAI when integrated with blockchain comprises multi-agent systems and intelligent expert systems. Also, it assists in decision-making, identifying credit worthy customers, and in providing business finance and startups.
2. *Medical Imaging.* The integration of XAI and blockchain implements a secure medical diagnosis framework for medical imaging. It uses block-wise encryption and histogram shifting to ensure secure transmission of patient data for providing trustable information about patients (how, who, when, where the data is created). It also assists radiologists for making decisions about critical patients [40]. Some of the countries that use the key data-driven smart city applications are shown in Table 16.2.

16.10 METAVERSE

The term "metaverse" comprises two words: "meta" and "verse." Meta denotes beyond and verse denotes universe. Metaverse has recently been discussed in many areas and is an integration of numerous cutting-edge technologies. For these developing

TABLE 16.2
Country-wise Data-Driven Smart City Applications

City, Country	Applications	Main Key Smart City Indicators
Amsterdam, the Netherlands	Open databases and citizen participation in all areas	SG, SEnv, SM, SP, SL
Groening, the Netherlands	Energy efficient systems	SEnv, SM
Nice, France	Smart lighting and environmental monitoring	SEnv, SM, SL
Padova, Italy	Environmental monitoring via smart sensors	SEnv
Turin, Italy	Digital portal for all services and smart meters	SG, SEnv, SP
Barcelona, Spain	Green technologies, smart water management, open databases	SG, SEnv, SM, SEcon, SP, SL
Malmo, Sweden	"Citizen-centric" portals and data platforms	SG, SL
Brno, Czech Republic	Support for entrepreneurship	SEcon
Porto, Portugal	Porto innovation hub	SEcon
Norfolk, UK	Environmental monitoring and local authority	SG, SEnv

SG smart governance, Senv smart environment, SM smart mobility, SL smart living, SP smart people, SEcon smart economy [41].

technologies to have an impact across all disciplines, apps utilizing them must have a more developed version. For instance, AI is essential for supervising information, otherwise harmful information might spread. Also, a blockchain framework is needed for securing all the authenticated information. Data computing for data analysis is also required, which includes cloud computing, edge computing. 3D rendering, a brain computer interface (BCI), robotics, and Extended Reality (XR) are the most important technologies for metaverse to work efficiently. Metaverse attempts to create a very immersive and digital space for promoting internet penetration and upgrading the digital economy. At present, metaverse technology is still in its infancy [42].

16.11   CONCLUSION

Conservation is a state of harmony between men and land.

Aldo Leopold

Technological contributions for the development of human beings have been mostly achieved in every sector. For this technological growth, the IoT plays an important role for satisfying human needs instantaneously. A major application of the IoT is the smart city, which is influencing human lives a great deal. Therefore, models of smart cities, such as the smart city wheel model, IBM model, and Hitachi smart city model, are being discussed. However, sustainability has to be considered for their constructive development without adversely affecting human lives. Hence, sustainability can be achieved through three key factors: infrastructure, economy, and society. For each key factor, the main attributes, such as smart parking, renewable energy, and the green smart city, have been explained. People are more aware of the recent pandemic situation which affected the health and education sectors in particular. With this consideration, the green smart city, which comprises the two vital categories of smart health and smart education, has been discussed. Moreover, sustainable development is also attained with the help of technologies such as AI, ML, DL, and big data; therefore the role of these three technologies has been described. Security issues in technological development are considered to be the main drawback. Therefore to address such security issues, blockchain has emerged as a technique to resolve such threats. Hence, two main roles of blockchain in smart cities, which helps in sustainable development, relate to the smart economy and digital citizen participation. Sustainable development is attained globally in transportation sectors through the various designs of automobiles, such as Mercedes-Benz, BMW Vision iNext, and Toyota Lexus LS600HL. Thus, this chapter should help designers and educationalists to gain knowledge about the technological developments in smart cities with respect to sustainability aspects.

16.12   FUTURE SCOPE

"Sustainability" and "Green Smart Cities" are two active terms which will spread globally. It will help people to make their lives more comfortable. From the pandemic perspective, it will favor people in the tracking of hotspot areas, quarantine regions in specific cities, and also help in the checking of transportation and vaccination centers. Therefore, this chapter has provided a detailed outline of sustainable and smart green cities as a whole.

REFERENCES

[1] J.I. Criado, and J.R. Gil-García, Electronic government, management and public policies current status and future trends in Latin America. *Manage. Public Policy* 22(2) (2013).

[2] Amir H. Alavi, Pengcheng Jiao, William G. Buttlar, and Nizar Lajnef, *Internet of Things-Enabled Smart Cities: State-of-the-Art and Future Trends, Measurement.* July 2018, doi: 10.1016/j.measurement.2018.07.067

[3] B. Cohen, "The top 10 smartest European cities," Internet: http://smartertrends.co.uk/articles/2012-11-14/the-top-10smartesteuropeancities/original/?original=www.fast coexist.com/1680856/the-top-10-smartest-european-cities [May29, 2013].

[4] S.F. DeAngelis, "A thought probe series on tomorrow's population, big data, and personalized predictive analytics: Part 3, Where things stand," Internet: http://en.paperblog.com/a-thought-probe-series-ontomorrow-s-population-big-data-and-personalized-predictive-analytics-part-3-where-things-stand524392/, 2013.

[5] "Hierarchy of needs," Internet: http://gerardkeegan.com/glossary/hierarchy-of-needs, [May 3, 2013].

[6] P. Neirotti, A. De Marco, A.C. Cagliano, G. Mangano, G. Scorrano, Current trends in smart city initiatives some stylized facts. *Cities* 38(2014):25–36.

[7] *Report of the World Commission on Environment and Development: Our Common Future.* Available online: http://www.un-documents.net/our-common-future.pdf (accessed on 29 October 2019).

[8] M. Gasco Living labs: Implementing open innovation in the public sector. *Gov Inf Q.* 34 (2017):90–98.

[9] A. Komeily, and R. Srinivasan Sustainability in smart cities: Balancing social, economic, environmental, and institutional aspects of urban life. In *Smart Cities: Foundations, Principles, and Applications.* John Wiley & Sons: Hoboken, NJ, 2018, pp. 503–534.

[10] R. Hiremath, P. Balachandra, B. Kumar, S. Bansode, and J. Murali, Indicator-based urban sustainability-A review. *Energy Sustain Dev* 17 (2013):555–563.

[11] S. Morse, and E.D. Frase, Making 'dirty' nations look clean? The nation state and the problem of selecting and weighting indices as tools for measuring progress towards sustainability. *Geoforum* 36 (2005):625–640.

[12] H. Ahvenniemi, A. Huovila, I. Pinto-Seppä, and M. Airaksinen, What are the differences between sustainable and smart cities? *Cities* 60 (2017):234–245.

[13] Adeoluwa Akandea, Pedro Cabrala, Paulo Gomesa, Sven Casteleyn, The Lisbon ranking for smart sustainable cities in Europe. *Sustainable Cities and Society* 44 (2019): 475–487.

[14] Dastan Bamwesigye, and Petra Hlavackova, Analysis of sustainable transport for smart cities. *Sustainability* 11 (2019):2140, doi:10.3390/su11072140

[15] Chandra Shekhar Mishra, R.K. Pandey and Shakti Suryanshi, http://www.iaeme.com/IJCIET/index.asp

[16] W. Ejaz, M. Naeem, A. Shahid, A. Anpalagan, and M. Jo, Efficient energy management for the Internet of Things in Smart Cities. *IEEE Commun. Mag.*, doi:10.1109/MCOM.2017.1600218CM

[17] K. Aono, N. Lajnef, F. Faridazar, and S. Chakrabartty, Infrastructural health monitoring using self-powered Internet-of-Things. In *IEEE International Symposium on Circuits and Systems.* Montreal, QC, Canada, 2016, doi: 10.1109/ISCAS.2016.7538983

[18] A.C. Şerban, and M.D. Lytras, Artificial intelligence for smart renewable energy sector in Europe smart energy infrastructures for next generation smart cities. *IEEE Access,* 2020:77364–77377.

[19] E. Mocanu, P. H. Nguyen, M. Gibescu, and W. L. Kling, Deep learning for estimating building energy consumption. *Sustain Energy, Grids Netw* 6(2016):91–99.

[20] Fayez Alqahtani, Zafer Al-Makhadmeh, and Amr Tolba Wael Said, Internet of Things-based urban waste management system for smart cities using a Cuckoo Search Algorithm. *Cluster Comput.* (2020), doi:10.1007/s10586-020-03126-x

[21] Sonali Dubey, Pushpa Singh, Piyush Yadav, and Krishna Kant Singh, Household waste management system using IoT and machine. *Procedia Comput.Sci.* 167 (2020):1950–1959.

[22] Valérian Guivarch, Valérie Camps, André Péninou, and Pierre Glize. Self-adaptationof a learnt behaviour by detecting and by managing user's implicit contradictions. In *Proceedings of the 2014 IEEE/WIC/ACM International Joint Conferences on Web Intelligence (WI) and Intelligent Agent Technologies (IAT)*, 03. IEEE Computer Society, 2014, pp.24–31.

[23] Riaz Shaik, Narendra Kumar Gudapati, Nikhil Kumar Balijepalli, and Harshitha Raj Medida, A survey on applications of Internet of Things. *Int. J. Civ. Eng. Technol.* 8(12) (2017):558–571.

[24] Everton Gomede, Fernando Henrique Gaffo, Gabriel Ulian Briganó, Rodolfo Miranda de Barros, and Leonardo de Souza Mendes, Application of computational intelligence to improve education in smart cities. *Sensors* 18 (2018):267, doi:10.3390/s18010267

[25] "Why Smart Cities are needed now," white paper, Hitachi Ltd, Sep. 2012, Internet: http://www.hitachi.com/products/smartcity/download/pdf/whitepaper.pdf [May 15, 2013].

[26] Ana Iolanda Voda, and Laura-Diana Radu, How can artificial intelligence respond to smart cities challenges. *Smart Cities: Issues and Challenges*, 199–213, doi:10.1016/B978-0-12-816639-0.00012-0

[27] Kun Guo, Yueming Lu, Hui Gao, and Ruohan Cao, Artificial intelligence-based semantic Internet of Things in a user-centric Smart City. *Sensors* 18 (2018):1341, doi:10.3390/s18051341

[28] S. Srivastava, and A. Bisht, Safety and security in smart cities using artificial intelligence-A review.(2017):4.

[29] Mehdi Mohammadi, and Ala Al-Fuqaha, Enabling cognitive smart cities using big data and machine learning: Approaches and challenges, emerging trends, issues, and challenges in Big Data and its implementation toward future. *IEEE Commun. Mag.* (2018): 94–101, doi: 10.1109/MCOM.2018.1700298

[30] I. M. Coelho, V. N. Coelho, E. J. D. S. Luz, L. S. Ochi, F. G. Guimarães, and E. Rios, A GPU deep learning Metaheuristic based model for time series forecasting. *Appl. Energy* 201 (2017): 412–418.

[31] K. Greff, R. K. Srivastava, J. Koutnik, B. R. Steunebrink, and J. Schmidhuber, LSTM: A search space odyssey, *IEEE Transactions on Neural Networks and Learning Systems*, 28(10) (2016): 2222–2232.

[32] Raja A. Alshawish, Salma A.M. Alfagih, Mohamed S. Musbah, Big data applications in smart cities, in: *International Conference on Engineering & MIS (ICEMIS)*, Agadir, Morocco, 2016.

[33] A. Corradi, G. Curatola, L. Foschini, R. Ianniello, and C.R. De Rolt. Automatic extraction of POIs in smart cities: Big Data processing in ParticipAct, in *2015 IFIP/IEEE International Symposium on Integrated Network Management (IM)*, Ottawa, Canada, 2015.

[34] Jianjun Sun, Jiaqi Yan, and Kem Z. K. Zhang, Blockchain-based sharing services: What blockchain technology can contribute to smart cities, *Financ. Innov.* 2(2016):26, doi:10.1186/s40854-016-0040-y

[35] Islam Bouzguenda, and Chaham Alalouch, Nadia Fava. Towards smart sustainable cities: A review of the role digital citizen participation could play in advancing social sustainability. *Sustain. Cities Soc.* 50 (2019):101627.

[36] Hong Xu, and Xuexian Geng, People-centric service intelligence for smart cities. *Smart Cities* 2(2019):135–152, doi:10.3390/smartcities2020010

[37] Ibrar Yaqoob, Latif U. Khan, S. M. Ahsan Kazmi, Muhammad Imran, Nadra Guizani, and Choong Seon Hong, Autonomous driving cars in smart cities: Recent advances, requirements, and challenges.*IEEE Network* (2019), doi:10.1109/MNET.2019.1900120

[38] Future effectual role of energy delivery: A comprehensive review of internet of things and smart grid. *Renew. Sust. Energ. Rev.* 91 (2018):90–108.

[39] Parvaneh Asghari, Amir Masoud Rahmani, and Hamid Haj Seyyed Javadi. Internet of Things applications: A systematic review. *Comput. Networks* 148 (2019):241–261.

[40] Sweta Bhattacharya, et al."A review on deep learning for future smart cities." *Internet Technol. Lett.* 5.1 (2022): e187.

[41] Yamuna Kaluarachchi. "Implementing data-driven smart city applications for future cities." *Smart Cities* 5.2 (2022): 455–474.

[42] Zaheer Allam, et al. "The metaverse as a virtual form of smart cities: Opportunities and challenges for environmental, economic, and social sustainability in urban futures." *Smart Cities* 5. 3 (2022): 771–801.

[43] Taher M. Ghazal, Mohammad Kamrul Hasan, Muhammad Turki Alshurideh, Haitham M. Alzoubi, Munir Ahmad, Syed Shehryar Akbar, Barween Al Kurdi, and Iman A. Akour. "IoT for smart cities: Machine Learning approaches in smart healthcare—A review." *Future Internet* 13.8 (2021): 218, doi: 10.3390/fi13080218

17 Smart Cities Using Green Computing
A Citizen Data Scientist Perspective

Prashant Pansare
Rubiscape Labs, Pune, Maharashtra, India

17.1 INTRODUCTION

A smart city is a city that implements modern technologies and electronic systems to connect, improve efficiency, communicate, and share information with its citizens and enhance the quality of life. The smart city basically creates a digital ecosystem and technology, public and private stakeholders, and citizens who are the parts of this ecosystem. Thus, digitalization is one of the most important aspects of smart city initiatives. In smart cities, several sensors collect data from different sources. These data are analysed to obtain meaningful insights which are used to manage efficiently different objects or facilities, such as roads, buildings, the environment, and the traffic of the ecosystem. These are the assets of the smart city, and sound asset management is required to reap benefits from them. Increasing asset life expectancy or decreasing their downtime are two measures of good asset management that translate into superior performance, lower management costs, enhanced user experience, and better sustainability.

Digitalization is the adoption and integration of electronic resources, tools, and technologies to improve processes and facilities, tackle issues, and satisfy the needs of people and society. Smart cities use digital intelligence to translate data into insights to solve public problems, make data-driven decisions, and improve the standard of life. Hence, smart cities need technologies with the capability to process huge amounts of data or 'big data'. High-volume, high-velocity, and high-variety information are three main characteristics of big data, and smart technologies need to be able to process enormous amounts of heterogeneous data quickly using algorithms.

The building blocks of a smart city are:

- A digital technology base;
- Applications that use data to draw meaningful insights;
- Use of these insights by various stakeholders: residents, companies, and government organizations (see Figure 17.1).

DOI: 10.1201/9781003388814-17

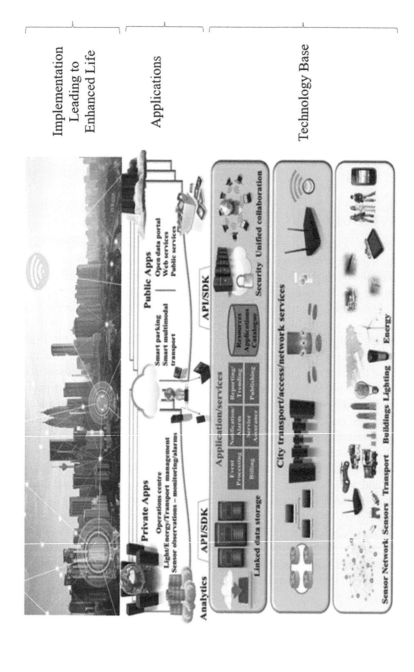

FIGURE 17.1   Building blocks of a smart city [1].

17.1.1 DIGITAL TECHNOLOGY BASE

Sensors and devices, communication networks, and open data portals form the digital technology base. Sensors record different parameters such as temperature, humidity, air quality, UV index, traffic, garbage production, energy consumption, water levels, and demand data. The data are constantly generated and uploaded to the cloud and are available online at every moment. This is like digitizing the behaviour of the city, which is feasible due to advancements in digital connectivity and the Internet of Things (IoT). Scalability, the need for upgrading, energy requirements, sustainability, and the involvement of citizens and stakeholders are the factors that a smart city should consider when adopting any innovative digital technology.

17.1.2 APPLICATIONS USING DATA

Data are the foundation for building applications which could be 'public apps' or 'private apps'. Applications use data generated by sensors to create alerts, insights, and design actions to solve or address specific public issues by generating sustainable solutions. Here, technology providers and app developers play an important role, and they require open data for building algorithms. For example, applications with eco-friendly solutions for transport can be designed to improve peoples' mobility or to improve energy efficiency through real-time monitoring and controlling heating and cooling. Applications can also be built as a bridge between government establishments and citizens where the latter can submit their ideas or suggestions on planning and administration, raise their concerns, and support certain initiatives, while government establishments can collect information, address complaints, create awareness, and digitize governmental functions such as tax filing, permits, licensing, approvals, and issuing certificates. Some other examples are apps for water-consumption tracking, telehealth, air quality tracking for a cleaner and more sustainable environment, and data management. These solutions can be classified into six categories: smart living, smart environment, smart mobility, smart government and governance, smart economy, and smart people [2]. These are the smart city indicators, and the smart city performs well in these six characteristics [3] (Figure 17.2).

- *Smart living*. These are the solutions that improve and enhance the standard of living in the smart city and include information and communication technology (ICT) enabled lifestyles, behaviours, and consumption by citizens of an urban area. Some of the actions or policies to support smart living provide cultural support and social cohesion, promoting telehealth, provision of emergency response facilities, smart utility metering, and 24/7 utility supply.
- *Smart environment*. These solutions aim at reducing the ecological footprint of a smart city and offer environmentally sustainable offerings and include smart energy solutions, pollution control and monitoring, smart buildings, and green urban planning. This can be ensured by reducing pollution, providing environment protection, having early warning systems in place, and the recycling of resources and solid waste.

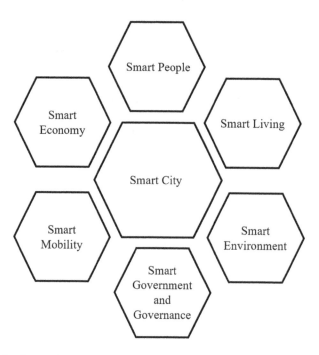

FIGURE 17.2 Smart city solutions.

- *Smart mobility.* Smart mobility is defining and developing sustainable mobility options and include ICT supported and integrated transport and logistics solutions. The actions required to provide smart mobility solutions include providing safe transportation systems, smart parking, smart traffic management, and the provision of safe and secure wireless connectivity in public places.
- *Smart government and governance.* These solutions seek to foster the involvement and contribution of citizens and to increase clarity, openness, and accountability in governance.
- *Smart economy.* This refers to making a city competitive through the implementation of e-commerce solutions, increased productivity, ICT enabled innovation, and the production and delivery of products and services.
- *Smart people.* This refers to protecting and increasing the knowledge, skills, abilities, education, and expertise of people in a city to help them develop potential and encourage their involvement in social welfare. These can be facilitated by providing digital education, collaboration between knowledge centres and industries, and promoting research, innovation, and development.

A single initiative of a smart city can come under multiple categories. For example, the Pune smart city initiative is working on intelligent lighting, sustainable transport, solid waste management, vehicle tracking, solar energy, and quality city management projects simultaneously. This involves many stakeholders and the mobilization of large and significant resources and hence is complicated.

17.1.3 Use of Insights by Various Stakeholders

The third building block is the application of these insights by residents, cities, organizations, and government establishments. Applications are successful when these stakeholders accept these solutions, use them, and change behaviour and actions to achieve the desired results. For example, applications providing live traffic conditions encourage people to take alternative, less congested routes. It is the travellers who are required to take suggested routes which in turn can reduce the traffic congestion. Energy management applications provide the consumption details and patterns and motivate users to use less energy. Thus, the adoption and usage of these applications often leads to better decision-making and positive behavioural change in society.

Most people within organizations use and manage data in some way through accessing databases and preparing spreadsheets, reports, or plotting charts and graphs for presentation to management. Thus, all individuals engage with data in some capacity. Imparting data science knowledge to these people to build a data-driven community, the citizen data scientist is the efficient and precise way to go.

The purpose of the research in this chapter is to explore the concept of the green smart city through the lens of green computing, data analytics, and the contribution of the citizen data scientist. After setting up the context above, Section 17.2 carries out a detailed literature review to understand the various aspects, applications, and challenges in smart city solutions, green computing, and the role of the citizen data scientists in smart city development. Section 17.3 presents smart-city technologies and aspects of quality of life, green computing, and AI and ML analytics for the green smart city and Citizen data scientist. Section 17.4 focuses on green computing.

17.2 LITERATURE REVIEW

There is research available on smart city applications, green computing, and citizen data scientists. The focus in this review is on use cases specifically relating to smart city solutions applying green computing methods and involvement by citizen data scientists in developing and implementing such solutions. This section examines the literature pertaining to smart cities, green computing, various technologies for smart city solutions, citizen data scientists, and analytics for green smart cities. At the end, the research objective pertaining to citizen data scientists and their contribution to sustainable smart city solutions will be stated.

17.2.1 Smart Cities

Smart cities apply modern technologies and solutions to enhance the standard of living and make efficient use of resources. A smart environment, smart people, smart livelihood, smart economy and economic policy, and smart mobility are the five 'smart city quality of life' domains [4]. Deriving meaningful information from the data and building models using data science technologies are driving automation in cities, making them intelligent [5]. The populations of cities are growing and only a transition to smart cities can meet the growing demand on resources by citizens and improve their quality of life. The authors of [6–7] studied the factors supporting and hindering smart city adoption and suggest that compatibility, service quality, and relative advantage are the main bolsters. The features and facilities desired by citizens decide the standard ranking and value of smart city services, and [8] suggests that enhancement in the standard of living of citizens is the major success parameter in smart city management. Smart city solutions are designed to address environmental issues and solve the problems caused by growing urbanization. These problems include traffic congestion, air pollution and garbage disposal [9], safety, security and protection against natural disaster, noise and air pollution, and sewage treatment [10].

17.2.2 GREEN COMPUTING

Green computing involves the study and application of computers and related devices to achieve a greener, healthier, and safer environment without compromising the technological requirements of the present and forthcoming generations [11]. There are several areas where efforts are made to achieve this by reducing energy consumption, greenhouse gas emissions, e-waste recycling, data centre consolidation and optimization, the virtualization of computer resources, and creating policies to support and encourage the eco-labelling of products and services [12].

17.2.3 TECHNOLOGIES FOR SMART CITIES

Smart cities make use of ICT to ingest data from heterogenous resources, to analyse these data to draw meaningful information, and to use this analysis to increase operational efficiency and enhance the standard of living. Thus, smart city solutions are based on data driven technologies such as the IoT, data analytics, AI, and machine learning (ML). Cities in different social structures and spatial contexts differ in development levels, and smart city technologies are selected to suit their local environments [13]. The most common applications in smart cities include smart lights and traffic management, smart power, smart energy management, city surveillance, smart buildings, and patient healthcare monitoring [14]. AI and data mining are the technologies used in developing and implementing the sustainable solutions [15]. The IoT, big data, 5G, the cloud, AI and edge computing are increasing smart city developments [16]. According to [17] smart cities implement AI, ML, and IoT-based systems such as smart traffic management systems, smart street lighting, and smart roads to collect, analyse, improve, and optimize logistics systems, utility services, various amenities, and applications that interact with different platforms such as smart buildings, telehealth, smart healthcare, connected vehicles, smart farming, and Industry 4.0.

17.2.4 CITIZEN DATA SCIENTISTS

A citizen data scientist is a role in an organization that involves using big data tools and technologies for data analysis and building data and business models for their companies. It is performed by people having knowledge of big data tools and technology [18]. Urbanization along with population ageing is transforming society. Citizen data science can involve older adults and stakeholders in co-creating recommendations for local and urban improvements [19]. A study by [20], which shows that citizen scientists can generate meaningful and accurate data on the ecological status and pollution of water streams due to the use of pesticides [21], suggests that citizen data scientists have the potential to study water quality [22] and conduct research to compare the quality of the data collected with the data collected by researchers and to determine if training citizen data scientists can improve the quality of the data. The author also studied whether investing in citizen data scientists would earn a good return on investment. The findings show that citizen scientists' involvement can scale up the reach, and operation, and increase the sampling power of various ecological surveys without hampering the quality of the data. Citizen scientists take part in various projects from various sectors and areas, from astronomy to air quality and from population ecology to biology. The findings of the research by [23] shows that citizen science can be used to analyse community health and gain insights in the health of citizen scientists. Hartmann [24] conducted text and image analysis research based on citizen science data and found the data to be large, powerful, and comprehensive in terms of users and spatial and temporal coverage.

17.2.5 ANALYTICS FOR THE GREEN SMART CITY

Data is an invaluable resource for cities. Cities use this data to perform the real-time monitoring and analysis of city life, and to use the insights for building data driven, efficient, open, transparent, sustainable, and competitive cities [25]. Transforming from an analogue city to a digital, smart city is tough and has significant challenges. Cities select different data and data analytics tools and techniques for this transformation based on the type of data available, the processes and resources available, and their priorities. Other important factors contributing to the adoption of data and data analytics include structures, strategy, leadership, culture, data infrastructure, data governance, skills, training, capacities, and budgets. ICT helps smart cities to solve city problems, improve infrastructure and facilities, and address public concerns. Cities can use big data to understand residents' travel patterns, to predict travel times, and to improve traffic management and enhance transportation infrastructure [26]. With the advent of location and activity detection and tracking technologies and location-based services, large volumes of location-based data are created, and these data can be used to understand city and human life and to find solutions to issues such as traffic congestion and air pollution [27]. IoT devices and embedded systems generate huge amounts of data, which can be explored to derive meaningful and actionable insights and improve residents' quality of life [28]. Cloud computing, fog computing, and edge computing are used for IoT analytics, and the implementation of AI and DL models can increase the efficiency and performance of IoT analytics. Cities can increase the use of data and data analytics by being skill driven, by hiring data experts and data analytics professionals, by being process driven by coupling strong leadership with set processes and training, or by being data infrastructure driven by employing strong leadership and sophisticated data infrastructure [29]. The IoT and optimization algorithms can be used to decrease the ecological footprint and improve energy efficiency and reduce pollution and greenhouse gas emissions.

The objectives of this study are:

1. To research important smart city technologies, green computing, and AL and ML analytics;
2. To determine the relevance of AI and ML techniques in developing green smart cities;
3. To study citizen data scientists and their contribution in developing smart cities.

17.3   SMART CITY TECHNOLOGIES, GREEN COMPUTING, AND AI AND ML ANALYTICS FOR THE GREEN SMART CITY

17.3.1   SMART CITY TECHNOLOGIES AND QUALITY OF LIFE

Safety and security, time and convenience, health, environmental quality, social connectedness and community involvement, employment, and the cost of living are the important dimensions of the quality of life [1]. Implementing smart technologies can help cities to improve upon these aspects.

17.3.1.1   Safety and Security

The goal of building smart safety and security solutions is to ensure an effective response to major emergency situations such as fire. Monitoring, communication, and mobility solutions are three main components of the public security system that are integrated with and controlled by a command-and-control centre.

17.3.1.2   Data Driven Monitoring

Video surveillance and analytics allows for remote monitoring, real time detection of certain events, interpretation, recording and creating alerts [30], or generating real-time responses on the occurrences of these events. The analysis can be for motion detection, object identification and tracking, person identification and re-identification, or crowd detection and management. For example, number plate detection can be used in traffic management solutions or for providing intelligence in the case of vehicle theft.

Data driven monitoring applications can be developed to track and fight crime. Applications for public safety solutions can enhance protection and offer critical incidence support in such events. Intelligent transportation systems and the IoT allow for the collection, processing, and analysis of data on vehicles, road infrastructures, and driver behaviour and can be used to design road safety solutions, build models to help prevent accidents, and make the driving experience safe for road users.

- *Communication.* Stable, reliable, and uninterrupted communication between surveillance devices, control centres, data centres, and Government establishments responsible for public safety and security is required to ensure integration and interoperability between them. To enable this, robust, reliable, and scalable networks are required.

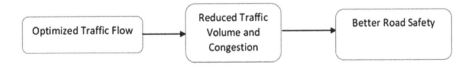

FIGURE 17.3   Mobility solutions.

- *Mobility solutions.* Mobility solutions enable the tracking, capturing, and streaming of real-time data on an incident. Some examples are smart mobility applications such as intelligent synchronizing of traffic signals, digital signage, or mobile apps for real-time tracking and monitoring, including smart-parking apps or IoT sensors, which prevent breakdowns and enable car sharing, bike sharing, and demand-responsive transport [31].

How citizens benefit (Figure 17.3):

- Quick response to calamities, disasters, and distress by concerned units;
- Early detection of potential distress and preventive action to avoid, minimize, or control it;
- Emergency response management;
- Enhanced security;
- Real-time tracking, monitoring, and information about traffic congestions, road closures or blocks, activities, events, or processions.

17.3.1.3   Time and Convenience

Time and convenience are critical aspects for improving quality of life. By 2025, cities deploying smart-mobility applications could reduce on average travelling times by 15–20% [32]. Smart parking solutions help people reduce the time needed for searching for available parking spaces. Smart traffic lights monitor and regulate traffic flow. Collecting people and vehicle movement data can be used to regulate traffic, track incidents, monitor traffic violations, and allow easy movement across the city.

17.3.1.4   Health

The smart city can make the health care system smarter with integrated and interoperable systems across health services. Smart healthcare technology solutions connect and interact with smart devices to collect data generated. These data are analysed by doctors, researchers, and health care professionals to derive insights, conduct personalized diagnoses, and offer customized treatment plans based on diagnosis. Thus, smart health care solutions can enable health care services to be accessed by all and to foster the social, economic, and environmental factors that influence citizens' overall health including physical, mental, social, and emotional health.

Smart healthcare implements new, data driven technologies like AI, the IoT, and big data to respond to the requirements of the healthcare ecosystem. AI and big data are used in drug discovery and vaccine distribution. The global IoT-based healthcare market is projected to increase by 29.9% and to reach $322.2 billion by 2025. Health applications can be developed to prevent, detect, track, treat, and monitor chronic health issues. Applications remotely monitoring patients, applications to identify elevated risk profiles (e.g. people with diabetes or heart disease), infectious-disease surveillance systems, and telemedicine are some examples that can help address health monitoring and use the disability-adjusted life years parameter to plan for a better healthcare system.

17.3.1.5   Environmental Quality

With urbanization and industrialization, the environmental footprint is also increasing. Smart cities can use smart technologies to reduce the impact and protect the environment.

Solutions like smart buildings, smart parking, traffic management solutions, smart energy solutions to optimize consumption, and mobility solutions like ridesharing can reduce carbon emissions by 10–15% [33]. In cities with higher water consumption, consumption tracking and metered water connections can encourage people to reduce usage and conserve water by about 15%. Installing sensors and analysing sensor data can be used to reduce water leakage from pipes. This can help cities conserve up to 80 litres of water per individual every day. Intelligent garbage solutions can be used to organize trash collection only when required, thus saving cost, and reducing carbon dioxide emissions. Smart trash cans powered by solar energy can automate pick-ups when full. Use of non-conventional energy sources for energy efficient lighting can reduce the use of conventional energy sources, lower pollution levels, and reduce the carbon footprint. IoT sensors for the prevention of leakage and breakage, deploying air quality sensors to monitor and improve air quality, as well as sound level monitoring are some of the applications for achieving a cleaner and sustainable environment, and for optimizing consumption (Figure 17.4).

17.3.1.6   Social Connectedness and Community Participation

Digital platforms can facilitate real-world interactions and social connections. Digital channels connecting government and community can be used to report concerns by citizens, collect data, gather input or suggestions on planning issues, and to digitize government functions such as tax filing, permit issuing, and licensing. This would also make government organizations more responsive and to disseminate information. Participation from citizens is the key to success and smart city transformation will only be successful if citizens are involved. There are various ways in which citizens can participate in smart city initiatives such as by using apps on mobile devices and providing data for improving the city's physical infrastructure and services. This can reduce facility costs and also improve sustainability.

FIGURE 17.4   Environmental quality enhancement.

17.3.1.7 Employment

One of the negative impacts of smart technologies is the reduction of certain types of jobs, such as jobs involving repetitive tasks and administrative jobs, but they also create others, such as data science, analysis, maintenance, and data engineering jobs. Online portals for business licensing, permits, and tax filing can simplify and quicken the processes contributing to a more entrepreneurial culture.

17.3.1.8 Cost of Living

Smart solutions help with saving by encouraging more efficient usage of water, energy, and other utilities and healthcare solutions. Smart streetlights and smart buildings can help with saving by reducing consumption and lowering repair and maintenance costs. Enterprises can save by employing smart transportation solutions and building smart manufacturing plants.

Innovation and new technologies such as AI and the IoT are steering the growth of smart cities around the world. AI can help in designing solutions to solve urban issues and the challenges of citizens, mobility, safety and security, or environmental concerns. It augments data handling, processing, and transformation abilities to convert data into meaningful insights and use these insights for designing and implementing solutions to build environmentally sustainable cities. Smart cities generate data in various formats, such as sensor data, text, images, videos, and social media data. AI can be used to analyse this large amount of data, build models, and employ cloud computing to reduce operational costs and optimize resources.

17.3.2 Artificial Intelligence

AI refers to modern technology where systems or machines are capable of simulating human intelligence to execute tasks, to reason, and to iteratively improve themselves based on learning experience. AI can efficiently handle and sift through massive amounts of data and can be used to build algorithms to forecast, make predictions, and build affordable, cost efficient solutions to drive sustainable smart city initiatives.

AI has two subsets: ML and neural networks. In ML computers learn from data and past experience, without being explicitly programmed and improve their performance on some tasks and decision-making practices.

17.3.3 Machine Learning

Supervised, unsupervised, semi-supervised, or reinforcement algorithms are subareas of ML:

- Supervised algorithms apply learning from past labelled data to new data. In supervised learning, the model is trained using the past data and target values to design a solution to the specific business problem. The trained AI model then executes predetermined actions. The model is continuously trained with the new data to explore new solutions and possibilities to provide better sustainable solutions.

- Unsupervised algorithms draw inferences from unlabelled data. In unsupervised learning, unlabelled and unclassified datasets are used to train AI algorithms, which are then used to find hidden patterns or trends in the data.
- Semi-supervised algorithms are trained with a small amount of labelled data and then work with new and unlabelled data.
- Reinforcement learning algorithms are trained with a specific goal or objective and are given a set of rules to achieve that goal.
- Machine learning algorithms strive to detect relationships among the parameters in a given dataset, to find patterns, and to make decisions with minimum human intervention.

17.3.4 DEEP LEARNING

Deep learning is a sub-field of ML where multi-layered artificial neural networks are trained to perform tasks like decision making, speech recognition, object location and identification, and natural language processing. Neural networks are composed of multiple layers of interconnected neurons with mathematical functions to process input data, transmit information, and predict an output. Deep learning algorithms are trained on large amounts of labelled data and hence require more time and large computational operations to train a model. Also, these models may not be able to identify causal relationships between variables explicitly. Deep learning has wide application in healthcare, manufacturing, finance, and retail industries where it is used to analyse huge amounts of data to derive insights and use them for better decision making.

17.3.5 THE INTERNET OF THINGS

The IoT is a huge network of digital devices that communicate and interact with each other. These devices include sensors, actuators, monitoring devices, and AI programs that can monitor, evaluate, and control particular factors or parameters.

A smart city deploys a number of actuators and sensors that collect relevant data for the betterment of the different systems. Data may be collected directly from purposefully installed sensors or from sensors positioned for some other uses, but they can only collect secondary albeit useful information. The component part is mainly categorized as solutions and services. Security, remote monitoring and surveillance, analytics, and network management come under the solutions category. Application segments are further classified into specific applications such as air quality monitoring, lighting, energy and utilities, traffic, public safety, and environmental monitoring. The end-user segment comprises government organizations, information and technology, telecommunications, manufacturing and automation, and energy.

Smart city applications depend heavily on a high-capacity, robust, low-latency communication network that can connect several devices. According to the report 'Smart Cities: Digital Solutions for a More Liveable Future' by the McKinsey Global Institute [34], cities could work to enhance some crucial quality of life parameters by 10–30% by using smart technologies. This is in terms of lives saved, reduced crime events and rates, shorter travel, reduced health burden, and reduced carbon emissions.

17.3.5.1   IoT Application in Smart Cities

Smart cities are effectively using IoT technology to improve the standard of living and to solve urban life problems. Data from IoT devices are collected and presented in real time, can uniquely identify 'things' or particular IoT devices, and can constantly collect, monitor, and control 'things' without human interference.

17.3.5.2   Traffic Management

Traffic tracking, monitoring, control, and management is one of the critical challenges cities need to deal with. Increasing usage of vehicles is causing traffic congestion and increased pollution. There is also a rise in the number of road accidents and related damages. As per the World Health Organization (WHO), about 1.35 million people lose their lives every year due to road traffic accidents. The report also states that about 50 million individuals suffer non-critical injuries. The IoT can provide the mechanism to manage and improve this scenario due to its capability to connect, monitor, and control things. Connected cars, safer roads, smart traffic lights, driver or vehicle monitoring, and smart parking are some of the ways.

17.3.5.3   Intelligent Transport System/Connected Vehicles

An intelligent transport system also known as connected vehicles is a system with integrated wireless networks for connecting and communicating with nearby devices and outside systems to improve safety, security, and transportation. These systems can include GPS, apps that can unlock cars, and cellular networks. Vehicle-to-vehicle communication is an example of a connected vehicle application. This allows a car to share the speed, location, traffic situation, road condition, and information on other vehicles on the road. Other examples of a connected vehicle application are vehicle-to-infrastructure communication and vehicle-to-roadside unit communications which allow cars to share information on traffic signals, road signs, toll booths, and other relevant information to improve traffic management.

Connected vehicles allows for two-way communications between different vehicles such as cars, buses, networking and mobility devices such as mobiles, and infrastructure through a network connection. Those communications can enable near-real-time information sharing, analysis, and decision making to trigger an automated response. GPS is used for location tracking, navigation, and surveying and supports intelligent transport systems. This satellite-based navigation system provides real-time location and time data that allow a vehicle to detect traffic jams and navigate the best route based on traffic conditions to avoid delays. Internally, an intelligent transport system can be connected to a smartphone to activate features like vehicle turn on or off with a smartphone, location tracking, and the sharing of the location of the vehicle in the event of theft.

The quality of the vehicle is assessed by analysing various parameters determining vehicle performance and by tracking the maintenance carried out. This is used to decide on the need for and to schedule maintenance. Real-time assistance, monitoring driving patterns, navigation assistance, a speeding alarm, and updates on the state of the roads can ensure a safe driving experience.

Connected cars can connect to the network and communicate with emergency services to send messages and warnings. This helps in planning a quick response and providing care to decrease the death toll (Figure 17.5).

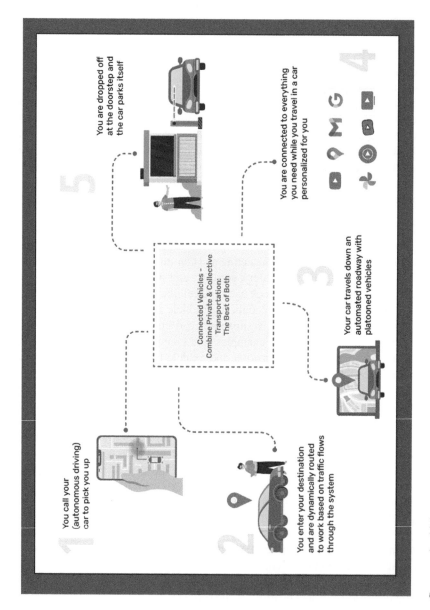

FIGURE 17.5 Connected vehicles.

17.3.5.4   Smart Traffic Signal Systems

These are the intelligent traffic signal systems that use advanced tools and technologies such as sensors and cameras to collect real-time data to adjust signal timings and improve the traffic conditions and safety, and to communicate with connected cars to increase the efficiency of the transportation system. The aim of the system is to collect real-time data on traffic conditions and use insights to adjust the timings of traffic signals. This can reduce accidents around traffic lights and also lessen their violations. Heat and moisture-sensitive sensors are installed in traffic lights for monitoring meteorological data, which is used to monitor weather and road visibility. Better traffic light control is offered by changing the brightness and intensity of the lights to improve visibility. The smart traffic lights duration can also be changed based on the number of cars on the road. For example, if there are a greater number of cars on the road than predicted then the duration of the green light is adjusted to reduce idling time and to avoid congestion. This lowers greenhouse gas emissions and improves the air quality index.

17.3.5.5   Driver or Vehicle Monitoring

Telematics uses telecommunications and informatics to create systems to understand, track, monitor, manage, and improve remote objects such as vehicles and drivers. Telematics use GPS technology, mobile phone networks, wireless communications, and IoT sensors to collect data on speed, location, condition, speeding and braking, and the movement of the vehicle. Collected data are communicated over the wireless networks to servers, where it is analysed and feedback is provided to the driver. This enables closer monitoring and encourages better driving practices. Telematics data can be used by insurance companies to design insurance plans that offer discounts to safe and efficient drivers. Telematics with AI, ML, and the IoT can derive new insights which can be used to optimize traffic management and reduce adverse environmental impacts in smart cities (Figure 17.6).

17.3.5.6   Smart Parking

Smart parking refers to the use of IoT technology to improve the efficiency and convenience of parking by automating it. This involves the application of sensors, cameras, and other smart devices that collect real-time data on available parking slots in the parking space in the vicinity and to direct drivers to these vacant spaces. There are many advantages of using smart parking solutions such as reduced congestion, reduce drivers' time and fuel costs spent in searching for a space, improved parking management, and better user experience. Smart parking systems can also increase the efficiency of the parking system by increasing the utilization of the existing parking spaces, thereby reducing the need for new places.

The driver/user checks the availability of the free parking spaces through the parking app. When the user enters the details of the desired spot for parking, such as a particular mall or street, the app sends the information to the service provider, which checks the availability of free parking spaces in the vicinity. If parking is available, it gives an appropriate response to the user (Figure 17.7).

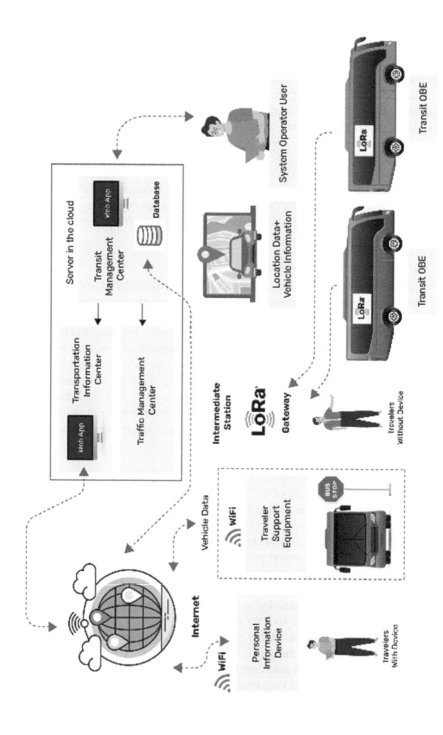

FIGURE 17.6 Vehicle tracking system.

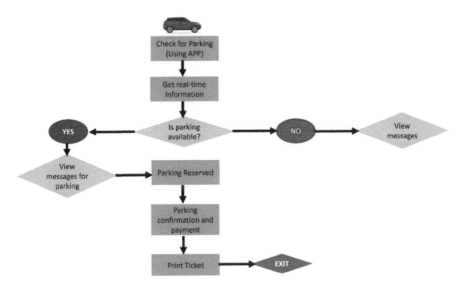

FIGURE 17.7   Smart parking.

17.3.5.7   Smart Utility Meters

Electricity, natural gas, water, transport, and communication are five key utilities in modern city life. Individuals and businesses need these utilities to function. In this data driven world, there is a need for a metering system that measures the consumption of these utilities and a mechanism that decides the charges that consumers should pay based on certain factors. A smart meter is very similar to the traditional meter at home and records utility consumption such as energy, gas, or water. For example, electricity is measured based on the input voltage, current, and power factors. Water is metered based on the volumetric usage. However, the smart meter is an electronic device that records electric energy usage and communicates that information with a utility at predetermined intervals and eliminates the need for a meter reader. The meter can also be controlled remotely (Figures 17.8 and 17.9).

In addition to reporting utility usage, the smart meter can inform the utility immediately whenever there is any outage or issue in the area. This helps utilities to increase efficiency and customer satisfaction. Smart meters are designed to collect a variety of data on consumption such as when and how much the resource is being used and what kind of resources are used. Utilities can use advanced analytics techniques to study this data to optimize operations and draw meaningful, actionable insights. Utility companies can determine costs and plan infrastructure needs based

FIGURE 17.8   Traditional utility meter.

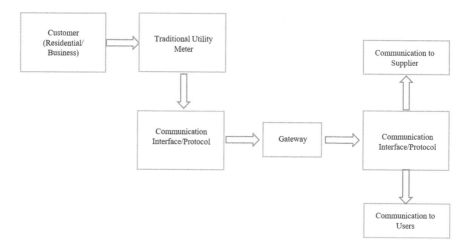

FIGURE 17.9 Smart meter.

on these insights. Smart meters also allow users to monitor and control their energy consumption. Consumers can be offered incentives by having time-of-day metering where they can shift their load to off-peak hours.

Smart utility meters can also be used for improving services to citizens. Most of the citizens are not interested in interacting with the municipal corporation or local government departments for providing inputs or suggestions, but the smart metering data can be used by municipalities to improve customer service and experience by proactively providing relevant information and offering efficient services. Capturing revenue losses, doing maintenance checks, and working proactively on leakage and failure alerts are some of the ways in which government departments can improve their operations, stay future ready, and offer a more sustainable environment.

17.4 GREEN COMPUTING

Due to increasing digitalization, carbon emission and energy demands have increased tremendously. There are 4.66 billion active internet users and 6.3 billion smartphone users worldwide, and this will reach 7.5 billion by 2026. This is having a severe impact on the environment. The increased manufacturing and networking require-ments associated with these proliferations will increase carbon emissions. Carbon emissions of digital technology companies were 3.7% in 2018, up from 2.5% in 2013. Data centres' energy consumption is around 3% of global electricity usage and is increasing significantly with the increased implementation of cloud solutions and will reach 8% by 2030.

Smart cities are consuming large amount of resources such as the IoT, computing resources, advance tools and technologies, security solutions, and communication technologies for development and hence the growing financial cost, environmental costs, and sustainability are becoming major concerns. Green computing is about using resources in environmentally sustainable ways and using the IoT, big data, and data analytics to fulfil user demand while reducing adverse environmental impacts.

Green computing also known as sustainable computing is the design, use, and disposal of computers, chips, computing systems, software, and other resources in an ecologically responsible way. The green computing objective is to create sustainable computing models that reduce greenhouse gas emissions resulting from energy use, increase energy efficiency, maximize renewable energy sources, and minimize environmental impact.

IoT-enabled devices are equipped with sensors and communication technologies to collect and share data for real time monitoring, analysis, and control. They require large amounts of energy to run and transmit data, resulting in increased carbon emissions in the environment. The production and disposal of IoT devices also causes carbon emissions and there is a need for resource conservation, environmental governance, and efforts to reduce carbon emissions. The IoT has potential to reduce greenhouse effects and ecological footprint by designing processes and policies to ensure sustainable use of the IoT.

17.4.1 Green IoT

The green IoT is the use of IoT devices and technologies to reduce the environmental impact of human actions and to achieve sustainable living. Examples of the green IoT include smart energy management, application of low energy consumption devices or electrical appliances, deploying carbon-free materials, using renewable energy resources and promoting reusability, smart water and smart waste management reducing the amount of waste, and conserving resources. AI, ML, and computing technologies are making the IoT more intuitive and user-friendly. Cloud computing can address the issues of high energy consumption and high resource usage. It offers virtualization, dynamic provisioning, multi-tenancy, and green data centre techniques which help reduce carbon emissions and reduce energy consumption. Organizations can host their on-premises applications on the cloud and thereby reduce their direct energy requirements by up to 90%. Green IoT solutions use green RFID, green cloud computing, green machine-to-machine technology, green wireless sensor networks, and green data centres.

- *Green RFID.* RFID tags are small electronic devices with a microchip to store data and an antenna for enabling radio frequency communication between a reader and a tag. An RFID tag can have built-in batteries for communication or may use energy from the reader. Green RFID applications include the use of RFID tags for recycling and waste management where they can be used to track and segregate recyclable materials, thereby reducing landfill waste for sustainable agriculture where the health of livestock can be monitored using RFIDs and for smart energy management in buildings and industrial facilities.
- *Green cloud computing.* Green cloud computing is employing cloud computing in an environmentally sustainable way. Green computing is offered as infrastructure as a service where in-cloud computing solution providers store, process, and manage data remotely thus reducing the requirement for on-premises equipment and servers, platform as a service, or software as

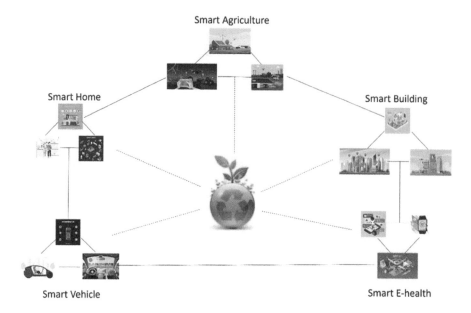

FIGURE 17.10   Green IoT.

a service models, where organizations can use resources on a needs basis, reducing the excess infrastructure requirement. Cloud solution providers can optimize the load requirements by consolidation, thus reducing the number of servers or other equipment requirements.

- *Green machine to machine (M2M).* Green M2M is using M2M in an environmentally sustainable way. Some of the examples are smart grids employing energy efficient and improved transmission, smart transportation and connected vehicles reducing carbon emissions though transport management, and improved power and communication through smart devices and technologies.

- *Green data centres.* Data centres are used for the storage, handling, and processing of all types of data. Green data centres employ resources in a sustainable and environmentally friendly way by reducing the impact of high energy consumption. Some of the examples of green data centres are those using renewable energy sources such as solar and wind, using energy efficient devices and tools, and reducing the number of devices used by handling the workloads optimally, or by green procurement, thus reducing energy consumption (Figure 17.10).

17.4.2   AI and ML Analytics for the Green Smart City

A smart city is a city that uses modern tools and technologies to improve the quality of life for its residents in an environmentally sustainable way by reducing waste and consumption across domains and functions. These data come from different sources such as geospatial data, traffic and vehicle data, crime data, and healthcare data and

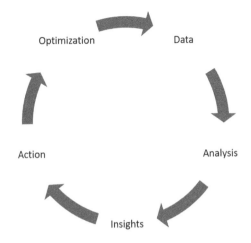

Optimization

Data

Action

Analysis

Insights

FIGURE 17.11 Data analytics for a smart city.

are used to build efficient and environmentally friendly cities for people. Increasing the use of devices generates a huge amount of data constantly. Traditional data analysis and data management systems need to be enhanced to fulfil modern data requirements. The real potential of the IoT lies in analysing these data, drawing insights, and using these business insights for a sustainable city. To make sense of all the data we need data analytics facilitated through IoT platforms. AI and ML algorithms are used to draw inferences and relationships among disparate sources to reveal meaningful and actionable information. An enormous amount of data is constantly generated in cities through different transactions, activities, and applications. Collaboration among various stakeholders such as the government, residents, organizations, and citizen data scientists is needed to share and analyse dataset laws and regulations to support the transformation from city to smart city and to decide how successful the transition is. The real-time handling and processing of data are required for automation, real-time monitoring, and the control of different systems and creating smart solutions. Hence, the availability of open data is the main axis in smart city solutions. IoT architecture and infrastructure needs to be customized to address needs in the application of specific domains of smart cities such as health, mobility solutions, traffic, and the environment (Figure 17.11).

17.4.3 Data Analytics for a Smart City

Data analytics is the tool used by smart cities for analysing the large amount of data in the most efficient manner and to draw meaningful information. Data comes in different formats and structures and can be classified. Based on how the data is classified, different tools and techniques are selected for analysis. Data analytics helps in deriving actionable insights from IoT data.

In IoT analytics, data is categorized as structured or unstructured, or as data in motion or data at rest. Structured data is organized in a particular format, schema, or model such as tables or csv files or spreadsheets. Many computing systems generate structured data, such as banking transactions data and log files. Traditional relational database management systems are able to store, process, and handle structured data. Data collected by IoT sensors such as weight, pressure, rain, temperature, and humidity are structured data. Unstructured data does not have a logical schema. Text, images, and video clips are examples of unstructured data. Around 80% of data generated by businesses is unstructured and hence data analytics methods to analyse such data are required. ML algorithms can be used to draw meaningful insights from images or video footages by decoding speeches, face recognition or object detection, and the identification algorithms. Smart IoT devices generate both structured and unstructured data.

Data can also be explained based on their state or location. Data in motion are the data that have actually been transmitted over the network, such as streaming videos, files sent over the internet, online transactions, Web browsing, data being exchanged between systems and devices, and emails. The data at rest are the data stored and which have not actively been transmitted, such as data in hard drives or files, or data in databases. The data from smart devices are an example of data in motion as they are transmitted over the network to its final destination. Data processing usually happens at the edge. Sensor readings, device status, data held by brokers or in a storage array at the data centre are examples of IoT data at rest.

Data analysis can be classified as descriptive analysis, diagnostic analysis, predictive analysis, or prescriptive analysis based on the results they produce (Figure 17.12).

Descriptive data analysis is used to summarize, describe, or explain the characteristics of data. Diagnostic analysis provides answers to why the particular problem or event happened. Predictive analytics tells us what would happen in the future, before it happens, while prescriptive analytics recommends solutions for future problems that may occur. Scalability and volatility of the data are major concerns in data analytics. As discussed earlier, ML is central to smart city solutions and analytics. Smart things collect a large amount of data constantly which are analysed to take smart

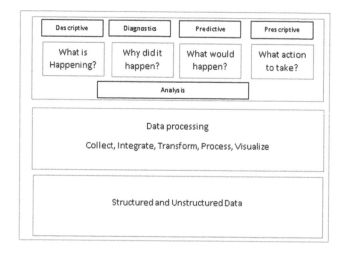

FIGURE 17.12 Types of data analysis.

actions. The manual processing of these data is complex and cumbersome, and AI and ML are used for such data. Smart parking is one such example. Let us see some examples of how analytics can be used in smart city solutions.

17.4.4 TRANSPORT

Monitoring traffic data and investigating trends can be used to design appropriate solutions to handle the traffic, resulting in reduced congestion and delays, reduced accidents, and assisting transport authorities in designing better ways to manage and monitor transportation inside the city. Data analytics can help make public transport efficient and smart by predicting demand for transport on specific occasions or days and adjusting the transport depending upon the prediction.

The city council of smart city Nanjing in China has installed sensors into 10,000 taxis, 7000 buses, and 1 million private cars. The Nanjing Information Centre collects daily traffic data, experts analyse this data, and updates are communicated to commuters on their smartphones. The government, instead of spending money on new roads, has created new traffic routes based on these insights which improved congestion.

17.4.5 BETTER ASSET MANAGEMENT

Data analytics facilitates better monitoring and management of city infrastructure. Predictive maintenance is used to improve equipment performance, reliability, and reduce downtime, thereby saving costs involved in reactive maintenance. Kansas City, Missouri is running a smart sewer upgrade project costing $4.5 billion and has managed to save $1 billion in project infrastructure costs by implementing the IoT, data analytics, and green infrastructure.

17.4.6 CRIME MONITORING AND PREVENTION

Data analytics can help to identify crime prone areas and times when crime rates are high. This information can be used to increase the security in such areas to prevent the crimes. For example, data analytics is used by many smart cities, such as London, Chicago, New York, and Los Angeles, to stop crimes being committed. New York has set up a Real-Time Crime Center that contains more than ten years data with records of criminal complaints, arrests, and 911 call records.

17.4.7 CITIZEN DATA SCIENTIST

Smart city solutions have the ability to address the issues arising from urbanization. Organizations worldwide and across industries strive to be more data-driven, and there needs to be a more skilled workforce which can design and implement such solutions effectively.

The citizen data scientist's role complements the role of data scientists. Citizen data scientists are valuable and affordable in addition to data scientists, which allows them to concentrate and work on more critical initiatives such as developing

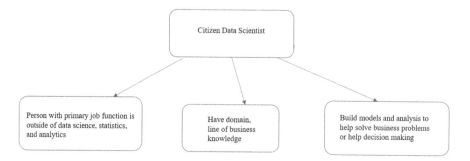

FIGURE 17.13   Citizen data scientist.

complex algorithms and advanced data models. According to Gartner [35], 'a citizen data scientist is a person who creates or generates models that leverage predictive or prescriptive analytics, but whose primary job function is outside of the field of statistics and analytics'. Thus, these people have other job titles within the organizational business group and acquire machine advance statistics and learning skills as required. They have strong domain knowledge and find solutions to business problems through data and analytics. Citizen data scientists make organizations analytics driven.

A lack of clarity of the role and responsibilities of citizen data scientists are the main challenges faced by organizations in developing such scientists and may result in creating conflicts among expert and citizen roles (Figure 17.13).

17.4.8   EMPOWERING CITIZEN DATA SCIENTISTS

To enable and promote the citizen data scientist role organizations need to build an ecosystem with people, processes, and technology as its components. Empowering citizen data scientists requires a data-driven culture, executive backing, a place for guidance, right tools, and training to put these tools to use.

17.4.9   PEOPLE

Business analysts, data engineers, business translators, ML architects, and developers work as members of the ecosystem. These members will work with citizen data to support them to learn the skills they lack. Reskilling or upskilling employees will help them to take on these enhanced data roles and companies can make better use of the data. It will also encourage healthy collaboration and communication among different stakeholders of the ecosystem.

17.4.10   PROCESSES

An organizational data culture sets the process to be used by data scientists and citizen data scientists for applying data and data analytics techniques. The use of big data demands programming knowledge, knowledge of statistics, ML, database management, data visualization, and domain knowledge. Hence training requires combining multiple skill sets. Citizen data scientists possess line of business expertise which data scientists may be lacking, and this is the greatest advantage they offer. Data democracy will allow the sharing of knowledge organization. In practice, data scientists work on advanced analytics and statistics, building workflows and modelling. Citizen data scientists run and test these workflows to ensure that they are ready to be deployed in the production environment. Gradually, over the period, citizen data scientists can perform additional tasks such as modifying workflows or creating new workflows of their own. Thus, organizations can empower citizen data scientists to learn and build ML models. These models can be used by relevant business users to make better decisions.

17.4.11 TECHNOLOGY

An easy to use, powerful, scalable, and self-service data analytics platform can empower data scientists to build models that can solve business problems or optimize products or processes. A good analytics platform allows citizen data scientists to connect, collect, and ingest all the relevant data; to prepare, clean, and transform data without exporting it; to create, train, and test ML models without any or only low coding requirements and share these models for implementation.

17.4.12 DATA COLLECTION

The data could be structured, unstructured, or semi-structured and can come from different data management systems. Those with ever increasing volumes of data are the main challenges while providing data ingestion capabilities. The data analytics platform requires the provision of seamless, direct, reliable, and secure access to multiple data sources with native interfaces and integration to source data.

17.4.13 DATA PREPARATION

The data in the original format is raw, rarely ready for analysis, and needs to be prepared for creating contextual data for analytics. The data may require missing value handling, outlier detection and handling, merging, sorting, filtering, or aggregation. Advance data preparation techniques like creating calculated fields, principal component analysis, or factor analysis are also applied as required. The data analytics platform with the ability to prepare and enrich data within the platform provides centralization trackability and better security.

17.4.14 MACHINE LEARNING MODELS

Every business is different and customized ML models are required to suit the specific business requirement. Citizen data scientists may need help to build models completely using coding, but they have the knowledge of the model that is suitable for a given business case. The analytics platform providing the drag and drop functionality to build models reduces the requirement of learning a programming language for coding.

17.4.15 SHARING MODELS

The trained models are shared with other business users where they can use these models on their data. Thus normal business users can use these models to derive insights from their data without citizen data scientists involvement and without any need to learn coding or ML tools and algorithms.

A data literacy program can help citizen data scientists to learn the skills required and make sense of the data, to learn and use dashboarding, data transformation, direct querying, and other such tools to complement their capabilities.

Automated machine learning (AutoML), self-service reporting software, augmented analytics and low code, and no code data science platforms can help organizations to democratize data science.

17.4.16 WHO CAN BE A CITIZEN DATA SCIENTIST?

A citizen data scientist's position is for employees who have the skills of the data scientist but do not have the data scientist's qualification. Any team member using external data to perform his or her job or in decision making can be a citizen data scientist. He or she may already be using tools like Excel for analysis and visualization. AutoML solutions are used for selecting the best model for the given business problem, to automate data pre-processing, feature engineering, and optimization. In auto-generated augmented analytics, an ML algorithm detects patterns or trends for generating insights. Low code, no code platforms provide drag and drop functionality which require minimum coding knowledge.

To be a citizen data scientist, the employee should be familiar with working with data, should have divergent thinking, strong analytical skills, the ability to assess the information meaningfully, and should have industry adjacency.

REFERENCES

1. Jonathan Woetzel, J. R. (2018). *Smart Cities: Digital Solutions for a More Livable Future*. McKinsey&Company. Retrieved from https://www.mckinsey.com/capabilities/operations/our-insights/smart-cities-digital-solutions-for-a-more-livable-future
2. Priano, F. H. and Cristina Fajardo Guerra. A framework for measuring smart cities. *Proceedings of the 15th Annual International Conference on Digital Government Research*, June 2014, pp. 44–54. https://doi.org/10.1145/2612733.2612741
3. Giffinger, R. C.-M. (2007). *Smart Cities: Ranking of European Medium-Sized Cities*. *Centre of Regional Science (SRF)*. https://www.smart-cities.eu/download/smart_cities_final_report.pdf
4. Chen, Z. and Chan, I.C.C. (2023). Smart cities and quality of life: a quantitative analysis of citizens' support for smart city development. *Information Technology & People*, 36(1), 263–285. https://doi.org/10.1108/ITP-07-2021-0577
5. Sarker, Iqbal H. (2022). Smart city data science: Towards data-driven smart cities with open research issues. *Internet of Things*, 19, 100528. ISSN 2542-6605, https://doi.org/10.1016/j.iot.2022.100528
6. Javed, Abdul Rehman, Zikria, Yousaf Bin, Rehman, Saifur, Shahzad, Faisal and Jalil, Zunera (2021). Future smart cities: Requirements, emerging technologies, applications, challenges, and future aspects. *TechRxiv*. Preprint. https://doi.org/10.36227/techrxiv.14722854.v1

7. Choi, Junho (2022). Enablers and inhibitors of smart city service adoption: A dual-factor approach based on the technology acceptance model. *Telematics and Informatics*, 75, 101911, ISSN 0736-5853, https://doi.org/10.1016/j.tele.2022.101911

8. Wirtz, B.W., Becker, M. and Schmidt, F.W.(2022). Smart city services: an empirical analysis of citizen preferences. *Public Organiz Rev* 22, 1063–1080. https://doi.org/10.1007/s11115-021-00562-0

9. Bhatta, B. (2010). *Analysis of Urban Growth and Sprawl from Remote Sensing Data.* Springer Science & Business Media. https://link.springer.com/chapter/10.1007/978-3-642-05299-6_1

10. Broere, W. (2012). Urban problems-underground solutions. In: Zhou, Y., Cai, J. and Sterling, R. (Eds), *Proceedings of the 13th World Conference of ACUUS: Underground Space Development-Opportunities and Challenges.* Research Publishing, Singapore, pp. 1528–1539.

11. Saha, B. (2014). Green Computing. *International Journal of Computer Trends and Technology (IJCTT)*, 14(2), 18–22.

12. Sarker, Iqbal H. (2022). Smart city data science: Towards data-driven smart cities with open research issues. *Internet of Things*, 19, 100528, ISSN 2542-6605. https://doi.org/10.1016/j.iot.2022.100528

13. Jiang, Huaxiong, Geertman, Stan and Witte, Patrick (2023). The contextualization of smart city technologies: An international comparison. *Journal of Urban Management*, 12(1), 33–43, ISSN 2226-5856, https://doi.org/10.1016/j.jum.2022.09.001

14. Heidari, A., Navimipour, Nima Jafari and Unal, Mehmet. (2022). Applications of ML/DL in the management of smart cities and societies based on new trends in information technologies: A systematic literature review. *Sustainable Cities and Society*, 85, 104089, ISSN 2210-6707. https://doi.org/10.1016/j.scs.2022.104089..

15. Rajput, Sudhir Kumar, Choudhury, Tanupriya, Sharma, Hitesh Kumar and Mahdi, Hussain Falih. (2022). Smart city driven by AI and data mining: The need of urbanization. https://doi.org/10.1007/978-981-19-4052-1_16

16. Razmjoo, Armin, Gandomi, Amir H., Pazhoohesh, Mehdi, Mirjalili, Seyedali and Rezaei, Mostafa. (2022) The key role of clean energy and technology in smart cities development. *Energy Strategy Reviews*, 44, 100943, ISSN 2211-467X, https://doi.org/10.1016/j.esr.2022.100943

17. Lilhore, Umesh Kumar, Lucky Imoize, Agbotiname, Li, Chun-Ta, Simaiya, Sarita, Pani, Subhendu Kumar, Goyal, Nitin, Kumar, Arun and Lee, Cheng-Chi. (2022). Design and implementation of an ML and IoT based adaptive traffic-management system for smart cities. *Sensors*, 22(8), 2908. https://doi.org/10.3390/s22082908

18. Arruda, H.M., Bavaresco, R.S., Kunst, R., Bugs, E.F., Pesenti, G.C., Barbosa, J.L.V. (2023). Data science methods and tools for industry 4.0: A systematic literature review and taxonomy. *Sensors (Basel)*, 23(11), 5010. https://doi.org/10.3390/s23115010

19. Wood, G.E.R., Pykett, J., Banchoff, A., King, A.C., Stathi, A. (2023) Improving your local area citizen scientists and community stakeholders. Employing citizen science to enhance active and healthy ageing in urban environments. *Health & Place*, 79, 102954. https://doi.org/10.1016/j.healthplace.2022.102954

20. von Gönner, J., Bowler, D.E., Gröning, J., Klauer, A.K., Liess, M., Neuer, L. and Bonn, A. (2023). Citizen science for assessing pesticide impacts in agricultural streams. *Science of the Total Environment*, 857(Pt 3), 159607. https://doi.org/10.1016/j.scitotenv.2022.159607

21. Scott, Andrew B. and Frost, Paul C. (2017). Monitoring water quality in Toronto's urban stormwater ponds: Assessing participation rates and data quality of water sampling by citizen scientists in the FreshWater Watch. *The Science of the Total Environment*, 592, 738–744.

22. Van der Velde, T., Milton, D.A., Lawson, T.J., Wilcox, C., Lansdell, M., Davis, G., Perkins, G. and Hardesty, B.D. (2017). Comparison of marine debris data collected by researchers and citizen scientists: Is citizen science data worth the effort? *Biological Conservation*, 208, 127–138. https://doi.org/10.1016/j.biocon.2016.05.025

23. Grootjans, S.J.M., Stijnen, M.M.N., Kroese, M.E.A.L., Ruwaard, D. and Jansen, M.W.J. (2022). Citizen science in the community: Gaining insight in community and participant health in four deprived neighbourhoods in the Netherlands. *Health & Place*, 75, 102798, ISSN 1353-8292. https://doi.org/10.1016/j.healthplace.2022.102798

24. Hartmann, M. C. (2022). Text and image analysis workflow using citizen science data to extract relevant social media records: Combining red kite observations from Flickr, eBird and iNaturalist. *Ecological Informatics*.

25. Rob, R. (2014). The real-time city? Big data and smart urbanism. *GeoJournal*, 79(1), 1–14. http://www.jstor.org/stable/24432611

26. Ushakov, Denis, Dudukalov, Egor, Mironenko, Ekaterina, Shatila, Khodor. (2022). Big data analytics in smart cities' transportation infrastructure modernization. *Transportation Research Procedia*, 63, 2385–2391, ISSN 2352-1465, https://doi.org/10.1016/j.trpro. 2022.06.274

27. Huang, Haosheng, Xiaobai Angela Yao, Jukka M. Krisp and Bin Jiang. (2021). Analytics of location-based big data for smart cities: Opportunities, challenges, and future directions. *Computers, Environment and Urban Systems*, 90. https://doi.org/10.1016/j.compen vurbsys.2021.101712

28. Atitallah, Safa Ben, Driss, Maha, Boulila, Wadii, Ghézala, Ben and Hajjami, Henda. (2020). Leveraging deep learning and IoT big data analytics to support the smart cities development: Review and future directions. *Computer Science Review*, 38, 100303.

29. Ruhlandt, Robert Wilhelm Siegfried, Levitt, Raymond, Jain, Rishee and Hall, Daniel. (2020). One approach does not fit all (smart) cities: Causal recipes for cities' use of "data and analytics". *Cities*, 104, 102800, ISSN 0264-2751, https://doi.org/10.1016/j.cities. 2020.102800

30. Elharrouss, Omar, Almaadeed, Noor, Al-Maadeed, Somaya. (2021). A review of video surveillance systems. *Journal of Visual Communication and Image Representation*, 77, 103116, ISSN 1047-3203, https://doi.org/10.1016/j.jvcir.2021.103116

31. Šurdonja, Sanja, Giuffrè, Tullio, Deluka-Tibljaš, Aleksandra. (2020). Smart mobility solutions – necessary precondition for a well-functioning smart city. *Transportation Research Procedia*, 45, 604–611, ISSN 2352-1465, https://doi.org/10.1016/j.trpro.2020. 03.051

32. Barkley, B. E. (2018). A long ride to work: Job access and the potential impact of ride-hailing in the Pittsburgh area. *Community Development Reports*.

33. World Economic Forum. (2019). Retrieved from webforum.org: https://www.weforum. org/agenda/2019/01/why-digitalization-is-the-key-to-exponential-climate-action/

34. Jonathan Woetzel, J. R. (2018). *Smart cities: Digital solutions for a more livable future*. McKinsey & Company. Retrieved from https://www.mckinsey.com/capabilities/ operations/our-insights/smart-cities-digital-solutions-for-a-more-livable-future

35. Gartner. (2018, May 12). *Gartner*. Retrieved from gartner.com: https://blogs.gartner. com/carlie-idoine/2018/05/13/citizen-data-scientists-and-why-they-matter/

Index

Pages in *italics* refer to figures and pages in **bold** refer to tables.